LEAD POISONING

Exposure, Abatement, Regulation

Edited by

Joseph J. Breen and *Cindy R. Stroup*

CRC Press
Taylor & Francis Group
Boca Raton London New York

CRC Press is an imprint of the
Taylor & Francis Group, an **informa** business

First published 1995 by Lewis Publishers

Published 2019 by CRC Press
Taylor & Francis Group
6000 Broken Sound Parkway NW, Suite 300
Boca Raton, FL 33487-2742

First issued in paperback 2020

ISBN 13: 978-0-367-57974-6 (pbk)
ISBN 13: 978-1-56670-113-6 (hbk)

Visit the Taylor & Francis Web site at
http://www.taylorandfrancis.com

and the CRC Press Web site at
http://www.crcpress.com

Library of Congress Cataloging-in-Publication Data

Lead poisoning : exposure, abatement, regulation / edited by Joseph J. Breen. Cindy R. Stroup.
    p. cm.
    "The Kenneth G. Hancock pollution prevention collection, ACS Division of Environmental Chemistry."
    This volume is based on the ACS Division of Environmental Chemistry symposium.
    Includes bibliographical references and index.
    ISBN 1-56670-113-9
    1. Lead—Toxicology.   2. Lead—Government policy—United States.   I. Breen, Joseph J.   II. Stroup, Cindy R.
III. American Chemical Society. Division of Environmental Chemistry.
  RA1231.L4L379   1995
  615.9/25688—dc20

                                        94-35436
                                           CIP

Library of Congress Card Number 94-35436

MEMORIAM

Kenneth G. Hancock, Ph.D., Director, Division of Chemistry, Mathematics and Physical Sciences Directorate, National Science Foundation (NSF), died suddenly and unexpectedly September 10, 1993 in Budapest, Hungary, while attending an international environmental chemistry workshop with colleagues from France, Hungary, and several Eastern European countries. Ken's death was a shock and loss not only to his family, but also to his colleagues and friends. Ken, an excellent chemist, administrator, teacher, and dedicated civil servant, was a graduate cum laude from Harvard University in 1963. His graduate studies were at the University of Wisconsin (Ph.D., 1968). A tenured associate professor of chemistry at the University of California, Davis, he joined the NSF staff in 1977. Ken played a leadership role in a number of areas at NSF, including international affairs and the fostering of environmentally benign chemical synthesis and processing.

Ken's commitment to green chemistry and pollution prevention prompted the establishment of **The Kenneth G. Hancock Pollution Prevention Fund** to be administered by the American Chemical Society's (ACS) Division of Environmental Chemistry. The Fund will *recognize student achievement in environmental chemistry, especially in environmentally benign chemical synthesis and processing* through scholarships to undergraduate and graduate students. Contributions may be made to the address listed below.

The Kenneth G. Hancock Pollution Prevention Collection has been established to provide an endowment for The Fund, through royalties earned from publications addressing current issues in pollution prevention. This Lewis Publishers volume, **Lead Poisoning Regulation, Exposure, Abatement,** is the first offering in the Collection. Other Division of Environmental Chemistry symposia volumes have been identified as future editions in the Collection. Appropriate volumes from other ACS divisions will be solicited to support The Fund as offerings in the Collection.

The Kenneth G. Hancock Pollution Prevention Fund
American Chemical Society
Division of Environmental Chemistry
1155 Sixteenth Street, NW
Washington, D.C. 20036

PREFACE

Lead poisoning in children has once again emerged as a major public and environmental health problem. Congress has put HUD, EPA, and CDC under intense pressure to combat childhood lead poisoning and to reduce environmental lead exposures. HUD is implementing the Lead Poisoning Prevention Act and has released national survey results indicating 74% (57 million) of all private residences in the U.S. contain some lead-based paint. ATSDR estimated in 1984 that 17% of all American preschool children had blood lead levels that exceed 15 µg/dl. CDC has revised its blood lead action level for young children from 25 to 10 µg/dl.

EPA, CDC, and HUD have each developed individual lead strategies and have jointly formulated a *National Implementation Plan for the Prevention of Childhood Lead Poisoning from Residential Exposure to Lead-Based Paint*. The implementation plan addresses six general areas: monitoring lead indicators, conducting blood-lead screening, enhancing public awareness and promoting action, developing testing and abatement capacity, testing housing units, and abating housing units.

The ACS Division of Environmental Chemistry symposium on which this volume is based provided an exciting forum for academic, research institute, local, state, and federal government researchers to exchange information on the important issue of lead poisoning in children. This volume, with 29 chapters in four parts and an epilogue, highlights issues on lead abatement, exposure, programs and policies, chemical measurement and sampling methods, and statistical analysis.

Part I, Lead Exposure and Abatement, provides insights from studies assessing lead exposures from paint, dust, soil, and lead battery recycling operations. It also documents the lessons learned from the HUD Abatement Program and the pilot comprehensive abatement study.

Part II, Program and Policy Issues, is a unique collection of strategic federal policy statements from EPA, HUD, and HEW-CDC. It includes the details for the National Implementation Plan, and a local government's efforts to provide low-cost effective risk communication and public outreach to the community. Alliance to End Childhood Lead Poisoning articulates a coordinated national action plan and promotes a prevention approach to meet national goals.

Part III, Chemical Measurement Methods, offers seven chapters on analytical issues in the measurement of lead in blood, paint, dust, and soils. The development of efforts to develop standard reference materials is reported by the National Institute of Standards and Technology.

Part IV, Sampling Methods and Statistical Issues, rounds out the technical portion of the volume. Sampling strategies as well as sampling methodologies are presented and discussed. The relationships among lead levels in biological and environmental media are investigated, and the interpretive problems are discussed. The use of multielement analysis of environmental samples, as an approach to investigate sources, is discussed.

The Epilogue, OPPT's **Check Our Kids for Lead** Program, is a most unusual offering. It presents the efforts of one organization to empower its employees to make a personal difference in confronting lead poisoning in children. The program serves as a model for other government organizations (federal, state, and local), university and community organizations (academic departments, churches, and public service), and corporations (small and large businesses) to educate themselves and take personal and corporate responsibility for addressing this important public and environmental health problem. Check our kids for lead. It's the right thing to do!

DISCLAIMER

This book was edited by Joseph J. Breen and Cindy R. Stroup in their private capacity. No official support or endorsement of the U.S. Environmental Protection Agency is intended or should be inferred.

Joseph J. Breen
Economics, Exposure and Technology Division
Cindy R. Stroup
Chemical Management Division

Office of Pollution Prevention and Toxics
U.S. Environmental Protection Agency
Washington, D.C. 20460

JOSEPH J. BREEN is Chief of the Design for the Environmental Program. Economics, Exposure and Technology Division, Office of Pollution Prevention and Toxics, U.S. Environmental Protection Agency, Washington, D.C. Major programmatic responsibilities for Dr. Breen (Ph.D., Chemistry, Duke University) have included measurement and exposure issues related to asbestos in schools and public buildings, the operation of the National Human Monitoring Program, the TSCA Section 4 Test Rule for dioxins and furans in commercial products, and TSCA Section 6 monitoring for PCBs. The reemergence of lead poisoning in children as a national public and environmental health issue focused his attention on assessing lead-abatement program efficacy and formulating a national laboratory proficiency testing and accreditation program for lead analysis. He has co-chaired previous ACS symposia and edited ACS Symposium Series volumes: *Environmental Applications of Chemometrics* and *Pollution Prevention in Industrial Processes: The Role of Process Analytical Chemistry.* More recent programmatic responsibilities include pollution prevention and the Design for the Environment Program's environmentally benign chemical synthesis and processing.

CINDY R. STROUP is Chief of the Technical Programs Branch in the Chemical Management Division, Office of Pollution Prevention and Toxics, U.S. Environmental Protection Agency, Washington, D.C. Major programmatic responsibilities for Ms. Stroup (MS, Statistics, Georgetown University) include the design of national surveys to determine the incidence of asbestos in schools and public buildings, the incidence and levels of selected toxic substances in the adipose tissue of the general U.S. population (National Adipose Tissue Survey), the incidence of leaking underground storage tanks, and the incidence of metal-shredding facilities with PCB-contaminated "fluff" in excess of 50 ppm. She has primary responsibility for technical support to the Lead-Based Paint Program in the Office of Pollution Prevention and Toxics. Current responsibilities include the development of guidelines for management of lead-based paint in repair and remodeling activities, the evaluation of exposures from repair and remodeling activities, evaluation of various lead paint abatement activities, and the field evaluation of lead testing methods, including XRF and test kits.

CONTRIBUTORS

Victoria A. Albright
Westat, Inc.
Rockville, Maryland

Vicki R. Anderson
Office of Pollution Prevention and Toxics
U.S. Environmental Protection Agency
Washington, D.C.

Edmond C. Baird, III
Price Associates, Inc.
Washington, D.C.

Susan L. Barnes
Industrial Testing Laboratories, Inc.
St. Louis, Missouri

K. Bauer
Midwest Research Institute
Kansas City, Missouri

Michael E. Beard
Atmospheric Research and Exposure Assessment
   Laboratory
U.S. Environmental Protection Agency
Research Triangle Park, North Carolina

Kent A. Benjamin
Office of Pollution Prevention and Toxics
U.S. Environmental Protection Agency
Washington, D.C.

Gershon H. Bergeison
Office of Emergency and Remedial Response
U.S. Environmental Protection Agency
Washington, D.C.

Sue Binder
Centers for Disease Control and Prevention
Atlanta, Georgia

David A. Binstock
Research Triangle Institute
Research Triangle Park, North Carolina

Casey Boudreau
Battelle Memorial Institute
Arlington, Virginia

Joseph J. Breen
Office of Pollution Prevention and Toxics
U.S. Environmental Protection Agency
Washington, D.C.

Samuel F. Brown
Office of Pollution Prevention and Toxics
U.S. Environmental Protection Agency
Washington, D.C.

David A. Burgoon
Battelle Memorial Institute
Columbus, Ohio

Bruce E. Buxton
Battelle Memorial Institute
Columbus, Ohio

E. Byrd
National Institute of Standards and Technology
U.S. Department of Commerce
Gaithersburg, Maryland

Joseph, S. Carra
Office of Pollution Prevention and Toxics
U.S. Environmental Protection Agency
Washington, D.C.

Gary S. Casuccio
RJ Lee Group, Inc.
Monroeville, Pennsylvania

Ivan Chang-Yen
University of the West Indies
St. Augustine, Trinidad, West Indies

Manie Chen
Office of Pollution Prevention and Toxics
U.S. Environmental Protection Agency
Washington, D.C.

Robert P. Clickner
Westat, Inc.
Rockville, Maryland

Tamara Collins
Battelle Memorial Institute
Columbus, Ohio

Paul Constant
Midwest Research Institute
Kansas City, Missouri

Brion T. Cook
Office of Pollution Prevention and Toxics
U.S. Environmental Protection Agency
Washington, D.C.

Georgene Cooper
Office of Pollution Prevention and Toxics
U.S. Environmental Protection Agency
Washington, D.C.

David C. Cox
QuanTech, Inc.
Arlington, Virginia

Randy J. Cramer
Office of Pollution Prevention and Toxics
U.S. Environmental Protection Agency
Washington, D.C.

Mark L. Demyanek
Radian Corporation
Tucker, Georgia

James R. DeVoe
National Institute of Standards and Technology
U.S. Department of Commerce
Gaithersburg, Maryland

Gary Dewalt
Midwest Research Institute
Kansas City, Missouri

V. Divljakovic
Industrial Testing Laboratories, Inc.
St. Louis, Missouri

George R. Dunmyre
RJ Lee Group, Inc.
Monroeville, Pennsylvania

R. Frederick Eberle
Dewberry & Davis
Fairfax, Virginia

C. Emrit
Macoya Health Centre
Macoya, Trinidad, West Indies

Michael S. Epstein
National Institute of Standards and Technology
U.S. Department of Commerce
Gaithersburg, Maryland

Eva D. Estes
Research Triangle Institute
Research Triangle Park, North Carolina

Mark R. Farfel
The Johns Hopkins University
Baltimore, Maryland

Carolyn Foster
QuanTech, Inc.
Arlington, Virginia

Arnold Greenland
Quan Tech, Inc.
Arlington, Virginia

P. M. Grohse
Research Triangle Institute
Research Triangle Park, North Carolina

Anne M. Guthrie
Alliance To End Childhood Lead Poisoning
Washington, D.C.

William F. Gutknecht
Research Triangle Institute
Research Triangle Park, North Carolina

S. L. Harper
Atmospheric Research and Exposure Assessment
    Laboratory
U.S. Environmental Protection Agency
Research Triangle Park, North Carolina

L. M. Harris
Office of Pollution Prevention and Toxics
U.S. Environmental Protection Agency
Washington, D.C.

Bradley C. Henderson
RJ Lee Group, Inc.
Monroeville, Pennsylvania

Robin Hertz
Battelle Memorial Institute
Arlington, Virginia

Laura L. Hodson
Atmospheric Research and Exposure Assessment
    Laboratory
U.S. Environmental Protection Agency
Research Triangle Park, North Carolina

A. Hosein-Rahaman
Macoya Health Centre
Macoya, Trinidad, West Indies

James Hogan
Connecticut Department of Health
Hartford, Connecticut

Karen A. Hogan
Office of Pollution Prevention and Toxics
U.S. Environmental Protection Agency
Washington, D.C.

J. S. Kane
National Institute of Standards and Technology
U.S. Department of Commerce
Gaithersburg, Maryland

John G. Kinateder
Battelle Memorial Institute
Columbus, Ohio

Joukko Koskinen
Melorex International
Oy, Espoo, Finland

Angela M. Krebs
Industrial Testing Laboratories, Inc.
St. Louis, Missouri

E. S. Lagergren
National Institute of Standards and Technology
U.S. Department of Commerce
Gaithersburg, Maryland

Bennett D. Lass
Metorex Inc.
Langhorne, Pennsylvania

Barbara A. Leczynski
Office of Pollution Prevention and Toxics
U.S. Environmental Protection Agency
Washington, D.C.

Benjamin S. Lim
Office of Pollution Prevention and Toxics
U.S. Environmental Protection Agency
Washington, D.C.

Robert A. Lordo
Battelle Memorial Institute
Columbus, Ohio

Kate K. Luk
Research Triangle Institute
Research Triangle Park, North Carolina

David G. Lynch
Office of Pollution Prevention and Toxics
U.S. Environmental Protection Agency
Washington, D.C.

A. F. Marlow
National Institute of Standards and Technology
U.S. Department of Commerce
Gaithersburg, Maryland

David Lee McAllister
New York City Department of Health
New York, New York

Mary McKnight
National Institute of Standards and Technology
U.S. Department of Commerce
Gaithersburg, Maryland

Ronald J. Morony
Office of Lead-Based Paint Abatement and
    Poisoning Prevention
U.S. Department of Housing and Urban
    Development
Washington, D.C.

K. E. Murphy
National Institute of Standards and Technology
U.S. Department of Commerce
Gaithersburg, Maryland

Larry E. Myers
Research Triangle Institute
Research Triangle Park, North Carolina

Barbara A. Myrick
Office of Pollution Prevention and Toxics
U.S. Environmental Protection Agency
Washington, D.C.

Roman J. Narconis
Industrial Testing Laboratories, Inc
St. Louis, Missouri

John D. Neefus
Research Triangle Institute
Research Triangle Park, North Carolina

James R. Pasmore
Metorex, Inc.
Langhorne, Pennsylvania

Peter A. Pella
National Institute of Standards and Technology
U.S. Department of Commerce
Gaithersburg, Maryland

Stanislaw Piorek
Metorex, Inc.
Langhorne, Pennsylvania

Bertram Price
Price Associates, Inc.
Washington, D.C.

K. W. James Rochow
Alliance to End Childhood Lead Poisoning
Washington, D.C.

John W. Rogers
Westat, Inc.
Rockville, Maryland

Charles A. Rohde
The Johns Hopkins University
Baltimore, Maryland

Steven W. Rust
Battelle Memorial Institute
Columbus, Ohio

Ferdinand Ruszala
Connecticut Department of Health
Hartford, Connecticut

Don Ryan
Alliance to End Childhood Lead Poisoning
Washington, D.C.

S. B. Schiller
National Institute of Standards and Technology
U.S. Department of Commerce
Washington, D.C.

Bradley D. Schultz
Office of Pollution Prevention and Toxics
U.S. Environmental Protection Agency
Washington, D.C.

John G. Schwemberger
Office of Pollution Prevention and Toxics
U.S. Environmental Protection Agency
Washington, D.C.

Heikki Sipila
Metorex International
Oy, Espoo, Finland

Sarah M. Smith
Yorktown High School
Arlington, Virginia

Ian M. Stewart
RJ Lee Group, Inc.
Monroeville, Pennsylvania

Cindy R. Stroup
Office of Pollution Prevention and Toxics
U.S. Environmental Protection Agency
Washington, D.C.

Frederick Todt
Battelle Memorial Institute
Columbus, Ohio

R. D. Vocke
National Institute of Standards and Technology
U.S. Department of Commerce
Gaithersburg, Maryland

Stevenson Weitz
Office of Lead-Based Paint Abatement and
    Poisoning Prevention
U.S. Department of Housing and Urban
    Development
Washington, D.C.

Emily E. Williams
Research Triangle Institute
Research Triangle Park, North Carolina

Bea M. Wilson
Atmospheric Research and Exposure Assessment
    Laboratory
U.S. Environmental·Protection Agency
Research Triangle Park, North Carolina

Sineta Wooten
Office of Pollution Prevention and Toxics
U.S. Environmental Protection Agency
Washington, D.C.

David Worsley
Connecticut Department of Health
Hartford, Connecticut

ACKNOWLEDGMENTS

A debt of gratitude is owed to all of the participants of the American Chemical Society (ACS) National Symposium, for their contributions to this document, and, in particular, to those who contributed material used in this volume. As editors we are particularly grateful for the support provided by Sineta Wooten, EPA's Chemical Management Division, in pulling together all the various materials that went into the document and diligently tracking the drafts, revisions, and final deliverables for the 29 chapters.

Credit is also due to the four ACS Symposium Session chairs: John G. Schwemberger, John V. Scalera, Benjamin S. Lim, and Brion T. Cook, from the EPA Office of Pollution Prevention and Toxics' Chemical Management Division, who organized and led each session. We appreciate the support received from the American Chemical Society's Division of Environmental Chemistry, especially from Dr. V. Dean Adams, Program Chair. Further, we are grateful for the personal interest of Brian A. Lewis in publishing this volume on lead poisoning in children.

CONTENTS

PART I: LEAD EXPOSURE AND ABATEMENT

PART II: PROGRAM AND POLICY ISSUES

PART III: CHEMICAL MEASUREMENT METHODS

Part I
Lead Exposure and Abatement

Chapter 1

The Prevalence of Lead-Based Paint in Housing: Findings from the National Survey

R. P. Clickner, V. A. Albright, and S. Weitz

CONTENTS

In the 1987 and 1988 amendments to the Lead-Based Paint Poisoning Prevention Act, Congress required the U.S. Department of Housing and Urban Development (HUD) to develop comprehensive and workable plans for the cost-effective inspection and abatement of lead-based paint in privately owned housing and public housing. The reports presenting the plans were required to contain estimates of the amount and characteristics of housing that contains lead-based paint, and of the cost of abatement. Congress also mandated the U.S. Environmental Protection Agency (EPA) to provide technical support to HUD in this effort. In response, HUD sponsored a national survey of lead-based paint in housing.

I. PURPOSE, DESIGN, AND METHODOLOGY OF THE NATIONAL SURVEY

This section presents a summary of the objectives, sample design, and survey methodology of the national survey of lead-based paint in housing, conducted by Westat, Inc., under contract to HUD.

Detailed descriptions of the methodology may be found in the reports to Congress on a "Comprehensive and Workable Plan for the Abatement of Lead-Based Paint in Privately Owned Housing," issued in December 1990, and on a "Comprehensive and Workable Plan for the Abatement of Lead-Based Paint in Public Housing," forthcoming.

The objective of the national survey of lead-based paint in housing was to obtain data for estimating (1) the number of housing units with lead-based paint; (2) the surface area of lead-based paint in housing, to develop an estimate of national abatement costs; (3) the condition of the paint; (4) the incidence of lead in dust in housing units, and in soil around the perimeter of residential structures; and (5) the characteristics of housing with varying levels of potential hazard, to examine possible priorities for abatement.

The study population consisted of nearly all occupied private and public housing constructed in the U.S. before 1980. Newer houses were presumed to be lead free because in 1978 the Consumer Product Safety Commission banned the sale of lead-based paint to consumers and the use of such paint in residences. The broad elements of the design included

- A national area probability multistage sample of lead-based paint in homes, to permit statistically valid estimates of the parameters required for the development of federal policy

- In-person visits to the sampled homes to identify, describe, and measure painted surfaces and to measure the lead content of the paint
- Analysis of dust and soil samples in and around each sampled home, to permit the analysis of the associations among lead in paint, dust, and soil

A. SURVEY DESIGN

The objectives of the study required in-person visits to all sampled dwelling units. The costs associated with in-person visits necessitated geographic clustering of the sampled homes. Therefore, a sample of 30 counties — stratified by region and selected with probability proportional to the 1980 population — was selected from the approximately 3000 counties in the U.S. In order to optimize the congressionally required estimates, the design was stratified on dwelling-unit age and type. Privately owned dwelling units were grouped into two types of housing and three age categories, as displayed in Table 1, which displays the national distribution of occupied housing across the six strata. The survey was conducted between December 1989 and March 1990. The final sample contained 284 privately owned dwelling units and 97 public housing units, for a total sample size of 381 dwelling units.

A survey sample requires a sampling frame; that is, a list of all dwelling units eligible for the survey. No such list exists nationally, or even in many localities, for private housing. However, HUD maintains a list of public housing projects suitable for use as a frame. Thus, private and public housing were sampled using different within-county designs. The different ownership structures between public and private housing also required different approaches to the owners and occupants. These two approaches are described below.

Privately Owned Housing

A multistage area probability design was developed and employed to sample private housing and approach the occupants. A description of the design and its within-county implementation follows.

- Five small geographic areas, called segments, were selected in each county. A segment is a block, or group of adjacent blocks, in an urban area; and a census enumeration district, or group of adjacent enumeration districts, in a rural area. To ensure that the full spectrum of income levels would be represented in the sample, a measure of wealth was computed for each segment. In each county the segments were sorted by this wealth measure, and five segments were selected using systematic random sampling.
- Field interviewers were sent to each of the 150 segments, to list every dwelling unit in the segment. This process created a frame for the sampling of dwelling units.
- Samples of dwelling units were selected from the lists, using systematic random sampling. A brief in-person screening interview was conducted with an adult occupant in each sampled dwelling unit, to determine if the unit was eligible for the survey and to which of the six age/type strata it belonged.
- A sample of dwelling units was randomly selected, within strata, from the eligible homes for inspection visits.

Public Housing

The public housing sample was designed in a somewhat different manner than the private housing sample. All housing projects in the 30 sampled counties were extracted from the public housing data tape HUD gave to Westat, Inc. Projects were then sampled for the study. Westat worked through the chain of command to contact the sampled projects and make appointments to inspect sampled units. Public housing tenants who consented to the inspection received the same $50 incentive as private housing occupants.

Table 1 National Distribution of Occupied, Privately Owned Dwelling Units Built Before 1980

| Type | Number of pre-1980 dwelling units (000) and construction year | | | Total |
	1960–1979	1940–1959	pre-1940	
Single family	29,137	18,782	18,499	66,418
Multifamily	6,548	1,690	2,521	10,759
Total	35,685	20,472	21,020	77,177

Source: 1987 American Housing Survey.

B. INSPECTION PROTOCOL

Resource and respondent burden limitations did not permit the complete identification, inspection, quantification, and testing of every painted surface in every inspected dwelling unit. Fortunately, the objectives of the study did not require such thorough inspections. It is possible in survey statistics to develop a good, clear picture of the aggregate population, with only limited information on each sampled individual. Consequently, the limited inspection protocol described below was followed in each dwelling unit. The inspection protocol was the same for both public and private housing units.

Lead-in-paint measurements were made with portable X-ray fluorescence (XRF) analyzers of the spectrum analyzer type, which the National Institute of Standards and Technology (NIST) had determined to be more accurate and more precise than the direct-reading XRFs used in earlier surveys.[1] While the spectrum analyzer XRF devices are an improvement over the earlier direct-reading XRFs, they still have limitations. In particular, spectrum analyzer XRF measurements made over brick or concrete are less accurate and less precise than those made over wood or plaster. These limitations notwithstanding, portable XRF technology was used because the survey included occupied dwellings where it was not feasible to take paint scrapings for laboratory analysis. Following the federal standard, paint was considered to be lead-based if the lead content was 1.0 mg/cm^2 or greater, as measured by XRF.

Interior Rooms

The rooms were inventoried and classified into wet and dry rooms, according to the presence or absence of plumbing in the room. One wet room and one dry room were randomly selected. All painted surfaces in each of these two rooms were identified and quantified; the substrate materials were identified; and the condition of the paint and substrate materials was noted.

Since not all rooms in a dwelling unit were inspected, it is possible to miss lead-based paint when it is really present somewhere else in the dwelling unit. To reduce the chances of misclassifying a dwelling unit with lead-based paint as "lead free," additional lead readings, termed "purposive" readings, were taken on surfaces that, in the opinion of the field technicians, were most likely to have lead-based paint. In some dwelling units these additional purposive samples did, indeed, find lead-based paint in dwelling units where no lead-based paint had been found via random selection. Outside, painted surfaces were inventoried and measured, and lead readings were taken of randomly and purposively selected exterior painted surfaces; soil samples were also taken.

Samples of dust were collected by vacuuming in three locations in each sampled room: the floor, a window sill, and a window well. In addition, a seventh dust sample was collected from the floor near the most-used entrance to the dwelling unit. The dust samples were sent to a laboratory where they were analyzed for their lead content, using ICP-AES.

Exterior

An exterior wall was randomly selected for inspection. All painted surfaces on the sampled wall were cataloged in the same manner as the interior rooms and subjected to XRF testing. Three exterior soil samples were collected: at the drip line along the sampled wall, at a remote location away from the building, and at the most-used entrance to the dwelling unit. The soil samples were sent to a laboratory for lead content analysis, using GFAA.

C. PRECISION AND ACCURACY OF THE SURVEY DATA

A complex survey with geographically clustered sampling, and differential probabilities of selection, typically has less precision than an unclustered sample with equal selection probabilities for all sampled units. The effect of the design on the precision of the data is called the "design effect." The design effect is the ratio of the actual-size sample to the size of a simple random sample with the same precision. For example, if the sample size is 750 and the design effect is 1.5, then the precision is the same as a simple random sample size of 500. The advantages gained by utilizing a complex design (which may be considerable) would be obtained at the cost of 250 units in the "effective" sample size. Approximate design effects were calculated for the national survey of lead-based paint in housing, and for selected subsets of the sample. The approximate design effect was 1.45 for the overall sample. Thus, overall, confidence interval widths increase by approximately 20% (the square root of the design effect). This analysis does not take XRF measurement error into account, or the effects of within-dwelling unit sampling.

II. PRIOR STUDIES

There have been four previous surveys of lead-based paint in housing. Three local surveys were conducted in the mid-1970s, and one national survey of public housing was carried out in the 1980s.

The Pittsburgh survey,[2] conducted in 1974 and 1975 by Pennsylvania's Allegheny County Health Department for the National Bureau of Standards (now the NIST) under HUD sponsorship, is by far the largest study of its type ever conducted. Approximately 3300 housing units were inspected out of a sample of 4000 units that represented the entire Pittsburgh urban area.

The Washington, D.C., survey,[3] conducted in 1973 by the National Bureau of Standards under HUD sponsorship, as a field test for the Pittsburgh survey, had a sample of 233 units (of which 115 were inspected), representing the city of Washington.

The Phoenix survey,[4] conducted in 1976 by the Arizona Department of Health Services, had a sample of 268 units, representing the census tract in Phoenix considered to be of highest priority because of the high number of both pre-1940 units and children under 5 years of age. One hundred and forty-six housing units were inspected.

"The Modernization Needs Study of Public Housing" included a survey of lead-based paint abatement needs in public housing, that was conducted in 1984–1985.[5] Two hundred and sixty-two public housing units, plus associated common areas, were inspected in 131 public housing projects in 34 cities. The 34 cities were selected because they had community lead-poisoning prevention programs that were willing to conduct the inspections according to a survey design prepared by Abt Associates, Inc., under HUD sponsorship. The results of the study were projected to the national stock of public housing.

As a basis for national estimates of the number of housing units with lead-based paint, these prior surveys are limited. The prior surveys also lack some of the information needed to analyze lead hazards in housing and to estimate the cost of abatement. In addition, percentages based on previous studies appear to have resulted in an underestimate of housing with lead problems. This finding underscored the need for a systematic national survey to generate estimates sufficiently reliable for analysis and policy development.

III. FINDINGS FROM THE NATIONAL SURVEY

The national survey of lead-based paint in housing produced a database with copious information on the incidence and characteristics of lead-based paint in housing, and its association with lead in dust and soil. This section presents the national survey findings that have been reported in the "Comprehensive and Workable Plan for the Abatement of Lead-Based Paint in Privately-Owned Housing: A Report to Congress." (Results concerning public housing are currently unavailable.)

A. PREVALENCE OF LEAD-BASED PAINT IN PRIVATELY OWNED HOMES

An estimated 57.4 million homes, 74% of all occupied housing units built before 1980, have lead-based paint somewhere in the building. As shown in Table 2, an estimated 9.9 million of these homes are occupied by families with children under 7 years of age. This is 71% of all pre-1980 housing units occupied by families with young children.

Older homes are more likely to have lead-based paint than are newer homes. An estimated 90% of dwelling units built before 1940 have lead-based paint in the interior or on the exterior, while 62% of homes built between 1960 and 1979 have lead-based paint. The age of the unit is the only recorded attribute for which the differences between categories are significant.

In particular, there are no significant differences in the incidence of lead-based paint, by the income of the household, the value of the home, or the rent. Although elevated blood-lead levels are more commonly found among poor children, well-to-do households are as likely to occupy homes with lead-based paint as are the poor. Similarly, there is no significant difference between single-family and multifamily housing units.

B. LEAD-BASED PAINT BY LOCATION

The survey also provides information on the location of lead-based paint within or outside the housing unit. Table 3 displays the number and percentage of occupied housing units with lead-based paint only on interior surfaces, only on exterior surfaces, and on both.

While most popular and public policy discussions have been concerned with lead-based paint on interior walls, and lead dust within the housing unit, the survey shows that lead-based paint is more common on

Table 2 Estimated Number of Privately Owned Occupied Housing Units Built Before 1980 with Lead-Based Paint, by Selected Characteristics[a]

Characteristic	Total occupied housing units (000)[b]	Housing units with lead-based paint anywhere in building		Number of housing units in sample
		Percent	Number (000)	
Total occupied housing units built before 1980	77,177	74 (6)[c]	57,370 (4,705)	284
Construction year				
1960–1979	35,681	62 (10%)	22,149 (3,407)	120
1940–1959	20,476	80 (9%)	16,381 (1,824)	87
Before 1940	21,018	90% (10%)	18,916 (2,056)	77
Housing type				
Single family	66,418	74% (7%)	49,476 (4,520)	227
Multifamily	10,759	73% (13%)	7,894 (1,358)	57
One or more children under age 7	13,912	71% (9%)	9,900 (1,302)	90
Census Region				
Northeast	16,963	93% (8%)	15,811 (1,379)	53
Midwest	19,848	76% (12%)	14,994 (2,416)	69
South	24,967	58% (11%)	14,558 (2,688)	116
West	15,399	80% (14%)	12,382 (2,120)	46
Owner-occupied	52,894	72% (8%)	38,251 (4,160)	179
Market value of home				
Less than $40,000	11,885	79% (15%)	9,399 (1,820)	39
$40,000 to $79,999	10,228	53% (17%)	5,442 (1,770)	46
$80,000 to $149,999	5,582	65% (17%)	3,641 (932)	45
$150,000 and up	7,405	87% (12%)	6,474 (891)	42
Renter-occupied	24,285	79% (9%)	19,120 (2,281)	105
Monthly rent payment				
Less than $400	16,339	69% (14%)	11,334 (2,314)	59
$400 and up	8,395	87% (12%)	7,324 (1,042)	40
Household income				
Less than $30,000	46,126	76% (7%)	35,124 (3,091)	156
$30,000 and up	31,048	72% (9%)	22,345 (2,642)	107

[a] Paint lead concentration ≥ 1.0 mg/cm^2.

[b] Total units data are from the 1987 American Housing Survey.

[c] Numbers in parentheses are approximate half-widths of 95% confidence intervals for the estimated percents and numbers. For example, the approximate 95% confidence interval for the percent of housing units with some lead-based paint is 74% ± 5% or 68% to 80%.

8

Table 3 Incidence of Lead-Based Paint (LBP) by Location in the Building

Location of LBP	Occupied housing units with lead-based paint	
	Number (000)	Percent[a]
Interior only	10,681	14%
Exterior only	17,967	23%
Both interior and exterior	28,718	37%
Anywhere in building	57,370	74%

[a] Base equals all 77,177,000 housing units built before 1980.

Table 4 Number and Percentage of Occupied Homes with Lead-Based Paint by Lead Concentration, Year of Construction, and Location of Lead-Based Paint

Location and construction year	Percentage of homes Paint lead concentration (mg/cm^2)			
	≥0.7	≥1.0	≥1.2	≥2.0
Interior	66%	51%	40%	22%
1960–1979	60%	41%	28%	7%
1940–1959	70%	59%	44%	20%
Built before 1940	73%	60%	57%	50%
Exterior	70%	60%	51%	36%
1960–1979	55%	42%	31%	12%
1940–1959	82%	76%	69%	46%
Built before 1940	83%	79%	69%	66%
Anywhere in Building	86%	74%	63%	43%
1960–1979	80%	62%	47%	18%
1940–1959	87%	80%	74%	52%
Built before 1940	94%	90%	79%	75%

Location and construction year	Number of homes (000) Paint lead Concentration (mg/cm^2)			
	≥0.7	≥1.0	≥1.2	≥2.0
Interior	51,008	39,401	31,024	17,239
1960–1979	21,409	14,768	9,991	2,498
1940–1959	14,333	12,058	9,009	4,095
Built before 1940	15,343	12,575	11,980	10,509
Exterior	53,674	46,686	39,641	27,562
1960–1979	19,625	15,058	11,061	4,282
1940–1959	16,790	15,474	14,128	9,419
Built before 1940	17,445	16,604	14,502	13,780
Anywhere in building	66,321	57,370	48,443	32,888
1960–1979	28,545	22,149	16,770	6,423
1940–1959	17,814	16,381	15,152	10,648
Built before 1940	19,661	18,916	16,604	15,693

the outside of the housing unit. An estimated 18.0 million occupied homes (23% of pre-1980 homes) have lead-based paint only on the exterior of the building, compared to an estimated 10.7 million homes (14%) with lead-based paint only in the interior. An estimated 28.7 million homes (37%) have lead-based paint both inside and outside the building. Table 4 shows that this pattern holds for virtually all standards in all time periods.

C. NONINTACT PAINT

Peeling or flaking paint constitutes a direct hazard to small children with pica (the habit of eating paint chips). This was the first hazard identified by research. Table 5 shows the incidence of nonintact paint, both in the aggregate and by the location of the paint. A dwelling unit is considered to have nonintact lead-based paint if at least 5 ft^2 of the lead-based paint in the dwelling unit is defective.

Table 5 Incidence of Nonintact Lead-Based Paint (LBP) by Location in the Building

Location of nonintact LBP[a]	Occupied housing units with nonintact lead-based paint	
	Number (000)	Percent[b]
Exterior only	8,577	11
Both interior and exterior	1,324	2
Anywhere in building[c]	13,820	18

[a] "Interior only" means the only nonintact LBP is in the interior; there may be intact LBP on the exterior. "Exterior only" has a similar meaning.

[b] Base equals all 77,177,000 housing units built before 1980.

[c] A housing unit has nonintact interior LBP if there are more than 5 ft² of damaged interior LBP. Similar definitions apply verbatim for exterior and any LBP. It is therefore possible for a housing unit to have nonintact "any" LBP without having either nonintact exterior LBP or nonintact interior LBP (for example, a house with 3 ft² of damaged interior LBP and 3 ft.² of damaged exterior LBP).

Some 13.8 million occupied units are estimated to have nonintact lead-based paint. This is 18% of the pre-1980 housing stock, and 24% of the pre-1980 stock with lead-based paint. The incidence of nonintact lead-based paint, just as the overall incidence of lead-based paint, is higher on the outside of housing units than on the inside. Moreover, there is a higher incidence of nonintact paint among units with exterior lead-based paint than among units with interior lead-based paint. The paint is damaged in 21% of the units with exterior lead-based paint, compared to 13% for units with interior lead-based paint.

D. ASSOCIATIONS BETWEEN LEAD IN PAINT AND DUST

There is evidence in the literature that lead-based paint is an important contributor to lead in dust. There is also evidence that lead-based paint, especially on the exterior, is an important source of lead in soil. The presence and condition of exterior lead-based paint are believed to be particularly important factors. This section presents the plan for the analysis of the association between lead in dust and soil and lead-based paint. These analyses of associations form a preamble to the analysis of the pathways of paint and dust lead in the following section.

Analyses were designed to test the hypotheses that defective lead-based paint, through flaking and chipping, contributed more dust lead to the environment than did sound, or intact, lead-based paint. The incidence of excessive dust lead between homes with or without lead-based paint was calculated by the location of the dust, by the location (interior or exterior) of the paint, and by its condition. (A dwelling unit was considered to have excessive dust lead if the lead content of the dust exceeded the HUD-recommended dust-lead clearance levels for lead-based paint abatement, which are 200 μg/ft² for floors, 500 μg/ft² for window sills, and 800 μg/ft² for window wells.)

However, it was necessary to exercise caution in assuming that a significant association would indicate the existence of a direct relationship between defective, or damaged, lead-based paint and excessive dust lead. In fact, the association may well be the reverse of that; the likelihood of finding excessive dust lead may be greater if lead-based paint is in sound condition than if it is defective. If so, one reason may be that defective paint is a condition found in most dwelling units from time to time. With periodic repainting, it is corrected. In the typical home it may be the repainting activity, with associated scraping and sanding, that generates the dust lead, not the defective paint condition itself.

The public health literature over the last few years has repeatedly implicated lead in house dust as the most common source of low-level childhood lead poisoning, within a dwelling unit. The national survey includes information on the presence and location of dust within the housing units that were sampled. Table 6 shows the number of units with dust-lead loadings in excess of federal guidelines for homes with or without lead-based paint, and also shows the incidence of dust according to the location of the lead-based paint. The HUD "Interim Guidelines" for the abatement of lead-based paint in housing contain recommended clearance levels for dust lead after lead-based paint abatement. The levels are 200 μg/ft² for floors, 500 μg/ft² for window sills, and 800 μg/ft² for window wells. The same clearance standards are used in Maryland and Massachusetts.

Some 17% of the occupied homes with lead-based paint had dust lead exceeding these guidelines, while only 4% of the dwelling units without any lead-based paint had excessive dust lead. Thus, it appears that over 80% of homes with lead-based paint may not be contaminated with high dust-lead levels. On the other hand, the chance of a unit having excessive dust lead is about four times greater if it has some

Table 6 Dust-Lead Loadings in Occupied Housing Units With or Without Interior or Exterior Lead-Based Paint (LBP)

Presence of LBP	Dust within guidelines[a]		Dust exceeding guidelines[a]	
	Number (000)	Percent	Number (000)	Percent
No LBP at all	19,084	96	723	4
Interior LBP only	10,013	94	671	6
Exterior LBP only	15,423	86	2,546	14
Both interior and exterior LBP	21,984	77	6,733	23
Any interior LBP	31,997	81	7,404	19
Any exterior LBP	37,407	80	9,279	20
Any LBP	47,420	83	9,950	17

[a] HUD Interim Guidelines.

Table 7 Rate of Occurrence of Occupied Housing Units with Dust Lead in Excess of the Federal Guidelines

	Interior surface–dust lead		
Location	Federal guideline[a] (μg/sq ft)	Number (000) of housing units above guideline[a]	Percent of housing units above guideline[a]
Anywhere	varies	10,674	14%
Window well	800	8,632	11%
Window sill	500	2,572	3%
Floor	200	986	1%
Window only[b]	varies	9,688	13%
Floor only	200	986	1%
Both floor and window[b]	varies	0[c]	0%

[a] HUD Interim Guidelines.
[b] Window includes window sill, window well, or both.
[c] There were no sampled housing units in this cell. Nationally there is some small number of housing units in this cell.

lead-based paint than if it does not have any. Table 6 also suggests that excessive dust lead is more likely to be generated by exterior lead-based paint than by interior paint. Evidence suggests that excessive dust-lead levels are reached typically when there is lead in both the interior and exterior paint. The incidence of units with excessive dust lead is highest for units with lead-based paint both inside and outside the house.

Table 7 offers an explanation for these findings. It shows the incidence of dust lead in different locations within the housing unit. Most of the dust is located around the windows, either in the window wells or on the window sills. Window wells and sills can easily receive dust from either the inside or the outside of the house. Fewer than 1 million units have floor dust with lead concentrations above the guidelines. There is likely to be more dust lead in the wells than on the sills or floor, because there is typically more dust there (the wells are cleaned less often) and probably because there is abrasion of paint caused by the opening and closing of the windows.

E. ASSOCIATION BETWEEN LEAD-BASED PAINT AND SOIL LEAD

The national survey also provides information on lead in the soil surrounding the housing unit. Lead in soil is a possible source of lead in house dust, as the soil is tracked or blown into the house. Soil lead can result from exterior lead-based paint, among a variety of environmental sources. Table 8 presents the estimated numbers of occupied dwelling units, nationwide, with soil lead, associated with the presence and condition of exterior lead-based paint. There is a strong statistical association. Table 8 indicates that the probability of excessive soil lead somewhere on the property (i.e., near the entrance, at the drip line, or at a remote location) is four to five times larger when exterior lead-based paint is present than when it is not. Soil lead is especially likely if the paint is defective. However, it is still true that 79% of the time that lead-based paint is present, the soil lead is within the guidelines.

The guidelines used in the survey are the interim guidance on soil lead cleanup levels at Superfund sites, recently issued by EPA.[6] Following a recommendation by the Centers for Disease Control,[7] the cleanup level is set at 500 to 1000 ppm total lead, "to be followed when the current or predicted land use

Table 8 Association Between Lead in Soil and Exterior Lead-Based Paint Condition for Privately Owned Housing Units

Presence and condition of exterior lead-based paint	Lead in soil anywhere			
	Within guideline[a]		Exceeding guideline[a]	
	Number (000)	Percent	Number (000)	Percent
No LBP	29,563	94	1,941	6
LBP present, intact	28,415	79	7,358	21
LBP present, not intact	5,145	52	4,756	48
Any exterior LBP	33,560	73	12,114	27
Total	63,123	82	14,055	18

[a] The guideline is 500 ppm. See EPA Interim Guidance.

is residential.'' When the soil lead is between 500 and 1000 ppm, site-specific conditions should be considered in determining the necessity of cleanup. In order to be conservative with respect to soil lead on residential property, this report uses the lower limit of the EPA range, 500 ppm, in all references to the federal guidelines for soil lead.

F. PATHWAYS

Of great concern to researchers are the pathways by which lead may be transported from lead-based paint to dust that may eventually be inadvertently ingested, particularly by young children. It may be hypothesized that exterior lead-based paint deteriorates, contaminates the soil, and finds its way into the dwelling, in the form of dust. Further, it is hypothesized that interior lead-based paint contributes in various ways to surface dust. The analyses of the survey data support these hypotheses.

Table 9 shows hypothesized major pathways of lead from paint to dust.* Some pathways are depicted as being possibly two way. For example, dust is shown to move back and forth between the floor at the entrance to the dwelling, and the soil near the entrance. Other pathways are depicted as being one way. For example, a pathway is shown from paint on the walls to dust on the floor, but not in the reverse direction. While it is possible that dust on the floor can be disturbed and, subsequently, can adhere to the wall, the amount of such dust is expected to be negligible. Thus, this particular pathway is depicted as being one way.

The statistics shown in Table 9 are the correlation coefficients between the natural logarithms of the pairs of survey measurements of lead associated with the pathways. The distribution of the measurements of lead are skewed to the right. The logarithmic transformation helps normalize the distribution and reduces the influence that a few large observations might have on our analysis.

All of the correlations shown in Table 9 are positive. All are statistically significant at the .05 level, and most are significant at the .001 level. This means that we must rule out chance in attempting to explain these associations.

Significant correlations do not in themselves imply cause and effect. However, in this case paint lead can be safely ruled out as being an effect of dust lead. And it is difficult to imagine a third factor that causes lead in both paint and dust. Thus, the most reasonable conclusion is that paint is one of the sources of lead in dust.

REFERENCES

1. McKnight, M.E., Byrd, W.E., and Roberts, W.E., *Measuring Lead Concentration in Paint Using a Portable Spectrum Analyzer X-Ray Fluorescence Device* (NISTIR W90–650), U.S. Department of Commerce, National Institute of Standards and Technology, Washington, D.C., 1990.
2. Shier, D.R. and Hall, W.G., Analysis of Housing Data Collected in a Lead-Based Paint Survey in Pittsburgh, Pennsylvania, Parts I and II (NBSIR 77–1250 and 77–1293), U.S. Department of Commerce, National Bureau of Standards, Washington, D.C., 1977.

---

* Separate statistics were derived based on data collected from ''wet'' rooms (rooms with plumbing) and ''dry'' rooms (rooms without plumbing). Separate analyses were done because of the possible error that could be introduced by the presence of lead pipe in ''wet'' rooms.

12

Table 9 Major Hypothesized Pathways of Lead from Paint to Dust, and Correlation Coefficients Between the Natural Logarithms of Survey Measurements of Lead for Each Pathway

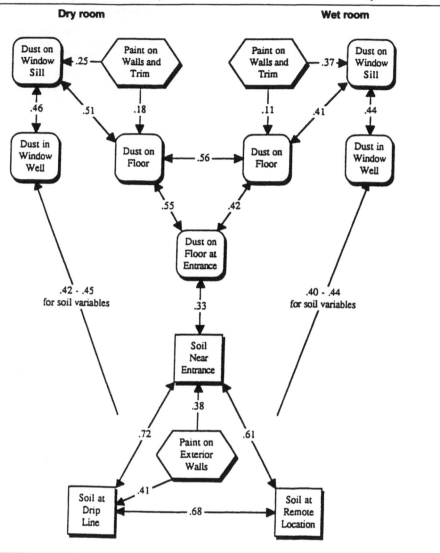

3. Hall, W. and Ayers, T., Survey Plans and Data Collection and Analysis Methodologies: Results of a Pre-Survey for the Magnitude and Extent of the Lead-Based Paint Hazard in Housing (NBSIR 74–426), U.S. Department of Commerce, National Bureau of Standards, Washington, D.C., 1974.
4. Lead-Based Paint: Report of Findings to the State Legislature (mimeo), Arizona Department of Health Services, Division of Environmental Health, Bureau of Sanitation, Phoenix, 1976.
5. Wallace, J.E., The Cost of Lead-Based Paint Abatement in Public Housing, U.S. Department of Housing and Urban Development, Washington, D.C., 1986.
6. Interim Guidance on Establishing Soil Lead Cleanup Levels at Superfund Sites (OSWER Directive #9355.4–02), U.S. Environmental Protection Agency, Washington, D.C., 1989.
7. Preventing Lead Poisoning in Children (99–2230), U.S. Department of Health and Human Services, Centers for Disease Control, 1985.

Chapter 2

Data Analysis of Lead in Soil
(HUD Survey Data)

S. F. Brown, B. D. Schultz, R. P. Clickner, and S. Weitz

CONTENTS

I. INTRODUCTION

The 1987 amendments to the Lead-Based Paint Poisoning Prevention Act required the Secretary of the Department of Housing and Urban Development (HUD) to "estimate the amount, characteristics, and regional distribution of housing in the United States that contains lead-based paint hazards at differing levels of contamination." In response to this act, HUD initiated the National Survey of Lead-Based Paint in Housing. The National Survey was a statistical sample of 381 dwelling units designed to represent the 77 million homes, in the 48 contiguous states, built before 1980. The National Survey produced a detailed, statistically sound national database on the nature and extent of lead-based paint in America.[1-3] Included in the study design, soil and house dust samples were collected from each home and analyzed for lead, to support the lead-in-paint data. Demographic information was also collected.

Because the relationships between soil lead, lead in paint, and house dust lead are not well known, the U.S. Environmental Protection Agency (EPA) was interested in thoroughly examining the soil lead data collected during the National Survey, to address the following objectives:

- To investigate the major sources of error in the National Survey soil data and to estimate the potential impact of error sources on classification bias in the estimates of the incidence of soil lead in housing;
- To statistically analyze the relationships among lead in soil, paint, and dust;
- To determine the suitability of the data for comparisons with historical, current, and future studies.

II. NATIONAL SURVEY METHODOLOGY

In the National Survey, **lead-in-paint** measurements were obtained on interior and exterior surfaces from each of the 381 sampled dwellings, using mobile X-ray fluorescence (XRF) spectrum analyzers. The analyzers measured milligrams of lead per square centimeter of painted surface (mg/cm^2).

Dust samples were collected with a vacuum sampler designed specially for this study, on floors, window wells, and window sills. Lead loading levels were determined — micrograms of lead per square foot of surface ($\mu g/ft^2$).

Measurement sites for paint and dust were selected both randomly and based on the judgment of the inspector. Dust samples were taken from two randomly selected rooms in the interior of each dwelling: one with plumbing ("wet" room) and one without plumbing ("dry" room). These rooms were also visually inspected, and the condition of painted surfaces was recorded. In addition, "purposive" XRF measurements were made on surfaces in other rooms deemed most likely to have lead-based paint, in the opinion of field technicians. Additional dust samples were collected near entryways.

Soil was collected at three locations close to each dwelling. The locations included (1) a "remote" sample on the property, intended to measure background lead from sources other than lead-based paint; (2) a "drip-line" sample near an exterior wall of each dwelling, potentially contaminated with lead from

Table 1 Percent Recovery for Soil Control Samples[a]

	Sample size	Averaged measured concentration (µg/g)	Spiked concentration (µg/g)	Recovery (percent) with 95% conf. interval	Comments
Casper	21	26.8 ± 2.6			
	18	300 ± 9	316.2	86.4% ± 3.0%	
	23	1720 ± 170	2097	80.7% ± 8.1%	Includes 1 low outlier
MRI	14	33.2 ± 1.7			
	15	336 ± 12	316.2	95.8% ± 3.8%	
	14	1985 ± 109	2097	93.1% ± 5.2%	

[a] Based on ICP-AES analysis of soil control samples.

Source: Analysis of Soil and Dust Samples for Lead (Pb), MRI, May 8, 1991.

lead-based paint; and (3) an "entrance" sample collected near the most commonly used entrance to the dwelling, to measure the potential associations with track-in lead. The soil lead measurements are reported as micrograms of lead per gram of soil (µg/g or ppm).

Composite soil samples were collected at each sampling location, made up from three soil subsamples combined into a plastic bag yielding one composite sample. Each subsample was collected by inserting a 2.9-cm-diameter metal tube-shaped corer into the ground and collecting the top 2 to 3 cm of soil. In order to consistently sample composite samples, the field technician first selected the sampling location and collected the first of three subsamples. The second subsample was taken 20 in. to the right, and the third, 20 in. to the left. Areas with run-off from potential lead sources were avoided (e.g., driveways), and samples were selected from the same side of the dwelling as the exterior XRF measurements.

Analyses of the samples were as follows: paint lead analysis was done *in situ* by XRF; soil and dust samples were sent to laboratories and analyzed by inductively coupled plasma-atomic emission spectrometry (ICP-AES) and by graphite furnace atomic absorption (GFAA) spectroscopy, respectively.

III. BIASES IN THE NATIONAL SURVEY SOIL MEASUREMENTS

In addition to the collected soil samples, spiked soils of known lead concentration were also sent to laboratories for analysis. These QA samples were "blinded" to chemists doing the analysis and served as ongoing method performance check samples (PCS). Table 1 shows the recoveries for the soil control samples. Recoveries are slightly, although significantly, below 100% and differ depending on the lab performing the measurement and on the spiking concentration. Because the soil lead concentrations were generally less than 300 µg/g, the mid-level spike was used to adjust the soil lead measurements to correct for recovery.

The soil measurements have a skewed distribution, with many low lead concentration measurements and few relatively high measurements. The distribution can be reasonably described by a lognormal distribution (i.e., the natural log of the measured lead concentrations have a normal or bell-shaped distribution). Figures 1, 2, and 3 show histograms of the log-transformed lead measurements for the entryway, drip-line, and remote soil samples, respectively. The histograms suggest that the data are slightly skewed, even after using the log transformation.

Table 2 shows descriptive statistics of the soil lead data. Three percent of the soil measurements are below the detection limit of analysis. In these cases half of the detection limit was used (less than or equal to 5 µg/g in all cases). This is expected to have no significant effect on the statistical analysis. Geometric means for the entrance, drip-line, and remote samples are 83, 72, and 47, respectively (arithmetic means are 295, 415, and 170). The remote sample is statistically different than the other two ($p < 0.001$); however, the entrance and drip-line samples are not significantly different from each other. The measurements at the three locations are all highly correlated with each other.

IV. ERROR IN THE SOIL DATA

Associated with each reported soil lead measurement is measurement variation (measurement error). As a result, the reported lead measurement will be different from the actual lead concentration in the soil sample and different from the average lead concentration in the vicinity of the sample. Although it is impossible

15

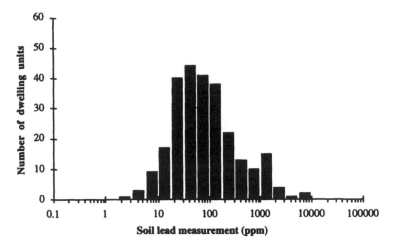

Figure 1 Distribution of the lead measurements in soil samples collected outside the dwelling unit entrance.

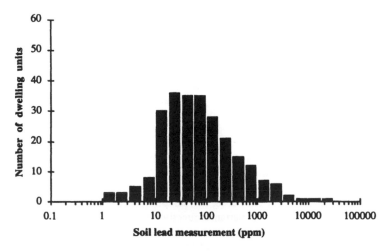

Figure 2 Distribution of the lead measurements in soil samples collected at the drip line.

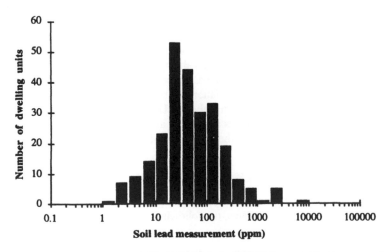

Figure 3 Distribution of the lead measurements in soil samples collected at remote locations away from the dwelling unit.

Table 2   Descriptive Statistics for Lead Measurements in Soil Samples

Set of data	Entrance samples	Drip line samples	Remote samples	Playground samples
Number of measurements	260	249	253	6
Arithmetic mean (ppm) (95% confidence interval)	295 (203–387)	415 (192–638)	170 (99–240)	19 (9–30)
Standard deviation (ppm)	753	1790	570	10
Coefficient of variation	2.55	4.31	3.35	.53
Percentiles (ppm)				
maximum	6,829	22,974	6,951	36
upper quartile	202	201	120	29
median	65	60	43	15
lower quartile	31	23	19	12
minimum	3	1	1	12
Geometric mean (ppm) (95% confidence interval)	83 (70–100)	72 (58–89)	47 (40–56)	18 (11–28)
Mean of the log-transformed measurements	4.42	4.27	3.85	2.87
Standard deviation of the log-transformed measurements	1.47	1.68	1.42	0.45

to know the difference between the measurement and the actual concentration for any one sample, it is possible to estimate the average magnitude of the measurement variation. Both the variance and the standard deviation measure the magnitude of the measurement variation.

To measure some of this variation, 10% of the soil samples collected in the field were split into two portions before being sent "blinded" to laboratories for analysis. From these measurements we can estimate the measurement variance associated with the sample preparation and measurement process at the lab (and the homogeneity of the splitting procedure). Two different laboratories analyzed soil samples in the National Survey. The variance from both labs in the log-transformed measurements are constant, independent of the lead concentration. For the lab that analyzed most of the soil samples, the variance contributed by the lab to the log-transformed measurements is 0.060. For the other lab the variance contributed by the lab to the log-transformed measurements is 0.132.

The contribution of the field sampling to the variance across an area is more difficult to estimate. Two factors contribute to the difficulty in estimating the sampling component of variance: (1) there are no measurements from which to estimate the variance directly, and (2) the variance will be a function of the size of the area over which the soil lead concentration is averaged.

Presumably, lead concentrations at locations that are close together are similar. Therefore, the variance of lead concentrations over small areas should be small. As the area over which the samples are taken increases in size, the variability of the lead concentration among randomly located samples increases. By making some assumptions we can use the ratio of the measurements at two different locations, to show the effect of distance between the samples on the variability of the measurements. To do this we must assume that (1) the sampling variability of the lead concentration is the same for the entrance, drip-line, and remote locations (an assumption often made for this type of data) and (2) that the ratio of the measurements from different locations is constant. The assumption of a constant ratio is consistent with scatter plots (Figures 4, 5, and 6) that show a linear relationship between the log-transformed soil measurements at different locations, with a slope of approximately 1.0. Using these assumptions the standard deviation of the log-transformed soil lead measurements around the average log-transformed soil lead concentration is 0.50 (the variance is therefore 0.707, and the variance of the difference between values is 1.41).

V. OUTLIERS IN THE SOIL DATA

Based on preliminary data analyses that suggested that the soil measurements differed among counties and among houses of different construction dates, the following three models were used to identify possible outliers:

- Fit a mean to the log-transformed entrance, drip-line, and remote measurements and assume that the residuals have a normal distribution (this is equivalent to assuming that the measurements have a log normal distribution).

- Fit a mean to the log-transformed entrance, drip-line, and remote measurements and assume that the residuals have a multivariate normal distribution. Assume that the cube root of the Mahalanobis distance for each dwelling unit has a normal distribution.
- Fit a two-way analysis of variance model with factors for vintage and county to the log-transformed entrance, drip-line, and remote measurements and assume that the residuals have a multivariate normal distribution. Assume that the cube root of the Mahalanobis distance for each dwelling unit has a normal distribution.

Although three dwelling units had somewhat unusual measurements based on visual inspection, no dwelling unit or individual measurement was classified as an outlier, using the Extreme Studentized Deviate

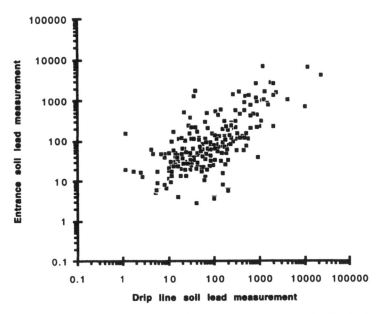

Figure 4 Plot of soil lead measurements at the entrance location vs. at the drip line location.

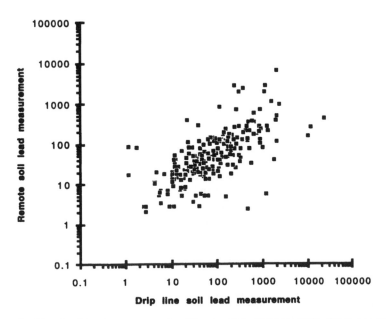

Figure 5 Plot of soil lead measurements at the remote location vs. at the drip line location.

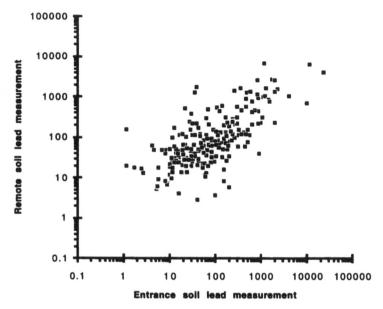

Figure 6 Plot of soil lead measurements at the remote location vs. at the entrance location.

Test. We checked the records for the dwelling unit with the largest Mahalanobis distance. The only indication of an unusual situation was that the soil samples were taken from frozen ground. Because frozen ground was encountered in other dwelling units in the same county and those dwelling units did not have unusual soil lead measurements, there was no rationale for questioning the data.

After review of the data we decided that no observations could be clearly identified as outliers and that the most extreme observations showed no identifiable problems. Therefore, we used all of the soil measurements in the subsequent analyses and found no reason to recommend excluding any soil measurements from other possible analyses.

Analysis of classification bias, using the soil measurements to classify a dwelling unit as having high soil levels ("high" is defined as greater than 500 ppm), looks at three scenarios: (1) classification based on one measurement; (2) classification based on the maximum of three measurements; and (3) classification based on the average of three measurements. Recommendations as to which procedure is appropriate depend primarily on what criteria are believed to be relevant. For example, if a child is assumed to play in the soil at different locations, with equal frequency, the average of soil lead concentrations across a site makes sense and would provide a relatively unbiased classification procedure. However, if interior dust is thought to be the major pathway for child lead exposure, then using the entrance sample for classification is more relevant, assuming contaminated soil is tracked into the dwelling. It follows that if the maximum exposure is thought to be important and avoidable, classification based on the maximum concentration is reasonable.

VI. CORRELATIONS BETWEEN SOIL, PAINT, AND DUST

Several variables are found to be statistically significant predictors of soil lead, in the analyses completed to date. Of these the average paint lead loading and the dwelling unit age stand out. Assuming that the model is correct, the parameters for the paint lead measurements suggest that paint lead has contributed 11 to 15% of the lead in the soil. The parameter estimates for dwelling unit age indicate that the soil lead concentrations increase significantly as the dwelling unit age increases. Soil lead concentrations are also correlated with local motor vehicle traffic volumes. The regression results do not help to identify other factors that might contribute to the lead in soil or to identify factors that might be correlated with dwelling unit age.

Correlations are often used to identify relationships between variables. If one variable causes, or directly affects, another in a linear manner, the two variables will be correlated. Although causation usually implies correlation, correlation does not imply a causal relationship. Even if there is a causal relationship, correlations cannot be used to determine which variable is the cause and which the effect. In many cases significant correlations are associated with a third, perhaps unmeasured, variable that affects the two correlated variables.

For example, if soil and dust lead concentrations are correlated, this may be due to either (1) dust lead comes from soil, (2) soil lead comes from dust, (3) the lead in soil and dust comes from a third source such as exterior paint or automobile exhaust, or (4) some combination of these factors.

Significant predictors for floor dust loading are soil lead measurements, paint lead measurements, percent paint damage, number of rooms with plumbing, and obviously total dust weight sampled. The parameter estimates are consistent with the hypothesis that soil contributes more lead to floor dust than does undamaged lead paint. Paint damage is associated with a significant and large increase in the dust lead loading. The parameter estimates are also consistent with the hypothesis that lead in dust at the entrance to the dwelling unit comes almost exclusively from exterior soil sources.

Parameter estimates for window sill dust suggest that soil contributes more lead than does lead paint, but do not support any one explanation as to which sources of paint lead contribute the most to window sill-dust lead loading. For window well loading, total dust weight was the only significant predictor. The other parameter estimates provide little insight into sources of lead in window wells.

VII. CONCLUSIONS

The National Survey of Lead-Based Paint in Housing generated a large database of soil, dust, and paint lead measurements from a representative snapshot of "average" U.S. housing. This report statistically defines the soil lead data by measuring errors, biases, correlations, and trends of the data within itself and among other parameters. The major findings are summarized below:

- Analytical recoveries for the lead in soil are significantly less than 100%.
- The soil lead measurements can be defined by a lognormal distribution.
- No observations can be clearly identified as outliers.
- Significant predictors of soil lead are average paint lead loading, dwelling age, and local traffic volume.
- Significant predictors of floor dust lead are soil lead, paint lead, percent paint damage, number of rooms with plumbing, and total dust weight collected.
- Significant predictors of window sill dust lead loading are soil and paint lead, and total dust collected.
- Significant predictor of window well dust lead loading is only total dust collected.

The EPA analysis of this database is still ongoing, and additional inputs (e.g., from census data) are being examined for possible further insight into the interpretation of results.[4]

REFERENCES

1. Comprehensive and Workable Plan for the Abatement of Lead-Based Paint in Privately Owned Housing. Report to Congress. U.S. Department of Housing and Urban Development, Washington, D.C., December 7, 1990.
2. Midwest Research Institute, Analysis of Soil and Dust Samples for Lead (Pb) (Final Report). EPA Contract No. 68–02–4252, U.S. Environmental Protection Agency and the U.S. Department of Housing and Urban Development, Washington, D.C., 1991.
3. Westat, Inc., Comprehensive Technical Report on the National Survey of Lead-Based Paint in Housing, EPA Contract No. 68–D9–0714, U.S. Environmental Protection Agency, Washington, D.C.
4. Data Analysis of Lead in Soil and Dust, U.S. EPA report, EPA 747-R-93-011, September 1993.

Chapter 3

National Survey of Lead Paint in Housing: Analysis of Error Sources

J. W. Rogers, R. P. Clickner, and M. Chen

CONTENTS

I. INTRODUCTION

The National Survey of Lead-Based Paint in Housing was conducted by Westat, Inc., under contract to the U.S. Department of Housing and Urban Development (HUD). In that sample survey of 381 housing units, a spectrum analyzer X-ray fluorescence (XRF) instrument was used to measure the lead content of painted surfaces in the sampled homes. One primary objective of the survey was to estimate the percentage of continental-U.S. dwelling units with lead-based paint. A dwelling unit is classified as having lead-based paint if any surface within the dwelling unit has an average paint lead concentration greater than 1.0 mg/cm^2. The determination of whether the average paint lead concentration across a surface is greater than 1.0 mg/cm^2 is subject to two sources of error: (1) XRF instrument bias, and (2) classification bias due to XRF instrument measurement error. The classification of a dwelling unit as having lead-based paint is subject to several additional error sources: (3) sampling error due to the random sampling of housing units, (4) sampling error due to the random selection of rooms and painted components within rooms, for the XRF measurements; and (5) classification bias due to the incomplete sampling of rooms. Failure to take these sources of error into account can produce biased prevalence estimates and underestimates of the confidence interval width. This paper focuses on the effects of XRF bias and measurement error on the classification of surfaces. The error sources affecting only the classification of dwelling units and the percentage of dwelling units with lead-based paint are discussed briefly. All of the sources of error are discussed in the complete report on the survey.

II. DEFINITIONS

The term "measurement error" or "error" refers to the difference between the estimate and the actual value being estimated: for example, (1) the difference between the XRF reading and the "true" or actual lead concentration on the surface being measured; or (2) the difference between the estimated proportion of surfaces with paint lead concentrations above 1.0 mg/cm^2 and the actual proportion. If the measurement error is greater than zero, then the estimate is greater than the value being estimated, and vice versa. The presence of error does not mean that the measurement was improperly made, but that there is a non-zero difference (usually assumed to be random) between an individual measurement and the quantity being estimated.

The term "bias" refers to the average measurement error or the average difference between the estimates of a quantity and the actual quantity being estimated. If the average measurement error is zero, then the estimate is said to be unbiased and individual estimates are sometimes greater than, and sometimes less than, the quantity being estimated. Using an archery analogy, an archer whose arrows all fall around the center of the target is unbiased. If the average measurement error is other than zero, the measurements are said to be biased. An archer whose arrows tend to fall on one edge of the target is biased. While the cause of biased measurements can come from several factors, an important factor relevant to this study is censoring. Censoring in this report refers to observations that are less than a minimum value and are recorded

1-56670-113-9/95/$0.00+$.50

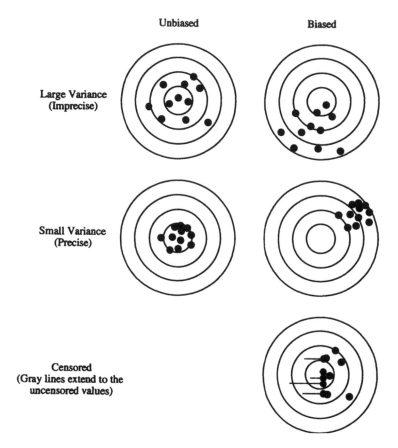

Figure 1 Illustrations of the definitions of bias, precision, and censoring.

as being at the minimum value. The XRF instruments used in the national survey never reported a reading below zero, even though readings below zero were possible due to instrument bias and measurement error. These XRF readings, which would have been below zero, but which were recorded as zero, are referred to as censored XRF readings and contribute to bias.

The measurement variance, standard deviation, and standard error measure the magnitude of the measurement errors around their average. Measurements with a small variance are said to be precise. A set of measurements can be both precise and biased. Using the archery analogy again, if all the arrows land in a tight pattern, the shots have a small variance, and the archer's aim is precise. Alternatively, if the arrows fall in a wide pattern, the shots have a large variance, and the archer's aim is imprecise. Figure 1 shows the concepts of precision, bias, and censoring using the archery analogy.

The percentage of dwelling units with lead-based paint depends on the definition by which a dwelling unit is classified as having lead-based paint. The definition used here classifies a dwelling unit as having lead-based paint if any surface has lead-based paint. A surface is classified as having lead-based paint if the average lead concentration across the surface is 1.0 mg/cm^2, or greater. Each sampled surface was classified based on one XRF reading using a spectrum analyzer XRF instrument calibrated to measure lead concentration as milligrams of lead per square centimeter (mg/cm^2).

Due to either bias or measurement error in the XRF reading, the classification of a surface as having, or not having, lead-based paint may not be correct. This is called "classification bias." Note that the estimated proportion of surfaces with lead-based paint may be biased, even though it is based on XRF measurements that are unbiased. Table 1 shows the qualitative effect of XRF bias and measurement error on the classification of surfaces. The difference between the XRF reading and the true lead concentrations will either (1) not affect the classification of a surface, or (2) result in one of two types of misclassification shown in Table 1.

Table 1 Effect of Measurement Error on the Classification of Surfaces as Having or Not Having Lead-Based Paint

Classification of the surface based on the XRF measurement	True classification of the surface	
	Has no lead-based paint	Has lead-based paint
Has no lead-based paint (XRF measurement is less than 1.0 mg/cm²)	Correct classification	Incorrect classification based on the XRF measurement. Results in too few surfaces or dwelling units classified as having lead-based paint
Has lead-based paint (XRF measurement is greater than or equal to 1.0 mg/cm²)	Incorrect classification based on the XRF measurement. Results in too many surfaces or dwelling units classified as having lead-based paint	Correct classification

III. XRF INSTRUMENT BIAS

The spectrum analyzer XRF instruments used in this study to measure paint lead concentrations tends to yield systematically biased measurements. The measurement bias was evaluated using XRF readings on surfaces with known paint lead concentrations. The bias was then subtracted from the XRF readings, to estimate the paint lead concentrations on the surfaces sampled in the national survey.

Validation measurements were taken throughout the survey field period, to routinely monitor the XRF performance. The validation measurements used cardboard shims painted with known lead concentrations. Two shims were used: one with 0.6 mg/cm² lead and the other with 2.99 mg/cm² lead. For simplicity, these two shims will be referred to as the 0.6 shim and the 2.99 shim. For each measurement the shims were placed on one of four substrate samples of either wood, steel, drywall, or concrete. In all, eight measurements, one for each combination of substrate and shim, were taken in the morning and evening on each day of field sampling. In addition, as part of the field data, the survey technicians recorded the substrate for each XRF reading taken in the survey dwelling units. After the field period these validation measurements were used to determine the bias in the XRF readings, and the factors that affected the bias. Using regression analysis on the data with less than 20% censored observations, it was determined that the bias in the XRF readings depended on the lead concentration in the shim, the substrate material, the instrument used to make the measurements, and time (there was a small drift in the XRF readings over time). For some instruments and substrates the XRF readings were generally greater than the paint lead concentrations; for others, they were less.

Based on the statistical analysis, it was decided to calculate the bias separately for each combination of instrument and substrate. It was also necessary to assume that there was a linear relationship between the paint lead concentration and the XRF reading. This assumption could have been evaluated if there had been at least three different shims used for the validation measurements. If the data had not been censored, regression analysis could have been used to estimate the bias or, equivalently, the relationship between the paint lead concentrations and the XRF readings. Because some of the data were censored, the procedure described below was used to model the relationship between the XRF readings and the paint lead concentration. This procedure includes substituting the median for the mean if more than 20% of the XRF readings were zero and adjusting for the slight drift in the readings over time.

For each instrument and substrate:

1. Determine the percent of the XRF readings, for the 0.6 shim and the 2.99 shim, that are zero (and possibly censored).
2. Calculate the expected XRF reading corresponding to February 1, 1990 (an arbitrary date toward the beginning of the field period) by adjusting for the drift over time. Call this quantity ModXRF. The constant in Equation 1 was determined from the regression analysis.

$$ModXRF = XRF \ reading + 0.00099 \times (days \ since \ 2/1/90) \qquad (Eq. \ 1)$$

3. For each shim, if the percentage of zeros (in the original data) is less the 20%, calculate the mean of ModXRF; otherwise calculate the median of ModXRF. This is the central tendency.

4. Calculate the slope (β) and intercept (α) for the line through the central tendency for the 0.6 and 2.99 shims. The resulting equation for the XRF reading in terms of the lead concentration is

$$\text{XRF reading} = \alpha + \beta \times \text{lead concentration} - 0.00099 \times (\text{days since } 2/1/90) \qquad \text{(Eq. 2)}$$

Using algebra, this equation can be rewritten to express the expected lead concentration corresponding to an XRF reading, as follows:

$$\text{lead concentration} = \frac{-\alpha}{\beta} + \frac{\text{XRF reading}}{\beta} + \frac{0.00099}{\beta} \times (\text{days since } 2/1/90) \qquad \text{(Eq. 3)}$$

Assuming that the same relationship between the XRF readings and the lead concentration holds for the validation data and the survey field data, Equation 3 was used to calculate the expected lead concentration corresponding to each XRF reading made with the same XRF instrument and on the same substrate. For simplicity here, the term "XRF reading" refers to the XRF reading recorded in the field. The term "XRF measurement" refers to the adjusted XRF reading after correcting for bias in the XRF instrument.

Because the true lead concentration cannot be less than zero, the equation for the XRF measurement becomes

$$(\text{XRF measurement}) = \text{Max} \begin{cases} \dfrac{-\alpha}{\beta} + \dfrac{\text{XRF reading}}{\beta} + \dfrac{0.00099}{\beta} \times (\text{days since } 2/1/90) \\ 0 \end{cases} \qquad \text{(Eq. 4)}$$

The bias determined from the validation data was used to correct the field measurements for each combination of XRF instrument and substrate. Figures 2 through 4 show example plots of the validation data vs. the shim lead concentration. The readings for each shim are shown as a sideways histogram with the base at the shim lead concentration. The relationship between the XRF reading (in the middle of the field period) and the lead concentration is shown as a solid line. The dotted line shows the relationship when there is no bias in the instrument.

Figure 2 shows the validation data for one instrument on a wood substrate. For these measurements the instrument has a small positive bias at lead concentrations above 2.0, and a small negative bias for lead concentrations below 2.0 mg/cm^2.

Figure 3 shows the validation data for one instrument on a steel substrate. For the 0.6 shim the XRF readings are biased, consistently greater than the lead concentration in the shim. For the 2.99 shim the XRF readings are very close to the lead concentration.

Figure 4 shows the validation data for one instrument on a concrete substrate. For the 0.6 shim the XRF readings are consistently less than the lead concentration in the shim. In fact, all but one XRF reading is zero. For the 2.99 shim the XRF readings are less than the lead concentration in the shim. The variance of the XRF readings on the 2.99 shim is greater on concrete than on the steel substrate. One would expect that the measurements on the 0.6 shim would vary around a reading of 0.6 mg/cm^2. However, because the XRF instrument is biased such that the reading is less than the lead concentration and because the instrument was programmed to present a zero reading in place of a negative reading, all but one of the XRF readings for the 0.6 shim are zero. Statisticians call this type of data censored. For these censored readings, we know that the reading would have been less than zero, but we do not know how much less.

These three examples illustrate the general finding that the spectrum analyzer XRF instruments are fairly accurate on wood and drywall, but tend to overestimate low concentrations on steel and underestimate low concentrations on concrete.

IV. XRF INSTRUMENT CENSORING

The XRF instrument uses a complicated formula (considered proprietary by the manufacturer) to calculate the XRF reading from the energy spectrum measured by the spectrum analyzer. Due to the random nature of the radiation source and detection, the XRF reading is subject to error. Even if there is no bias in the instrument, multiple XRF readings will sometimes be greater than the paint lead concentration and sometimes less. On surfaces with low paint lead concentrations, the XRF readings that might be calculated to be less than zero were reported as zero by the XRF instrument, thus resulting in censored data.

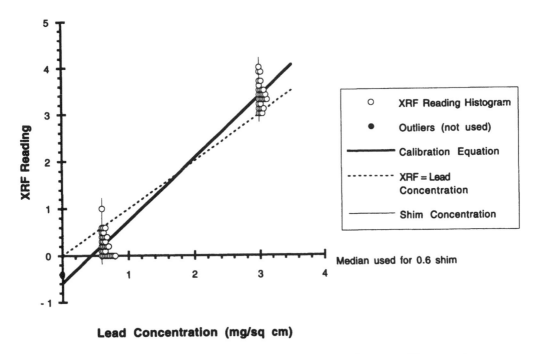

Figure 2 Validation XRF readings vs. lead concentration for instrument #39 on wood.

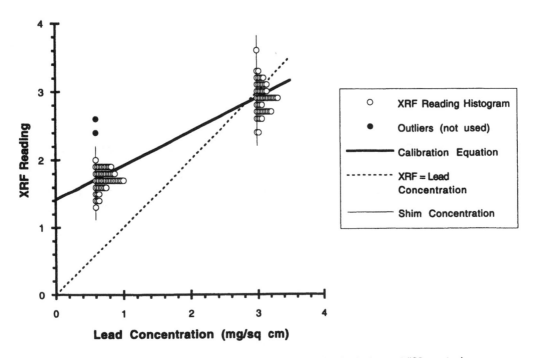

Figure 3 Validation XRF readings vs. lead concentration for instrument #38 on steel.

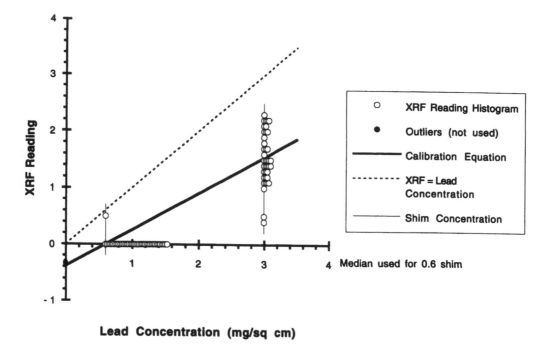

Figure 4   Validation XRF readings vs. lead concentration for instrument #36 on concrete.

In the survey, censoring affected primarily the low readings on concrete substrates. Since this group of readings comprised only a small proportion of all XRF readings, the affect of censoring on the national estimates is expected to be small. However, it should be noted that the spectrum analyzer XRF instrument can give biased or incorrect XRF readings at low lead concentrations, due to censoring. Adjusting for the XRF bias does not correct for the effect of censoring.

Although not used in the national survey, one approach to correct for censoring in the field is the following:

1. If the XRF reading is zero, insert a shim with a known lead concentration between the instrument and the surface being measured and take a second reading at the same location;
2. Estimate the bias in the XRF instrument, using samples of substrates as in the national survey;
3. Correct the second XRF reading, which used the shim for the instrument bias;
4. Subtract the known lead concentration in the shim from the corrected XRF measurement.

V. CLASSIFICATION BIAS DUE TO XRF MEASUREMENT VARIATION

The XRF instrument measurement error can induce a bias in the estimated prevalence of surfaces and housing units with lead-based paint. Due to this measurement error, the XRF reading on a surface with paint lead concentration less than 1.0 mg/cm^2 may be greater than 1.0, or vice versa, resulting in a misclassification (see Table 1). If the number of surfaces that are misclassified as having lead-based paint is different than the number of surfaces misclassified as not having lead-based paint, the estimated percentage of surfaces in a dwelling unit with lead-based paint will be biased. The amount of the bias depends on the variance of the measurement errors relative to the variance of the measurements.

An analysis of the data suggested that the XRF measurements can be reasonably described by a log-normal distribution. The following model for the XRF measurements was used to calculate the approximate classification bias on the estimated proportion of surfaces with lead-based paint:

$$Y = X + e \qquad \text{(Eq. 5)}$$

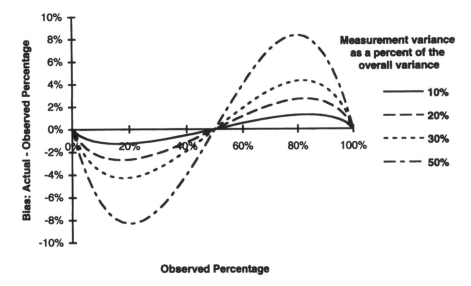

Figure 5 Misclassification due to measurement error: bias in the observed percent as a function of both the observed percent and the relative measurement variance.

where:

Y = log-transformed XRF measurement at a randomly selected location on the measured surface;
X = log-transformed geometric mean paint lead concentration across the surface, assumed to be normally-distributed with variance σ_x^2;
e = the measurement error in the log-transformed scale, assumed to be normally distributed with mean zero and variance σ_e^2.

For ease of calculation, the model uses the geometric mean, rather than the arithmetic mean, across the measured surface. Unless the measurement variance (σ_e^2) is larger than about 0.50, use of either the geometric or arithmetic average will produce similar results. Figure 5 shows the bias in the percentage of surfaces with lead-based paint, as a function of both the observed percentage and the relative magnitude of the measurement error (i.e., $\sigma_e^2 / (\sigma_e^2 + \sigma_x^2)$).

When estimating the average lead paint concentration across a surface, the measurement error σ_e^2 is the sum of the random measurement error in the XRF instrument and the error due to random selection of the location where the XRF reading was taken within the surface. Combining both of these sources of error, the standard deviation of the measurement error in the log-transformed scale was estimated to be roughly 0.34. The relative magnitude of the measurement variance to the variance of the XRF measurements depends on the set of surfaces being classified. If the objective is to estimate the percentage of surfaces within a dwelling unit with lead-based paint, the variance of the measurement error may be from 20 to 50% the variance of the XRF measurements. Therefore, for estimating the proportion of surfaces in a dwelling unit with lead-based paint, the estimate can be either low or high, by as much as 8%, due to the measurement error of the XRF instrument and the variation in the lead concentration within a surface (see Figure 5). If the objective is to estimate the proportion of surfaces nationally that have lead-based paint, the measurement error variance may only be 5% of the variance of the XRF measurements across surfaces nationally; in which case the bias in the estimate is less than 1%.

For the national survey, dwelling units were classified as having lead-based paint if any surface in the dwelling unit had lead-based paint (i.e., had an average paint lead concentration greater than or equal to 1.0 mg/cm^2). For classifying dwelling units instead of surfaces, calculating the bias in the national estimate is more difficult because the number of surfaces and the concentrations on the surfaces are different in each dwelling unit. Simulations used to estimate the proportion of dwelling units with lead-based paint, after adjusting for the measurement variance, are discussed briefly in the next section. Table 2 summarizes the bias for different estimates.

Table 2 Summary of the Effect of Measurement Error on the Bias of the Estimated Percentage of Surfaces and Dwelling Units Classified as Having Lead-Based Paint

Bias = estimated percentage − true percentage	Estimation of:	
	Percentage of surfaces	Percentage of dwelling units
Classification within: The continental U.S.	−1 to 1 Depends on the percentage	Estimated for the national survey using simulations. Results are being reviewed.
One dwelling unit	−8 to 8 Depends on the percentage	Not applicable

VI. CLASSIFICATION BIAS DUE TO WITHIN-DWELLING-UNIT SAMPLING PROCEDURES

Within each dwelling unit, XRF measurements were made on randomly selected surfaces within one randomly selected wet room and one randomly selected dry room (i.e., room with plumbing and room without plumbing, respectively). However, the classification of a dwelling unit as having lead-based paint depends on the lead concentration on all surfaces within a dwelling unit. Using only the XRF measurements on the measured surfaces, a dwelling unit might be misclassified as not having lead-based paint when it actually did, because the surfaces with lead-based paint were not among those that were randomly selected for measurement.

Although the survey procedures attempted to identify and measure those surfaces with the highest lead concentrations, the effort was not completely effective. As a result it was necessary to develop a mathematical simulation model, based on data from the survey and other sources, to predict the prevalence of housing units with lead-based paint. The model was used to correct for the classification bias due to incomplete sampling of rooms and surfaces within a dwelling unit and due to measurement variation discussed above. Extending the results from the measured surfaces to all surfaces in the unit was based on (1) data on the number of rooms in the unit, (2) data on the number of surfaces per room, and (3) assumptions about the relationship of the lead concentrations on unmeasured surfaces to those on the sampled and measured surfaces. The classification of dwelling units as having, or not having, lead-based paint can be quite biased when only a few surfaces within the dwelling unit are measured.

While it is not the purpose of this paper to present the details of the model, important aspects of the model include

1. With the exception of those dwelling units with very low paint lead concentrations, the geometric mean paint lead concentration in a dwelling unit can be described by a lognormal distribution. The characteristics of the distribution depend on the age (or vintage) of the dwelling unit. As might be expected among older dwelling units, the mean paint lead concentrations are both higher and more variable than in newer dwelling units.
2. The lead concentrations on surfaces within the dwelling unit are assumed to have a lognormal distribution around the geometric mean paint lead concentration for the home.

Figure 6 shows the distribution of the exterior paint lead concentrations for the survey homes, by the age (or vintage) of the home. As can be seen in Figure 6, lead concentrations increase as the age of the dwelling unit increases.

VII. SUMMARY AND RECOMMENDATIONS

Both XRF instrument bias and measurement error can result in biased estimates of the percentage of surfaces or dwelling units with lead-based paint. The direction and magnitude of the bias depends on (1) the measurement bias of the XRF instrument, (2) the relative variance of the measurement error and the XRF measurements, and (3) the actual proportion of surfaces or dwelling units with lead-based paint being measured. The spectrum analyzer XRF instrument has measurement bias that depends on the substrate below the paint surface, the XRF instrument, and the lead concentration in the paint. Correcting for the bias does not adequately correct for the censoring at low lead concentrations on some substrates. When estimating the proportion of surfaces or dwelling units with lead-based paint, variation associated with the measurement instrument, and variation in lead concentration across the surfaces being measured, cannot

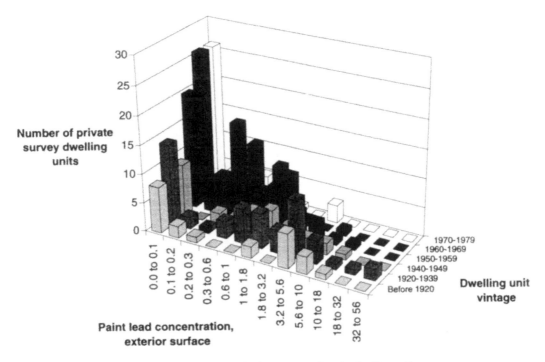

Figure 6 Exterior lead-based paint concentrations by dwelling unit age.

be ignored. All of these factors affect the estimates of the incidence of lead-based paint in housing. Adjustments for these sources of bias have been made for the National Survey of Lead Paint in Housing and are currently being reviewed.

Recommended procedures for similar surveys include

1. Use shims with several known paint lead concentrations to (1) monitor the operation of the XRF instrument, (2) adjust for the bias in the instrument, and (3) estimate the measurement variance associated with the XRF instrument.
2. Develop procedures for correcting for the censoring when it occurs, by possibly making a second measurement with a lead-containing shim between the instrument and the surface being measured.
3. Take multiple XRF readings at random locations on several surfaces, to estimate the measurement error variance associated with differences across the surface being measured.

Chapter 4

The HUD Lead-Based Paint Abatement Demonstration in Public Housing

R. F. Eberle

CONTENTS

I. BACKGROUND

The Lead-Based Paint Abatement Demonstration was designed by the Department of Housing and Urban Development (HUD) to have two components. The first component, the Federal Housing Administration Demonstration (FHA Demonstration), was a lead-based paint abatement demonstration in 172 vacant single-family housing units owned by the FHA as a result of foreclosure action. Findings from this component of the demonstration were published in August 1991.[1]

The FHA Demonstration utilized six methods of lead-based paint abatement: (1) encapsulation, (2) enclosure, (3) abrasive removal, (4) hand-scraping with a heat gun, (5) chemical removal, and (6) removal and replacement. This FHA Demonstration was designed to achieve three major objectives[2] with regard to lead-based paint abatement:

- To estimate the comparative costs of alternative methods of abatement
- To assess the efficacy of alternative methods of abatement
- To confirm the adequacy of worker protection safeguards during abatement

The findings[3] of the FHA Demonstration, as they relate to the objectives, were that encapsulation (sealing surfaces with durable coatings) was the least expensive method of lead-based paint abatement and generated the lowest levels of airborne lead dust during abatement. Enclosure (covering surfaces with durable materials, e.g., gypsum board) was also found to be a low lead-dust generation method, although it is more costly than encapsulation and cannot be used as extensively. The report does caution that more study is necessary because it is not known how long or effective these methods will remain over time.

Of the methods where lead-based paint was removed, hand-scraping with a heat gun generated more airborne lead dust than did the other abatement methods. Chemical removal was typically more costly than were other methods, required more worker protection than did other methods, and generated more hazardous wastes than did other methods. However, depending on worker skills, chemical removal can be used on a wide variety of substrates.

Replacement (removal of lead-containing substrate, e.g.,. baseboard) was found to be the most promising of the removal methods in almost all circumstances. This method generated relatively little airborne lead dust, although it did produce bulk hazardous waste.

II. THE PUBLIC HOUSING AUTHORITY DEMONSTRATION

The nature of the FHA Demonstration did not create many opportunities for abrasive (blasting, grinding, or sanding methods) removal. This was due to the random selection of the abatement method and the fact that many building substrates containing lead-based paint were not compatible to this type of removal.

The second component to be described in this presentation is the Public Housing Authority demonstration (PHA Demonstration) of lead-based paint abatement in 109 units of multifamily public housing. These units are owned by PHAs in Albany, NY, Cambridge, MA, and Omaha, NE. In each case the units are scheduled for modernization under the Comprehensive Improvement Assistance Program (CIAP), and they will all have been vacated prior to abatement. The CIAP is a funding mechanism by which HUD grants PHAs money to upgrade existing public housing. Each of the three PHAs has entered into a contract with

Dewberry and Davis: (1) to perform field testing for lead hazards; (2) to develop an abatement plan that is consistent with modernization requirements, but research oriented; (3) to assist in contractor solicitation and training; (4) to monitor and collect data during construction; and (5) to report on the findings of the research on each site.

The PHA Demonstration field activities are currently in progress. Reports will be issued to HUD on a city-by-city basis as the data is collected and analyzed. The Omaha report was released in spring of 1993. The other two city reports and an overall project report were completed in 1994.

The FHA Demonstration examined strictly the abatement activities mentioned above. No effort or work, except to secure a structure, was made to repair or upgrade a structure, beyond abating lead-based paint. In contrast, the PHA demonstration allowed us to examine the problems, and opportunities, that arise when lead-based-paint abatement and modernization are carried out together. In particular, the PHA Demonstration provided much more opportunity to examine the problems associated with removal and demolition of building components that would be replaced under modernization. On each site the PHAs elected to separate the work into two phases: one protected and one unprotected. The protected phase was the actual abatement where workers were trained in lead-based-paint hazards and where full worker protection against lead-based paint hazards was employed, while the unprotected phase was the construction that followed. When abatement was complete, final clearance wipes were taken in accordance with HUD "Interim Guidelines for Hazard Identification and Abatement in Public and Indian Housing"[4] (Guidelines), and when the work area passed final clearance, construction contractors completed the modernization work (unprotected phase).

Final clearance wipes are wipe samples taken on floors, window sills, and window wells of an abated area. These wipe samples are sent to a laboratory for analysis of lead content. Wipe samples must be taken in a known area, and results are reported in micrograms (of lead) per square foot. The Guidelines recommend that an area has passed final clearance if the wipe results are less than, or equal to, 200 $\mu g/ft^2$ for floors, 500 $\mu g/ft^2$ for window sills, and 800 $\mu g/ft^2$ for window wells.[5] It should be understood that this clearance procedure only documents how clean an area is or has been made after abatement; it has no direct relationship to the quality of the abatement work or any remaining intact lead-based paint. A thorough visual examination is required to assure these factors.

Housing authorities often are required to reconfigure interior partition walls during modernization. Removal of lead-based paint on walls must be performed during the protected phase, to allow demolition of any remaining walls during the unprotected phase. Otherwise, all demolition must be performed with worker protection in place. The primary method used to remove this paint during the PHA Demonstration was to chisel paint from plaster walls. Generally, a layer of the plaster came off with the paint. Wet methods were always used, and air monitoring indicated lead dust was below the action level. Air monitoring of both workers in their breathing zones and adjacent work areas was performed during the PHA Demonstration. Air monitoring involves drawing or sucking a known volume (liters) of ambient air through a filter. The filter is then analyzed for lead content by a laboratory, and a result is reported in micrograms per cubic meter.

HUD has adopted the following Occupational Safety and Health Administrations (OSHA) regulations for occupational exposure to lead.[6] These regulations cite an action level of 30 $\mu g/m^3$ of air and a permissible exposure limit (PEL) of 50 $\mu g/m^3$ of air, both averaged over an 8-h period. The action level is that level where employees must be enrolled in a medical surveillance program (blood-lead testing). The PEL is that level where respirators must be worn. If engineering controls cannot reduce levels below the PEL, then they must be supplemented by the use of respiratory protection. It should be noted that the PHA Demonstration by specification required the full use of worker protection at all times, to include respirators and engineering controls, without regard to air-monitoring data.

Toxic Characteristic Leaching Procedure[7] (TCLP) testing was performed on the resultant plaster debris. The TCLP test always determined that this debris was below the hazardous waste threshold of five parts per million (ppm); therefore, the debris was disposed of in a construction debris–approved landfill. (The TCLP test will be discussed later in this paper.)

Many PHA units have metal door frames that were fitted into openings during wall construction. This type of construction makes it nearly impossible to remove the frames without cutting through the wall. Although other methods were available, the demonstration in Cambridge used the needlegun to remove paint from door frames in a building. The FHA Demonstration did not present many opportunities to test the needlegun, due to limitations of the substrates (as the needlegun seems to work best on metal surfaces). These metal door frames presented an excellent opportunity to use this method in the demonstration.

A needlegun looks similar to a machine gun. It is pneumatically operated at approximately 110 psi, and about 30 3-mm rods simply hammer the surface. A local high-efficiency particulate air (HEPA) vacuum is attached to the needlegun, which collects lead dust. Worker protection for noise is critical, as well as for eye and respiratory protection. Powered air-purifying respirators were used for respiratory protection and are recommended. Benefits of this system include a reduced waste stream (only paint chips or dust) and a surface that is slightly roughened and ready for painting. The significance of a reduced waste stream is that the resultant paint chip and dust debris must almost always be disposed of as hazardous waste. Less volume or weight equates to less disposal cost.

The PHA Demonstration also looked at the integration of lead-based paint abatement with planned renovation activity under CIAP. This involved the phased sequencing of various construction activities, to include asbestos abatement. One housing authority utilized in-house labor (force account), and the others used abatement contractors. The use of force account labor allowed the abatement activity or method to change without the need for change orders. This allowed the opportunity for the PHA to adjust a sequencing for abatement based on field conditions, at minimal cost and delay. The PHA did have to accept more liability and purchase abatement equipment; however, the work force was easy to train and quickly became skilled at lead-based paint abatement.

The "force account" work force employed by the Housing Authority was obtained from local construction trade unions. This work force was under the direct supervision of full-time housing authority staff. These workers were trained in lead hazard awareness, as per the Guidelines,[8] by an industrial hygienist. In addition, a Dewberry and Davis field inspector provided additional lead-based paint abatement training. The commitment of resources by the Housing Authority and quality work performed by the work force were key to the undertaking and success of this project.

Deleading contractors possessed different advantages/disadvantages. They were, for the most part, already trained, owned this equipment, and carried their own liability insurance. Changes in the work process or sequence had to be more carefully analyzed in light of changed order costs. This is significant with respect to the recommended pilot abatement project,[9] as stated in the Guidelines, and the fact that data observed during the pilot abatement can be used to change the scope of the project.

Other elements of the demonstration include the extensive collection and analysis of air and clearance dust wipes samples, an examination of the extent to which lead hazards can be contained within dwelling units, and an assessment of the feasibility of lead-based paint abatement of lead hazards to the 0.06% by weight standard (or 600 ppm) level in selected units.

The 0.06% by weight standard was established, by the Consumer Product Safety Commission (CPSC), as the allowable maximum lead level in new paint. The Stewart B. McKinney Homeless Amendments Act of 1989 (PL100–628, November 7, 1988) required HUD to determine whether testing and abatement to such a standard is scientifically and practically feasible.[10]

The guidelines require abatement of lead-based paint hazards at a level of 1.0 mg/cm^2 and prescribe an abatement at 0.5% by weight (5000 ppm).[11] The machines used to detect levels of lead paint, X-ray fluorescence (XRF) detectors, measure paint in milligrams per square centimeter. Paint chip analysis performed in a laboratory by atomic absorption spectroscopy (AAS) can be reported in either milligrams per square centimeter (with area of paint chip) or percent by weight. It is important to understand that milligrams per square centimeter and percent by weight (ppm) are not directly comparable. Therefore, the XRF cannot be used for the 0.06% feasibility study, nor can the 1.0-mg/cm^2 abatement threshold be directly compared to the 0.06%-by-weight CPSC standard or the 5000%-by-weight abatement threshold. This work must be performed by laboratory analysis, with results reported in percent by weight.

Throughout the abatement process, personal air samples were obtained on all workers in all three cities. These data, for the most part, were full shift worker exposure. Data on worker tasks were collected hourly, which permits examination of the relationship between worker exposure and worker tasks. The results and data analysis are ongoing and will be provided in the final report to Congress.

Another objective of the demonstration was to examine the relationships between abatement methods, as compared to pass/fail rates of clearance wipe tests. Preliminary data analysis indicates that wipe cleanup on the floors was more difficult to achieve where caustic paint removal was performed than where other types of abatement methods were performed.

Another major component of the demonstration was an initial experiment in all three cities, to determine if migration of dust to adjacent units occurs as a result of abatement activity in a single unit. This was performed by abating a single unit per building or floor and performing air and wipe data in adjacent units.

While preliminary indications are that dust spread was minimal, further analysis of the data obtained from this experiment is necessary to draw definitive conclusions.

The demonstration scope also required a study of the feasibility of abatement of lead hazards to the 0.06%-by-weight standard. From the above discussion, this value is a low abatement threshold, especially when compared to the prescribed 0.5%-by-weight abatement threshold. This experiment is still evolving; however, it essentially involves the retesting of abated substrates in already abated units, to determine if they are below the 0.06%-by-weight standard. This experiment will be performed by core drilling abated substrates and analyzing the cuttings. In addition, previously tested lead-free paint will be applied to surfaces, and after a "curing period", this paint will be sampled by AAS to determine if lead is present above the 0.06%-by-weight standard. This data will then be analyzed to determine if abatement to this standard is feasible and practical.

The demonstration also looked at two additional topics: the application of the Guidelines in testing and abatement, and in the hazardous waste process. In all three cities the Guidelines were followed. The final report will comment on the application of the Guidelines to actual testing and abatement.

The waste disposal issue became an interesting component of the PHA Demonstration. The Toxic Characteristic Leaching Procedure[12] (TCLP) test for hazardous-waste determination was used for the PHA Demonstration. It produced a striking contrast to the Extraction Procedure (EP) Toxicity Test[13] that was used during the FHA Demonstration, since many more of the TCLP analyses were positive for lead. Essentially, two different laboratories were employed to perform TCLP test analysis of removed substrates. One laboratory reported that the substrate was below the hazardous threshold, and the other reported results far above the hazardous waste threshold. Upon investigation it was determined that each laboratory interpreted the TCLP procedure differently; so EPA's Office of Solid Waste (OSW) was consulted. EPA indicated the samples needed to fit through a 9 5-mm sieve, and they recommended dicing the sample into cubes or dowels, as part of the sample preparation protocol. When both laboratories performed this procedure, the results were above the hazardous waste threshold, but considerably less than the extremely high results of the first test.

The TCLP test is designed to simulate the leaching of lead from a substrate in a landfill. The EP Toxicity Test was an earlier test to simulate this same leaching process. While there are several differences between the two analytical methods, a significant difference is that the EP Toxicity Test included a "structural integrity procedure." This procedure can be performed on a solid component such as a piece of door frame, baseboard, or other such component. If the component "passed" this procedure, the solid "chunk" would be analyzed by the EP toxicity method, instead of cutting it to fit a 9.5-mm sieve. The TCLP test does not contain this provision.[14]

The issue with the TCLP test is not the results obtained; it is essentially the preparation of the sample for testing, in particular a solid waste such as a lead-based paint-containing door, window, or other such component. The guidelines require that all removed components be wrapped in 6-mil poly in manageable lengths.[15] It is not clear, however, that waste disposed in a landfill breaks into 9.5-mm cubes or is shed from the 6-mil poly. The laboratory that initially reported the waste below the hazardous threshold followed all parts of the TCLP test except for cutting the sample into 9.5-mm cubes. Instead, they used a solid piece of component to simulate what would actually be disposed of in the landfill.

No one associated with the PHA Demonstration wants to be a party to contaminating our Earth. However, it is not evident that the preparation of the sample for TCLP analysis is simulating what occurs to components in a landfill. Due to the expensive disposal costs of hazardous waste, it is feared that lead-based paint abatement decisions will become based on the potential disposal costs, not on what is the best abatement solution for lead-based paint. Hence, situations could occur where a decision has to be to manage the lead-based paint in place, instead of removing it. With the proper conditions, management in place should be the first option considered; however, it is a decision that must be made after carefully analyzing other options. More research is needed on testing methods and sample preparation, for determining disposal criteria of lead-based paint-containing components.

III. RESULTS

The entire PHA Demonstration report should be submitted to HUD by the Spring of 1995. Individual city reports will be submitted to HUD during this time period as well. These reports, when released for public distribution, can be obtained by placing an order through HUD USER at (301) 251–5154.

REFERENCES

1. The HUD Lead-Based Paint Abatement Demonstration (FHA), HUD-1316(1)-PDR, U.S. Department of Housing and Urban Development, U.S. Government Printing Office, Washington, D.C., 1991.

2. The HUD Lead-Based Paint Abatement Demonstration (FHA), HUD-1316(1)-PDR, U.S. Department of Housing and Urban Development, U.S. Government Printing Office, Washington, D.C., 1991, p. ii.

3. The HUD Lead-Based Paint Abatement Demonstration (FHA), HUD-1316(1)-PDR, U.S. Department of Housing and Urban Development, U.S. Government Printing Office, Washington, D.C., 1991, pp. 1x-1–1x-10.

4. Lead-Based Paint: Interim Guidelines for Hazard Identification and Abatement in Public and Indian Housing, 281–930/44406, U.S. Department of Housing and Urban Development, U.S. Government Printing Office, Washington, D.C., 1990, revised 1991.

5. Lead-Based Paint: Interim Guidelines for Hazard Identification and Abatement in Public and Indian Housing, sections 10.4.2–10.4.3, 281–930/44406, U.S. Department of Housing and Urban Development, U.S. Government Printing Office, Washington, D.C., 1990, revised 1991, 122–25.

6. Occupational Safety and Health Administration Regulation, 29 Code of Federal Regulation 1910.1025, Office of the Federal Register National Archives and Records Administration, Washington, D.C., 1988, chap. 17.

7. U.S. Environmental Protection Agency, 40 Code of Federal Regulation, Part 261 Appendix II, Office of the Federal Register National Archives and Records Administration, Washington, D.C., 1991, chap. 1.

8. Lead-Based Paint: Interim Guidelines for Hazard Identification and Abatement in Public and Indian Housing, Section 8.8, 281–930/44406, U.S. Department of Housing and Urban Development, U.S. Government Printing Office, Washington, D.C., 1991, 91.

9. Lead-Based Paint: Interim Guidelines for Hazard Identification and Abatement in Public and Indian Housing, Section 6.3.2, 281–930/44406, U.S. Department of Housing and Urban Development, U.S. Government Printing Office, Washington, D.C., 1991, 61.

10. Lead-Based Paint: Interim Guidelines for Hazard Identification and Abatement in Public and Indian Housing, 281–930/44406, U.S. Department of Housing and Urban Development, U.S. Government Printing Office, Washington, D.C., 1991, chap. 1, p. 2.

11. Lead-Based Paint: Interim Guidelines for Hazard Identification and Abatement in Public and Indian Housing, 281–930/44406, U.S. Department of Housing and Urban Development, U.S. Government Printing Office, Washington, D.C., 1991, chap. 4, p. 25.

12. U.S. Environmental Protection Agency, 40 Code of Federal Regulation, Part 261 Appendix II, Office of the Federal Register National Archives and Records Administration, Washington, D.C., 1991, chap. 1.

13. Test Methods for Evaluating Solid Waste, SW-846, 3rd ed., U.S. Environmental Protection Agency, Office of Waste and Emergency Response, Washington, D.C., 1986.

14. Barrows, B. A., Personal communication, 1992.

15. Lead-Based Paint: Interim Guidelines for Hazard Identification and Abatement in Public and Indian Housing, Section 10.2.2.1, 281–930/44406, U.S. Department of Housing and Urban Development, U.S. Government Printing Office, Washington, D.C., 1991, 117.

Chapter 5

Information Collected in the HUD Abatement Demonstration Program and Its Application in Planning a Follow-on Study

R. A. Lordo and M. Chen

CONTENTS

I. INTRODUCTION

Lead contamination of paint, and its contribution to lead in soil and dust, is regarded as a primary source of childhood lead poisoning in residential dwellings. In 1987 and 1988, amendments to the Lead-Based Paint Poisoning Prevention Act of 1971 (LBPPPA) required the U.S. Department of Housing and Urban Development (HUD), with assistance from the U.S. Environmental Protection Agency (EPA), to direct a broad-based program of research, demonstration, and policy actions aimed at identifying and abating lead-based paint hazards in privately owned and public housing. Included among the objectives of this program was to identify the most cost-efficient methods for abatement of lead-based paint hazards. This objective was satisfied by conducting a multicity demonstration program addressing the following program requirements:

- To estimate the comparative costs of a series of abatement methods
- To determine the efficacy of these methods
- To confirm the adequacy of worker protection safeguards for these methods

This program was known as the HUD Abatement Demonstration. This demonstration evaluated the short-term effectiveness of a variety of abatement strategies, by implementing these abatement methods on a set of HUD-owned (i.e., FHA-foreclosed) vacant single-family units in varying regions of the U.S.

The HUD Abatement Demonstration program provided useful information on the presence of lead in painted surfaces prior to abatement, and how lead-based paint abatement influences the lead content in affected media such as soil and dust. This latter information allowed the HUD Abatement Demonstration program to focus on short-term effectiveness of abatement methods. The EPA Office of Pollution Prevention and Toxics has used the information obtained from the HUD Abatement Demonstration program in planning the Comprehensive Abatement Performance (CAP) study, a study focusing on the long-term effectiveness of these same abatement methods. The primary objective of the CAP study is to determine the efficacy of the abatement methods over time, relative to lead levels in key media.

II. THE HUD ABATEMENT DEMONSTRATION

The project team for the HUD Abatement Demonstration were professionals in research design, field sampling, and laboratory analysis. The participants included

- Dewberry and Davis: project manager
- Speedwell, Inc.: developed the research design, performed all data analysis, and reported study results
- KTA-Tator, Inc.: performed field testing and laboratory analysis of *in situ* paint and paint samples
- Tracor Technology Resources, Inc.: performed soil and dust field sampling, analyzed soil and dust samples in the laboratory, and addressed worker protection issues
- The Marcor Group: assisted in developing abatement methods and provided abatement for pilot units

1-56670-113-9/95/$0.00+$.50
© 1995 by CRC Press, Inc.

The HUD Abatement Demonstration considered single-family units in seven cities: Baltimore, Birmingham, Denver, Indianapolis, Seattle, Tacoma, and Washington, D.C. These cities were selected based on their regional diversity and sufficient number of single-family properties of a certain age, with potential lead-based paint hazards.

A total of 304 units likely to contain lead-based paint were selected from the above cities for the abatement demonstration. To characterize the extent of lead-based paint in these units, painted surfaces in the interior and exterior were tested by portable X-ray fluorescence (XRF) analyzers, for the presence of lead-based paint. Units included in the demonstration were those found to have a large number of structural components covered by paint with a high concentration of lead, according to the XRF survey. When XRF readings yielded inconclusive results for a given surface, a second round of XRF readings on the surface were taken, or paint samples were taken for more conclusive laboratory analysis using atomic absorption spectrometry (AAS) methods.

Of the 304 units participating in the XRF-testing survey, 172 of them were found to have sufficiently large numbers of painted substrates exceeding the XRF-testing standards. Thus, lead-based paint abatement procedures proceeded in these 172 units. Three of these units had only pilot abatements performed, while the other 169 units were completely abated.

Exterior soil samples were collected prior to, and following, abatement at 130 of the 172 abated units. These soil samples were analyzed in the laboratory for lead content, using AAS methods. An average of 3.5 paired soil samples were collected along the exterior walls for each of these units. Comparison of pre- and post-abatement soil sample testing determined the extent to which the abatement process affected the lead concentration in the exterior soil.

Dust samples were also collected throughout each abated unit, following abatement, to note when the unit met regulatory clearance standards. The presence of lead in dust indicates the need to remove such dust regardless of whether it was the result of abatement efforts. Clearance standards of 200 $\mu g/ft^2$ for floors, 500 $\mu g/ft^2$ for window sills, and 800 $\mu g/ft^2$ for window wells were used to determine whether the unit met clearance standards. Dust samples were taken using premoistened commercial wipes. The dust samples were analyzed in the laboratory, using AAS methods. The cleaning process was repeated when dust sample analysis indicated clearance standards were not met. The results of laboratory analysis of dust wipe samples were used to determine failure rates of clearance for each abatement method.

Summaries of the sampling methods for interior dust and exterior soil are found in Table 1. Analytical methods are summarized in Table 2.

III. THE HUD DEMONSTRATION DATABASE

Information on lead concentrations in various key media (paint, soil, dust), and the type of lead-based paint abatement performed, was collected on each of the 304 HUD Abatement Demonstration units. These data originated from several sources among the contractors used in the HUD demonstration; thus, the information was expressed in a variety of formats. To incorporate all data into one central repository, the lead concentration and abatement information was merged into a unified SAS® database. The data residing in this database are as follows:

- Lead concentration in painted surfaces resulting from XRF testing
- Results of AAS analysis of paint samples when XRF testing was indeterminate or impossible for a given painted component
- Results of AAS analysis of soil samples taken at abated units both before and after abatement
- Results of AAS analysis of dust samples taken at abated units primarily after abatement
- The abatement method used on each abated component, and the amount of abatement (in square or linear feet) performed on each component

The SAS datasets constituting the HUD demonstration database are linked by the unit ID to make database information available to all datasets. The structure of the database allows easy access to all types of information from the demonstration and facilitates the linking of various types of information. Complete documentation has been prepared on this database, to indicate what the variables in the database represent and how the coded values of variables are interpreted.

HUD has recently provided additional pre-abatement information characterizing the units considered in the demonstration, including age, location, and physical condition of the unit. These additional data have been documented and incorporated into the HUD demonstration database.

Table 1 Sampling Methods for the HUD Demonstration

	Interior dust	Exterior soil
Sampling device	Chubs Thick Baby Wipes with Aloe (5 3/4 × 8 in.)	0.75-in. ID tube (0.5 in.^{2} surface area)
Sampling area	One square foot (floors, window sills, window wells)	Top 0.5 in. of soil is taken
Samples collected	One window sill per abated area	4 samples each taken 1 ft from foundation; one
	One window well per abated area	sample on each side of the unit
	One floor per abated area	
Compositing	None	All samples are a composite of 5 uniformly spaced cores along the length of the wall

Table 2 Analytical Methods for the HUD Demonstration

	Interior dust	Exterior soil
Sample preparation summary	Wipe ashed at 550–600°C for 2 h. Acid digested in HNO$_3$/H$_2$O$_2$. Diluted to 10 ml.	Oven dry, sieve. Oven dry at 105°C for 24 h. 1 g digested in HNO$_3$. Dilute to 100 ml.
Instrumental technique	Flame atomic absorption	Flame atomic absorption
Estimated LOQ	2 µg/sample	6 µg/g
Data reporting	µg/ft^2	µg/g dry weight
QA/QC notes	No reference material used. Used side-by-side sampling for duplicates.	Reference material not specified

IV. ABATEMENT METHODS

The HUD demonstration of single-family units originally considered six different abatement methods of lead-based paint:

- Encapsulation
- Enclosure
- Chemical stripping
- Abrasive stripping (sanding, blasting, grinding)
- Heat-gun stripping
- Complete removal and/or replacement of the painted components

The latter four methods are considered removal methods. Because of the diversity of housing components containing lead-based paint, it was generally true that no single abatement method could be used uniformly throughout a given housing unit. However, the plan for the follow-on CAP study was to classify each unit by one of two categories of abatement:

- Encapsulation/enclosure
- Removal methods

The rationale for this approach is that these two categories of abatement are likely to result in significantly different performance. In addition, these two categories of abatement have been adopted by HUD for the PHA component (multifamily dwellings) of the demonstration program, and a sampling program designed to assess all six abatement methods separately would prove too costly.

One important consideration in the current CAP study is the appropriate way in which to summarize and classify the abatement activities conducted at each unit. Detailed information was collected by HUD, which lists each type of interior and exterior structural component abated in the demonstration, along with the linear or square footage abated and the abatement method used. Each unit is classified according to the abatement category (i.e., encapsulation/enclosure vs. removal methods) accounting for the largest square footage of interior abatement. A summary of the resulting classification for all 169 abated units in the demonstration is presented in Table 3. Note in this table that encapsulation/enclosure and removal methods were used in the demonstration in roughly the same number of units, although encapsulation/enclosure was used somewhat more often.

The predominant interior abatement method is used as the primary means for classifying abated units in the CAP study, because the main objectives of this study are to assess abatement performance in terms

Table 3 Number of Units Abated in the HUD Demonstration

| City | Interior abatement category[a] | | Exterior abatement only | | Total |
	Encap/enclos	Removal	Encap/enclos	Removal	
Baltimore	11	9	—	—	20
Birmingham	8	12	2	1	23
Denver	33	18	5	1	57
Indianapolis	17	10	3	4	34
Seattle/Tacoma	12	10	1	3	26
Washington	6	3	—	—	9
Total	87	62	11	9	169

[a] Each unit is classified according to the abatement category accounting for the largest square footage of *interior* abatement.

Table 4 Number of Units Abated by Each Type of Method for Both Interior and Exterior Abatement

| Predominant exterior abatement method | Predominant interior abatement method | | Total |
	Encapsulation/enclosure	Removal methods	
Encapsulate/enclosure	63	25	88
Removal methods	22	37	59
Total	85	62	147[a]

[a] There are 20 units where only exterior abatement was performed, and 2 units where only interior abatement was performed.

Table 5 Summary of Square Footage Abated by Each Method[a]

| | Encapsulation/enclosure | | | Removal Methods | | |
	n[b]	Total ft² abated	Ft² abated by encap/enclos	n	Total ft² abated	Ft² abated by removal
Interior abatement	87	718	618	62	382	286
Exterior abatement	99	1046	933	68	925	778

[a] Square footage figures represent averages across those units for which given type of abatement was the predominant method used.

[b] n = Number of units for which given type of abatement was the predominant method used.

of indoor dust lead levels, and to investigate the relationship between indoor dust lead levels and lead levels from a variety of other sources. However, at many HUD demonstration units, a great deal of exterior abatement was also performed, and another objective of the CAP study is to assess the performance of this exterior abatement. Therefore, the data interpretation must consider which specific methods were used on both the interior and exterior of the unit. Table 4 shows that for 100 of 147 units where both interior and exterior abatement were performed, the same type of abatement was predominantly used both inside and outside in the demonstration.

In contrast, for 47 units in the demonstration, a different type of abatement was performed on the interior and exterior of the unit. In addition, there are 20 units where only exterior abatement was performed, and two units where only interior abatement was performed. The statistical model developed to analyze the sampling results of the CAP study explicitly considers the square footage abated indoors and outdoors, by each abatement method at each individual unit, in order to assess the performance of each method.

Two other important considerations for the data interpretation are the sometimes widely different square footages abated at different units, and the different mix of methods used. Table 5 summarizes the average square footage abated both indoors and outside in the demonstration, for each abatement category, and the average percentage of that square footage that was abated by each type of abatement. For example, Table 5 indicates that of the 149 units where some interior abatement was performed, encapsulation/ enclosure was the predominant type of abatement used indoors at 87 of those units, and in those units it accounted for approximately 85% of the interior square footage abated. Table 5 also indicates that a greater square footage was typically abated outdoors than indoors, and a greater square footage was abated by encapsulation/enclosure than by removal methods. However, Table 5 does not describe the wide variations in these results at different demonstration units. For example, the total square footage abated at different

Table 6 Number of Unabated Units Tested by XRF in the HUD Demonstration

City	Number of LBP components[a]				
	0	**1–2**	**3–9**	**10 or more**	**Total**
Baltimore	1	6	3	10	20
Birmingham	4	5	—	5	14
Denver	13	10	14	3	40
Indianapolis	5	9	5	—	19
Seattle/Tacoma	10	3	2	5	20
Washington	4	2	4	9	19
Total	37	35	28	32	132

[a] Number of structural components for which XRF testing identified the presence of lead-based paint.

units ranges from over 4200 ft^2 down to less than 100 ft^2. And the percentage of abatement associated with a single abatement strategy (either encapsulation/enclosure or removal methods) ranges from 100% at some units to just over 50% at others.

Of the 304 units considered in the HUD demonstration, 132 were not abated, due to their minimal lead-based paint contamination, according to XRF testing. Table 6 indicates the number of components in each of these units, for which XRF testing identified the presence of lead-based paint. Note that 37 of these units had no potentially contaminated components. These units satisfy the criteria for classification as control units.

V. SUMMARIES OF HUD DEMONSTRATION DATA

The HUD demonstration database contains lead concentrations found in paint, soil, and dust samples, as well as lead concentrations recorded in the XRF-testing survey. These data can be summarized to characterize the extent of lead-based paint in the sampled units and to observe the short-term effect of abatement on lead levels in the various sampled media.

Over 16,800 interior and exterior substrates[5] were tested by XRF analyzers or AAS laboratory methods in the demonstration. When performing the XRF tests, HUD took three replicate XRF readings at each sampling location and based their decisions, at each location, on the average of those three readings. The XRF/AAS testing procedure identified approximately 3600 substrates (21% of all tested substrates)[5] whose XRF/AAS readings for lead concentration exceeded the abatement standard of 1.0 mg/cm^2 (0.7 mg/cm^2 in Maryland).

Figure 1 indicates the percentage of tested substrates with lead concentrations at, or above, the standard, according to location in the unit. This figure shows that painted surfaces on the exterior of the unit had a higher proportion of XRF readings at, or above, the standard, compared to surfaces inside the unit. Among the rooms within the unit, higher proportions of XRF readings at, or above, the standard were found in kitchens and bathrooms, so-called "wet rooms" due to access to interior plumbing, compared to other living areas such as bedrooms, dining rooms, hallways, and living/family rooms. These findings are consistent with the fact that paint grades with higher lead content have historically been applied to exteriors and wet rooms, prior to lead-based paint regulations.

Figure 2 presents the proportion of XRF/AAS readings, for lead concentration, that are at, or above, regulatory standards within the five regions of the country represented in the demonstration. All components tested by XRF or AAS were represented in determining these proportions. This figure shows that the proportion of XRF/AAS readings at, or above, regulatory standards were relatively consistent among the five regions.

Figure 3 illustrates levels of preabatement soil lead concentrations according to the proportion of exterior painted components found to exceed lead concentration regulations, according to XRF testing. Thus, this figure illustrates the extent of lead contamination of exterior paint and soil, prior to abatement. This plot does not show any evidence that extensive contamination of exterior painted components affects levels of soil lead contamination prior to abatement.

The amount of influence that abatement methods have on soil lead concentrations can be determined by considering preabatement vs. postabatement soil concentrations. Figure 4 displays a plot of post- vs. preabatement soil concentrations for abated units, averaged across the soil samples taken close to the

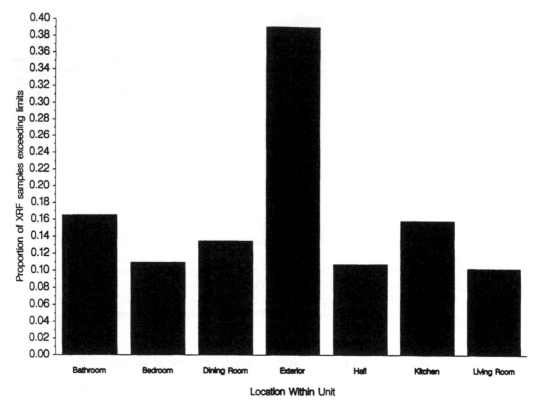

Figure 1 Proportion of XRF samples that exceed regulatory limits for lead concentration, according to location within unit.

foundation of the unit. The solid line in this plot indicates when pre- and postabated soil concentrations are equal. As indicated in the plot, the majority of abated units had average postabatement soil concentrations exceeding the preabatement averages. Across all pairs of soil samples, a difference of 112.5 ppm[5] was observed between the average postabatement and preabatement soil concentrations.

Increases in soil lead concentrations following abatement appear to imply that the abatement processes may cause additional lead contamination to soil in the immediate vicinity of the unit. This is because no soil lead abatement was performed in the demonstration. Figure 5 displays the relationship between the total square footage of abatement performed in a unit vs. the average postabatement soil lead concentration. While both low and high values of soil concentrations are associated with lower square footage of abatement performed (less than 1000 ft²), the proportion of units having higher soil concentrations increases as the square footage of abatement performed increases. Similar relationships with soil lead concentrations are seen when considering only square footage of exterior abatement. Thus, these findings indicate that a short-term effect of abatement is an increase in soil lead concentration, and this effect is related to the amount of abatement performed.

Initially, the demonstration collected preabatement dust samples, to indicate the presence of lead in dust prior to abatement. However, budgetary constraints prevented collecting preabatement dust wipe samples from all abated units. Thus, a comparison of post- vs. preabatement dust lead concentrations is limited by the number of units involved. Dust lead samples were taken on floors, window sills, and window wells in abated areas of the unit. Final postabatement dust wipe concentrations vs. preabatement concentrations are plotted in Figure 6 for floors, Figure 7 for window sills, and Figure 8 for window wells. Each point in these three logarithmic plots indicates an abated area of a unit. The solid lines in these plots indicate when pre- and postabatement concentrations are equal. The final postabatement dust concentrations represent the result of up to three cleaning iterations. Thus, the majority of postabatement dust concentrations are lower than preabatement concentrations, for a given area. While some postabatement concentrations remained above clearance standards, all units were eventually cleaned to meet regulatory standards.

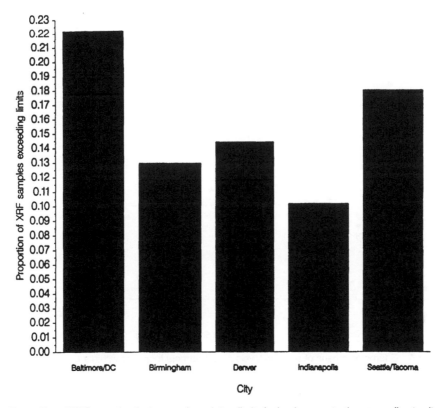

Figure 2 Proportion of XRF samples that exceed regulatory limits for lead concentration, according to city of survey.

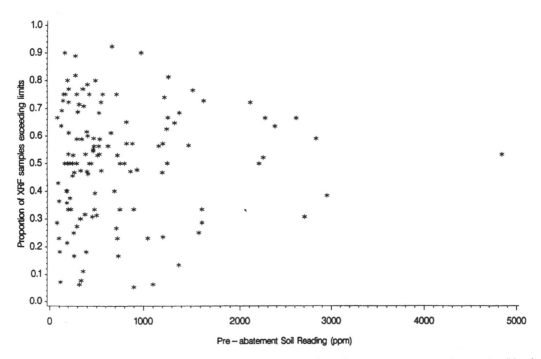

Figure 3 Proportion of exterior XRF samples exceeding lead regulatory limits, vs. average preabatement soil lead levels (ppm).

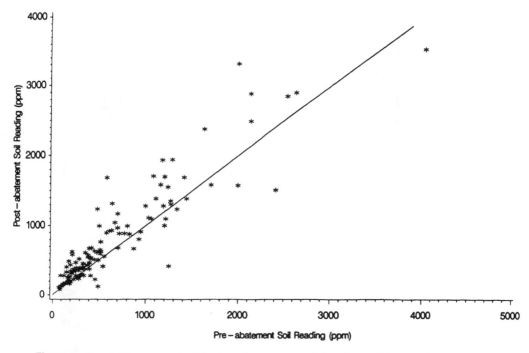

Figure 4 Geometric means of soil lead levels across sample locations within a unit (units = ppm).

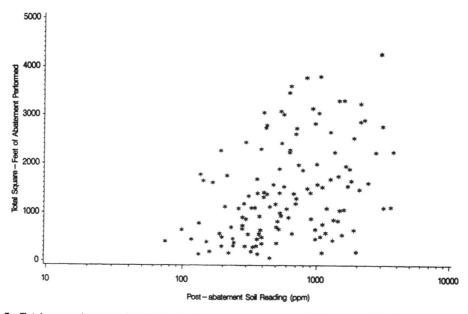

Figure 5 Total square footage of abatement performed vs. average postabatement soil lead concentration (ppm).

VI. THE HUD DATABASE USED TO PLAN THE CAP PILOT STUDY

The HUD demonstration results were used to design a pilot program for the follow-on CAP study. Among the objectives of the pilot program, a precursor to the full CAP study, was the estimation of the portion of total variability in interior dust lead concentrations that are attributable to the following sources:

- Among units abated with different methods
- Among units abated by the same method

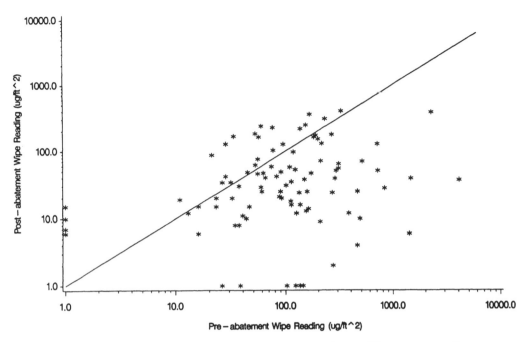

Figure 6 Final wipe lead levels, post- vs. preabatement (units = μg/ft²): component = floor.

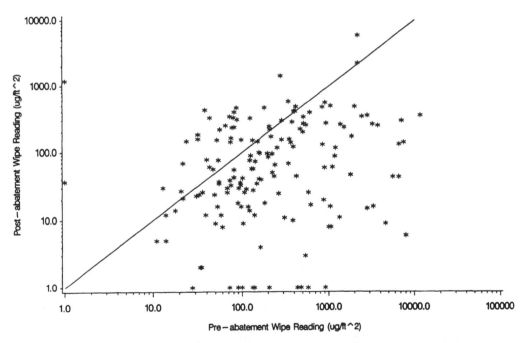

Figure 7 Final wipe lead levels, post- vs. preabatement (units = μg/ft²): component = window sill.

- Among rooms within a unit abated by the same method
- Among sampling locations within a room

Whereas the HUD demonstration is intended to focus on short-term cost effectiveness of abatement methods, the CAP study will provide important information about the longer-term effectiveness of these same methods. Although clearance testing of lead levels in dust was done immediately after abatement in the HUD demonstration, the longer-term performance of the abatement methods after these units have been

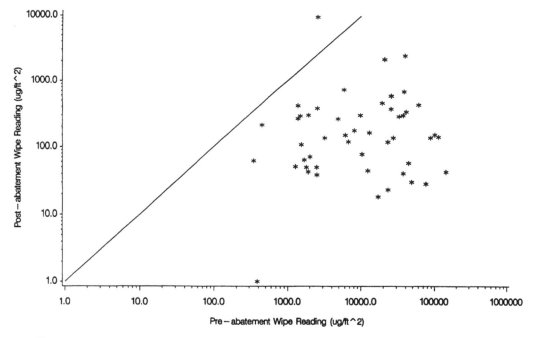

Figure 8 Final wipe lead levels, post- vs. preabatement (units = μg/ft²): component = window well.

reoccupied is not known. The CAP study is therefore necessary to preclude spending large sums of money abating lead-based paint, using methods that may prove, in the long term, to be ineffective at maintaining low lead levels in household dust.

Results from the CAP study will be compared with earlier results from the HUD demonstration. For this reason the sampling and analytical methods for the CAP study have been selected to be comparable to the methods used in the HUD demonstration.

Identifying variability due to different abatement methods required the pilot program to include units with differing predominant abatement methods (encapsulation/enclosure or removal methods). However, to also determine variability associated with different units abated by the same method, the pilot program needed to include units with the same predominant method on both interior and exterior components. Thus, one third of the units in the pilot program represented encapsulation/enclosure methods, and one third represented removal methods. The remaining one third consisted of control units. To access variability from different rooms within a unit, the units contained two rooms abated by the same predominant method used for the unit as a whole. The square footage of abatement performed by each abatement category was taken from the HUD demonstration database and used to select units for the CAP pilot study.

Since dust measurements are a primary means of assessing the performance of an abatement method in the CAP study, it was necessary to compare dust measurements from the CAP study, with the dust measurements obtained from the wipe/ashing protocol used in the HUD Abatement Demonstration program. However, a vacuum/total digestion measurement is to be used in the CAP study. Thus, a means of comparing vacuum vs. wipe sampling results was necessary in the CAP pilot program. Side-by-side vacuum and wipe samples were obtained in an additional room of the CAP pilot study units. As a result, comparisons in dust results could be made between the HUD Abatement Demonstration and the CAP pilot program, and between the merged dust results from both programs and XRF/AAS results in the HUD Abatement Demonstration.

Initial plans for the CAP study included selecting housing units from all seven urban areas in the FHA portion of the HUD demonstration. However, after conducting the pilot sampling and analysis program,[1] and subsequently developing a cost estimate for the CAP study, it was decided that the CAP study will only be conducted in Denver, where 57 of the 169 abated demonstration units are located (Table 3). Because the number of abated units in Denver is limited, all reoccupied units were included for recruitment in the CAP study. Results from a preliminary statistical power analysis indicated that approximately 40 abated

Table 7 Typical XRF Results for HUD Demonstration Units in Denver[a]

Type of housing unit	Interior surfaces			Exterior surfaces		
	Number LBP components[b]	Percentage LBP components	Average XRF result (mg/cm²)	Number LBP components[b]	Percentage LBP components[c]	Average XRF result (mg/cm²)
Typical abated unit	13.7	14	0.80	6.0	48	3.31
Typical E/E unit	14.2	15	0.84	6.2	49	3.29
Typical removal unit	11.2	14	0.59	5.1	46	3.38
Typical control unit	3.4	3	0.17	1.5	13	0.71

[a] Average results for 33 encapsulation/enclosure (E/E), 7 removal, and 20 control units selected for the CAP Study.
[b] Average number of components testing positive (i.e., XRF greater than 1.0 mg/cm²) for lead-based paint.
[c] Average percentage of components testing positive for lead-based paint.

units will be sufficient to detect meaningful differences between dust lead levels in abated and control units. Given the initial set of 57 abated units in Denver, 70% of these units must be successfully recruited into the study.

Control units for the CAP study were recruited from the set of unabated Denver units tested by XRF in the HUD demonstration and found to be essentially free of structural components covered with lead-based paint. For the purposes of identifying control units, the detailed XRF results supplied by HUD were used under the assumption that they provide an accurate and current assessment of these units. If unusually high dust lead levels are found at the CAP study control units, additional confirmatory testing at these units will be considered to ensure that no lead-based paint is present. Using a criterion that equally weights (1) the percentage of housing components testing positive by XRF for lead-based paint, and (2) the average XRF testing result, unabated units with the lowest XRF results are prioritized and selected as control units for the CAP study. Although these control units may not be completely free of lead-based paint, Table 7 demonstrates that they are expected to be relatively free of lead-based paint, when compared with abated units in Denver. If not all of the initial control units are successfully recruited, then those unabated units with the next lowest XRF results are selected from the prioritized list.

REFERENCES

1. Buxton, B., Draft Final Report for the Comprehensive Abatement Performance Pilot Study. Report from Battelle for U.S. Environmental Protection Agency, Office of Toxic Substances, Washington, D.C., Contract 68-D0–0126, 1991.
2. Buxton, B. and Dewalt, G., Detailed Design Document for the Comprehensive Abatement Performance Study. Report from Battelle and MRI for U.S. Environmental Protection Agency, Office of Toxic Substances, Washington, D.C., Contracts 68-D0–0126 and 68-D0–0137, 1992.
3. Buxton, B. and Dewalt, G., Quality Assurance Project Plan for the Abatement Performance Pilot Study. Report from Battelle and MRI for U.S. Environmental Protection Agency, Office of Toxic Substances, Washington, D.C., Contracts 68-D0–0126 and 68-D0–0137, 1991.
4. Buxton, B. and Dewalt, G., Quality Assurance Project Plan for the Comprehensive Abatement Performance Study. Report from Battelle and MRI for U.S. Environmental Protection Agency, Office of Toxic Substances, Washington, D.C., Contracts 68-D0–0126 and 68-D0–0137, 1992.
5. Dewberry and Davis, The HUD Lead-Based Paint Abatement Demonstration (FHA), HUD5845, HUD Office of Policy Development and Research, Washington, D.C., 1991.
6. Dewberry and Davis, The HUD Lead-Based Paint Abatement Demonstration (FHA), Vol. I, Appendices A–H, HUD5846, HUD Office of Policy Development and Research, Washington, D.C., 1991.
7. Dewberry and Davis, The HUD Lead-Based Paint Abatement Demonstration (FHA), Vol. II, Appendices I–P, HUD5847, HUD Office of Policy Development and Research, Washington, D.C., 1991.
8. Lordo, R., HUD Demonstration Lead-Based Paint Database. Draft Documentation for U.S. Environmental Protection Agency, Office of Toxic Substances, Washington, D.C., Contract 68-D0–0126, 1992.

Chapter 6

Results from the Pilot Comprehensive Abatement Performance Study

B. E. Buxton, S. W. Rust, F. Todt, T. Collins, C. Boudreau, R. Hertz, P. Constant,
G. Dewalt, J. G. Schwemberger, and B. S. Lim

CONTENTS

I. INTRODUCTION AND SUMMARY

This chapter presents results from the Comprehensive Abatement Performance Pilot Study, conducted in 1991 by Battelle Memorial Institute and Midwest Research Institute (MRI) for the U.S. Environmental Protection Agency's Office of Pollution Prevention and Toxics (OPPT). The objectives, approach, design, and results of this study, although briefly summarized here, are completely described in other documents.[1,2]

Under an interagency Memorandum of Understanding the Environmental Protection Agency (EPA) is providing technical support to the Department of Housing and Urban Development (HUD) with respect to the abatement of lead-based paint hazards in public and private housing. As part of its lead-based paint research activities, HUD is carrying out a demonstration program in 10 cities to assess the costs and short-term efficacy of alternative methods of lead-based paint abatement. A variety of abatement methods are being tested in approximately 120 multifamily public housing units in three cities — Omaha, Cambridge, and Albany — and in 172 single-family housing units in the FHA inventory in seven metropolitan areas — Baltimore, Birmingham, Denver, Indianapolis, Seattle, Tacoma, and Washington, D.C. The FHA portion of the demonstration has now been completed, and EPA's Office of Pollution Prevention and Toxics is conducting a follow-up study (referred to as the Comprehensive Abatement Performance [CAP] Study) of these housing units, with the following objectives:

1. Characterize levels of lead in household dust and exterior soil over time, for HUD demonstration and control homes.
2. Compare abatement methods or combination of methods relative to performance over time. Assess whether there are differences in performance.
3. Investigate the relationship between lead in household dust and lead from other sources; in particular, exterior soil, rugs, upholstered furniture, and air ducts.

As a precursor to the full CAP Study, OPPT conducted a pilot study in the Denver area, to investigate the field and laboratory procedures that are planned. This chapter describes some of the results from the pilot study.

II. STUDY APPROACH

The pilot study was intended to investigate the field, laboratory, and statistical analysis procedures planned for the full CAP Study. In particular, the objectives of the pilot study were as follows:

1-56670-113-9/95/$0.00+$.50
© 1995 by CRC Press, Inc.

50

- Test the sampling and analysis protocols
- Evaluate the questionnaires and other field data forms
- Provide variance estimates to help determine the final design of the full CAP Study
- Assess the performance (i.e., sensitivity, accuracy, and precision) of the sampling and analysis methods
- Compare analytical results for the MRI laboratory with those of another laboratory
- Compare the vacuum/total digestion protocol planned for the full CAP Study with the wipe/ashing protocol previously used in the HUD Demonstration Study

The first five objectives are all necessary precursors that will help to refine the study design and methods for the full CAP Study. The final objective is intended to further enhance our ability to assess the HUD abatement methods, by providing a bridge between earlier dust measurements from the HUD Demonstration study, obtained with a wipe sampling method, and our current dust measurements obtained with vacuum sampling.

Our data analysis approach for the pilot study focused on three statistical study objectives: (1) variance component estimation, (2) comparison of vacuum and wipe protocols, and (3) assessment of the performance of the sampling and analysis methods.

The field sampling design for the pilot study included samples to address the variance component estimation and comparison of vacuum and wipe sampling. A summary of the most important design considerations for the pilot study is contained in the following points:

- To assess variability associated with different housing units abated by the same method, we sampled from six Denver housing units abated by the same method on both interior and exterior components. Two units were selected from those predominantly abated by encapsulation/enclosure methods, two units were selected from those predominantly abated by removal methods, and two units were selected from those control houses already tested by HUD and found relatively free of lead-based paint.
- Sampling was performed in two different rooms of each house. To assess variability from different sources, a total of 18 regular vacuum dust samples were collected. These samples were collected from the following: floors (two per room), window stools (two per room), window channels (two per room), rug/upholstery (one per room), air ducts (one per room), and entryway samples (front and back). When selecting two abated rooms, rooms were chosen that were both predominantly abated by the same method used for the house in general.
- Soil samples were collected in the pilot study, to help assess potential nonpaint sources of lead contamination in interior dust. For two sides of each house, soil samples were collected both at the foundation of the house and at the property boundary. In addition, soil samples were collected immediately outside the front and rear entryways.
- For each of the six housing units included in the pilot study, one room was selected for comparative vacuum and wipe sampling. This room was a third room added to the two sampled rooms discussed above. Within each room selected for comparative sampling, a pair of side-by-side floor samples was collected immediately next to each other by both the vacuum and wipe sampling methods. In addition, side-by-side vacuum and wipe samples were collected on both the stool and channel of two windows in the room.
- Seven quality control samples (i.e., field side-by-sides, field blanks, and interlaboratory comparison samples) were collected to assess variability introduced by the sampling method, sample handling, and laboratory effects.

III. SUMMARY OF RESULTS

The CAP pilot study, although primarily intended to test sampling and analysis methods and to obtain variance estimates, did provide some interesting initial results concerning abated and control houses. It should be noted, however, that this pilot study included only a relatively small number of samples, and therefore, more definitive results regarding many of these issues can be expected from the full CAP Study. The results highlighted in this chapter are the following:

1. Although the design of this study did not consider renovation, two of the six houses were undergoing renovation at the time of sampling. Partial renovation was being performed in a control home, while an

encapsulation/enclosure home was undergoing full renovation. Units being renovated had high interior lead loadings on surfaces such as entryways, floors, and window stools; floor lead loadings in the units undergoing full renovation were estimated to be 70 times higher than those in control units; both higher lead concentrations in the dust (5 times higher) and higher dust loadings (14 times higher) appear to contribute to the higher lead loadings.

2. There is some evidence that abated units have higher interior lead loadings, due primarily to higher lead concentrations.

3. For floor lead loadings, abated rooms in abated units have lead levels that are comparable to those in control units; however, lead loadings in unabated rooms in abated units may be 10 times higher than in abated rooms in the same unit; higher dust loading appears to be the primary cause.

4. Soil lead concentrations for the three types of samples collected (boundary, entryway, and foundation) are very highly correlated from unit to unit both before and after correcting for renovation and abatement effects. In contrast, interior dust lead concentrations generally are not very highly correlated. However, interior dust lead concentrations are generally correlated with soil lead concentrations.

5. Based on paired data for the two sampling procedures, the wipe sampling procedure appears to produce lead loadings on the order of 5 to 10 times higher than the vacuum method; this would be consistent with a sampling efficiency of approximately 20 to 25% for the vacuum sampler.

A. RENOVATION AND ABATEMENT EFFECTS

Most of the data interpretation results for the CAP pilot study were obtained from log-linear regression analyses. All of the statistical models for dust samples contain an overall geometric mean. The models can contain random effects for unit-to-unit, room-to-room, sampling location-to-sampling location, and duplicate-to-duplicate variability. At the unit level there can be fixed effects for renovation (no renovation vs. some renovation vs. complete renovation) and abatement (no abatement vs. some abatement); and at the room level there can be a fixed effect for abatement. The statistical model for soil samples is similar to the model for dust samples. However, side-by-side replaces room-to-room as the within-unit variability source. Also, since exterior abatement information is not available by the side of the unit, a fixed effect for abatement is included only at the unit level.

Estimates of the geometric mean, fixed effects for abatement and renovation, and variance components are reported in Table 1 for lead loading, in Table 2 for lead concentration, and in Table 3 for dust loading. Rather than representing an overall mean for all units, the geometric mean represents the expected value of the dependent variable (i.e., lead loading, lead concentration, or dust loading) for unrenovated control units. The estimated geometric mean is reported as the top value in the fourth column, with the logarithmic standard error reported in parentheses below.

For renovation, house abatement, and room abatement effects, estimated effects are reported as the top value in the fifth through seventh columns of Tables 1 to 3. The estimate can be interpreted as the multiplicative effect of the presence of that condition. For example, to determine an estimate of the geometric average lead loading (Table 1) for vacuum window stool samples in abated, unrenovated houses, multiply the geometric mean for unrenovated control houses by the estimate for house abatement:

$$6.70 \times 5.47 = 36.65 \ \mu g/ft^2$$

Below each of these estimates, the logarithmic standard error of the estimate is reported in parentheses. The bottom value reported in these columns is the observed significance level of the test that the true multiplicative effect is equal to one (i.e., no multiplicative effect). Observed significance levels of less than .05 indicate significant results, while those between .05 and .10 indicate marginally significant results. Each fixed effect was tested for significance when added last among the fixed effects in the model, but before all the random effects in the model. The denominator mean square used in each test was the proper linear combination of the estimated variance components, as determined by expected mean square equations.

The room abatement effect was significant only for floor lead loading. In fitting lead loading and lead concentration to various models for components other than floors, the room abatement effect was never even marginally significant (i.e., the significance level was never even below 0.20). Therefore, room abatement is only included in models for the floor sampling results.

Table 1 Estimated Renovation Effects, Estimated Abatement Effects, and Variance Component Estimates from Mixed Model ANOVA: Lead Loading (μg/ft²)

Sample type	Component	Sample size	Fixed effects[a]				Random effects standard deviation[b]				
			Geometric mean for unrenovated control houses	Renovation	House abatement	Room abatement	Total	Unit	Room	Sampling location	Error
Vacuum	Air duct	10	649 (1.76)	0.01 (5.26) 0.43	0.49 (2.03) 0.76		1.95 (3.12)	1.54 (1.24) 0.07	1.19 (5)		0.47 (11)
Vacuum	Entryway	12	6.62 (0.52)	40.9 (0.74) 0.02	1.57 (0.60) 0.51		1.30 (8.90)	0.00 (1.66) 0.76	1.56 (6)		0.32 (2)
Vacuum	Floor	39	3.76 (0.49) 0.08	70.0 (0.70) 0.01	9.93 (0.99) 0.06	0.13 (0.90) 0.05	0.99 (12.79)	0.43 (0.55) 0.21	0.64 (4) 0.08	0.39 (1.48) 0.18	0.35 (11)
Vacuum	Window stool	25	6.70 (0.95) 0.13	6.11 (1.34) 0.26	5.47 (1.09) 0.21		1.86 (11.77)	0.69 (0.28) 0.27	1.66 (11.27) 0.01	0.35 (0.89) 0.33	0.32 (2)
Vacuum	Window channel	11	2873 (0.69) 0.00	0.29	0.59 (0.64) 0.45		1.02 (4.07)	0.00 (2.89) 0.98	1.37 (0.82) 0.42	1.06 (0.81) 0.19	0.35 (1)
Wipe	Floor	12	7.63 (0.41) 0.02	69.4 (0.57) 0.01	3.53 (0.47) 0.07		0.59 (4.15)	0.48 (1.96) 0.04			0.33 (6)
Wipe	Window stool	12	100 (0.82) 0.01	28.2 (1.00) 0.09	0.40 (0.94) 0.41		1.07 (3.88)	0.94 (2.11) 0.59	0.00 (0.87) 0.65	0.79 (1.39) 0.08	0.40 (3)

[a] Top value is multiplicative estimate, middle value is logarithmic standard error of estimate, and bottom value is observed significance level.
[b] Top value is estimated logarithmic standard deviation, middle value is estimated degrees of freedom, and bottom value (when present) is observed significance level.

Table 2 Estimated Renovation Effects, Estimated Abatement Effects, and Variance Component Estimates from Mixed Model ANOVA: Lead Concentration (μg/g)

			Fixed effects[a]				Random effects standard deviation[b]				
Sample type	Component	Sample size	Geometric mean for unrenovated control houses	Renovation	House abatement	Room abatement	Total	Unit	Room (dust) or side of house (soil)	Sampling location	Error
Vacuum	Air duct	10	875 (0.33) 0.00	0.51 (1.34) 0.63	0.84 (0.37) 0.67		0.58 (6.94)	0.00 (1.22) 0.68	0.67 (5)		
Vacuum	Entryway	12	96.3 (0.15)	4.85 (0.21) 0.00	3.25 (0.17) 0.01		0.47 (8.06)	0.00 (3.46) 0.89	0.60 (6)		
Vacuum	Floor	39	106 (0.35)	4.89 (0.50) 0.05	2.86 (0.71) 0.18	0.73 (0.65) 0.64	0.70 (12.14)	0.31 (0.55) 0.21	0.45 (3.54) 0.08	0.36 (4.64) 0.03	0.25 (11)
Vacuum	Window stool	25	245 (0.95)	1.42 (1.35) 0.81	4.06 (1.10) 0.29		1.56 (6.98)	1.06 (1.53) 0.04	1.02 (7.89) 0.06	0.44 (1.80) 0.23	0.29 (2)
Vacuum	Window channel	11	2150 (0.83)	0.33 (0.98) 0.38	0.95 (0.92) 0.96		1.12 (4.94)	0.35 (0.03) 0.45	1.06 (2.95) 0.19	0.00 (0.28) 0.63	0.17 (1)
Soil	Boundary	15	53.6 (0.61)	2.14 (0.85) 0.44	2.41 (0.70) 0.29		0.89 (4.77)	0.65 (1.34) 0.11	0.61 (5.93) 0.00		0.05 (3)
Soil	Entryway	16	65 (0.30)	1.92 (0.45) 0.23	4.71 (0.34) 0.02		0.64 (11.11)	0.00 (0.14) 0.56	0.45 (1.02) 0.27		0.51 (4)
Soil	Foundation	17	109 (0.70)	2.39 (0.98) 0.44	1.97 (0.80) 0.46		0.99 (4.11)	0.81 (1.87) 0.04	0.54 (5.09) 0.01		0.17 (5)

[a] Top value is multiplicative estimate, middle value is logarithmic standard error of estimate, and bottom value is observed significance level.

[b] Top value is estimated logarithmic standard deviation, middle value is estimated degrees of freedom, and bottom value (when present) is observed significance level.

Table 3 Estimated Renovation Effects, Estimated Abatement Effects, and Variance Component Estimates from Mixed Model ANOVA: Dust Loading (mg/ft²)

Sample type	Component	Sample size	Fixed effects[a]				Random effects standard deviation[b]				
			Geometric mean for unrenovated control houses	Renovation	House abatement	Room abatement	Total	Unit	Room	Sampling location	Error
Vacuum	Air duct	10	**742**	**0.02**	**0.58**		**1.65**	**1.31**	**1.00**		
			(1.49)	(4.45)	(1.72)		(3.11)	(1.24)	(5)		
				0.43	0.78			0.07			
Vacuum	Entryway	12	**69**	**8.33**	**0.48**		**1.09**	**0.00**	**1.17**		
			(0.54)	(0.77)	(0.62)		(8.64)	(0.17)	(6)		
				0.07	0.32			0.56			
Vacuum	Floor	39	**36**	**14.31**	**3.47**	**0.18**	**1.09**	**0.70**	**0.70**	**0.00**	**0.51**
			(0.64)	(0.90)	(1.19)	(1.03)	(7.94)	(1.41)	(7.31)	(0.30)	(11)
				0.06	0.34	0.13		0.06	0.03	0.64	
Vacuum	Window stool	25	**27**	**4.32**	**1.35**		**0.88**	**0.00**	**0.75**	**0.57**	**0.26**
			(0.27)	(0.37)	(0.30)		(18.54)	(3.28)	(3.52)	(2.67)	(2)
				0.02	0.38			0.78	0.18	0.14	
Vacuum	Window channel	11	**1336**	**0.87**	**0.62**		**1.06**	**0.00**	**0.63**	**1.21**	**0.18**
			(0.75)	(0.29)	(0.79)		(6.50)	(0.97)	(0.06)	(0.96)	(1)
				0.95	0.57			0.75	0.56	0.09	

[a] Top value is multiplicative estimate, middle value is logarithmic standard error of estimate, and bottom value is observed significance level.

[b] Top value is estimated logarithmic standard deviation, middle value is estimated degrees of freedom, and bottom value (when present) is observed significance level.

Lead Loadings

The geometric average lead loadings in unrenovated control houses for different sample types, in decreasing order are

- Window channel (vacuum 2873 $\mu g/ft^2$)
- Air duct (649 $\mu g/ft^2$)
- Window stool (vacuum 6.70 $\mu g/ft^2$, wipe 100 $\mu g/ft^2$)
- Floor (vacuum 3.76 $\mu g/ft^2$, wipe 7.63 $\mu g/ft^2$)
- Entryway (6.62 $\mu g/ft^2$)

The renovation effect was only statistically significant in explaining the results for entryway samples, and both vacuum and wipe floor samples. For all sample types except air ducts and vacuum window channels, the estimated effect of renovation was to increase lead loadings. The effect was strongest for both vacuum and wipe floor samples.

In general, abatement history was found to be less significant than renovation for lead loading. For no component was this effect strongly significant. In the cases of vacuum and wipe floor samples, a marginal significance was observed. For all sample types except air ducts, vacuum window channels, and wipe window stools, houses that have been abated in the past have higher lead loadings.

The effect of room abatement was found to be significant only for floor vacuum samples. Abated rooms were observed to have lower floor lead loadings than unabated rooms in abated houses. By also controlling for the floor substrate (e.g., wood, tile, etc.), the observed significance of the room and house abatement effects was changed by only about 1%. Thus, it does not appear as though the floor substrate had a strong impact on the sampling results.

Lead Concentrations

The geometric average lead concentrations estimated for unrenovated control houses for different sample types in decreasing order are

- Window channel dust (2150 $\mu g/g$)
- Air duct dust (875 $\mu g/g$)
- Window stool dust (245 $\mu g/g$)
- Foundation soil (109 $\mu g/g$)
- Floor dust (106 $\mu g/g$)
- Entryway dust (96.3 $\mu g/g$)
- Entryway soil (65 $\mu g/g$)
- Boundary soil (53.6 $\mu g/g$)

The three dust sample types with the highest concentrations are in the same order as for lead loadings. Note that none of the soil lead concentrations is very high, but foundation soil levels are very close to the floor and entryway dust levels.

For lead concentrations the renovation effect was only statistically significant in explaining the results for entryway samples and floor samples. As was the case for lead loadings for all sample types except air ducts and window channels, the estimated effect of renovation was to increase lead concentrations. Also consistent with the results for lead loadings, the effect was seen to be strongest in floor samples.

The house abatement effect was only found to be significant in both types of entryway samples (vacuum and soil). However, for all sample types except air ducts and window channels, houses that have been abated in the past have higher lead concentrations, although the differences are not always statistically significant. The component with the strongest estimated abatement effect was soil entryway samples.

Room abatement history was only included in the analysis for floor lead concentrations. It was not observed as significant; however, as was the case for lead loadings, abated rooms were observed to have lower floor lead concentrations than unabated rooms in abated houses.

Dust Loadings

The geometric average dust loadings in unrenovated control houses for different sample types in decreasing order are

- Window channel (1336 mg/ft^2)
- Air duct (742 mg/ft^2)
- Entryway (69 mg/ft^2)
- Floor (36 mg/ft^2)
- Window stool (27 mg/ft^2)

Table 4 Unit-to-Unit Correlations Among Sample Types After Correction for Renovation and Abatement Effects: Lead Loading

		Vacuum						Wipe		
		Air duct	Bed/rug uph	Entryway	Floor	Window stool	Window channel	Floor	Window stool	Window channel
Vacuum	Air duct		0.99ª	−0.63	−0.16	0.12	0.71	−0.02	0.21	
			0.09ª	0.57	0.90	0.92	0.50	0.99	0.87	
			(2)ª	(2)	(2)	(2)	(2)	(2)	(2)	
	Bed/rug/uph			−0.60	−0.19	0.09	0.72	−0.01	0.14	
				0.59	0.88	0.94	0.49	0.99	0.91	
				(2)	(2)	(2)	(2)	(2)	(2)	
	Entryway				−0.38	−0.81	−0.75	−0.22	−0.76	
					0.62	0.19	0.25	0.78	0.24	
					(3)	(3)	(3)	(3)	(3)	
	Floor					0.36	−0.27	−0.53	0.11	
						0.64	0.73	0.47	0.89	
						(3)	(3)	(3)	(3)	
	Window stool						0.63	0.55	0.91	
							0.37	0.45	0.09	
							(3)	(3)	(3)	
	Window channel							0.66	0.67	
								0.34	0.33	
								(3)	(3)	
Wipe	Floor								0.65	
									0.35	
									(3)	
	Window stool									
	Window channel									

ª Top value is estimated correlation coefficient, middle value is observed significance level, and bottom value is degrees of freedom.

The renovation effect was only statistically significant for dust loadings in window stool samples. For all sample types except air ducts and window channels, the estimated effect of renovation was to increase dust loadings. The effect was strongest for floor samples.

Abatement was not found to be significant for any of the components; however, the strongest estimated effect was for floors. For floors and window stools, abated houses have higher dust loadings.

The effect of room abatement was not found to be statistically significant for floor samples. However, abated rooms were observed to have lower floor dust loadings than did unabated rooms in abated houses.

B. CORRELATIONS AMONG SAMPLE TYPES

As noted earlier, one objective of the CAP Study is to investigate the possible sources of lead in household dust, such as exterior soil and previously contaminated rugs, upholstered furniture, and air ducts. This objective was addressed by examining the statistical correlation found in lead loadings and concentrations between samples taken at different locations (e.g., entryway floors vs. window stools) and in different media (e.g., interior dust samples vs. exterior soil samples). These correlations were calculated using the geometric means for each housing unit and sample type.

Lead Loading

The correlation matrix for lead loading unit means is presented as Table 4. To locate a correlation of interest, locate the row corresponding to the first sample type, and the column corresponding to the second sample type. Correlation information for the two sample types is presented in the corresponding box where the following three values are presented:

- Top value: correlation coefficient between the logarithms of the geometric unit means
- Middle value: observed significance level of the test of the hypothesis of no correlation (correlation coefficient equal to zero)

• Bottom value: degrees of freedom associated with the variance estimates used in calculating the correlation coefficient

The lead loading unit means are presented graphically in Figure 1. This figure is a scatterplot matrix, or a collection of bivariate plots organized into matrix form. As with the correlation matrix, to locate a plot of interest, identify the row associated with one sample type, and the column associated with the other sample type. The plot is presented in the corresponding box. Within each box the horizontal axis represents increasing values of the column variable, on a logarithmic scale. Similarly, the vertical axis represents increasing values of the row variable, on a logarithmic scale. The abbreviations employed on the diagonal to identify the different sample types are defined in Table 5.

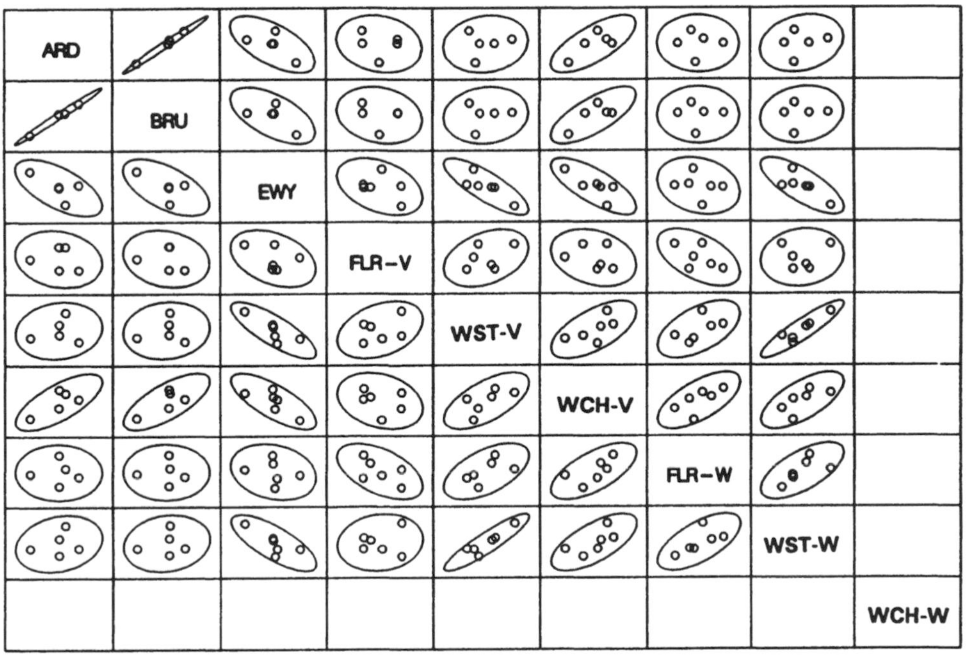

Figure 1 Scatterplot matrix of geometric unit means for different sample types after correction for renovation and abatement effects: lead loading ($\mu g/ft^2$).

Table 5 Symbols Used to Denote Sample Types in Tables and Figures

Sample type	Symbol	Description
Air duct dust	ARD	Dust samples from an *air duct*
Bed cover-rug-upholstery dust	BRU	Dust samples from a *bed cover, rug, or upholstered furniture*
Entryway dust (interior)	EWY-I	Dust samples from *inside* an *entryway*
Floor dust	FLR	Dust samples from the *floor*
	FLR-V	*Vacuum* dust samples from the *floor*
	FLR-W	*Wipe* dust samples from the *floor*
Window stool dust	WST	Dust samples from a *window stool*
	WST(1/2)	Dust samples from a *split window stool*
	WST-V	*Vacuum* dust samples from a *window stool*
	WST-W	*Wipe* dust samples from a *window stool*
Window channel dust	WCH	Dust samples from a *window channel*
	WCH(1/2)	Dust samples from a *split window channel*
	WCH-V	*Vacuum* dust samples from a *window channel*
	WCH-W	*Wipe* dust samples from a *window channel*
Soil	BDY	Soil samples from the *boundary* of the property
	EWY-O	Soil samples from *outside* an *entryway*
	FDN	Soil samples near the *foundation* of the unit

The ellipse plotted in each box of Figure 1 is the ellipse that contains 95% of the probability associated with the estimated bivariate normal distribution for the plotted data. The narrower the ellipse, the stronger the correlation between the two sample types. If the ellipse is oriented from the lower left-hand corner of the box to the upper right-hand corner of the box, the sample types are positively correlated. If, on the other hand, the ellipse is oriented from the upper left-hand corner of the box to the lower right-hand corner of the box, the sample types are negatively correlated.

It may be possible that the correlation present in the lead loading data, or conversely the lack of correlation, is due to nonrandom factors such as renovation or abatement. For example, if all units that were abated have very high lead loadings on both floors and window sills, and unabated units have very low levels for both of these sample types, then floor loadings and window sill loadings will be very highly correlated when there may be no correlation at all beyond the effect of abatement history. Therefore, the correlation matrix and scatterplot matrix were created after controlling for fixed renovation and abatement effects.

Specifically for each sample type, the residuals from the log-linear models were averaged to produce average residuals for each unit. These average unit residuals were then used in calculating the correlation coefficients in Table 4. The average unit residuals were also plotted in a scatterplot matrix in Figure 1. When controlling for the fixed effects, one must realize that some degrees of freedom for estimation of correlation are sacrificed to estimate the fixed effects. This is accounted for in the significance levels and degrees of freedom provided in Table 4.

After correcting for renovation and abatement affects, none of the correlation estimates was observed to be significant. (Since only six houses were sampled in the pilot study, and two house-level fixed effects were found to be important, the statistical power to detect non-zero correlations is relatively low.) However, there are several relationships worth noting. Lead loadings for entryway, floor (vacuum), and floor (wipe) samples were all found to be negatively correlated with each other after correcting for the fixed effects. This result is counter intuitive (i.e., positive correlations would be expected) and may suggest that the effects of renovation and abatement are stronger than any house-to-house relationship between these sample types. Lead loadings for air ducts and bed/rug/upholstery samples had the highest correlation coefficient. Lead loadings for window stool vacuum and wipe samples were also found to be positively correlated. In addition, lead loadings for entryway samples were found to be negatively correlated with those for every other sample type, after correction for the fixed effects.

Lead Concentration

Table 6 contains unit-to-unit correlation coefficients for the lead concentration data after correcting for renovation and abatement effects. This table is analogous to Table 4, but is for lead concentrations rather than for lead loadings. The mean lead concentration residuals are plotted in scatterplot matrix form in Figure 2; this figure is analogous to Figure 1.

After correction for renovation and abatement effects, there are many positive relationships exhibited in the data, and some were statistically significant. Lead concentrations for floor samples are significantly correlated with soil samples taken at the boundary, entryway (marginal), and foundation. The lead concentrations among soil samples are still strongly correlated after controlling for the fixed effects. This may indicate that it is not the fixed effect of renovation or abatement that causes the data for these soil sample types to be correlated.

C. COMPARISON OF VACUUM AND WIPE SAMPLING PROCEDURES

One of the objectives of the pilot study was to compare the CAP vacuum sampling protocol with the HUD demonstration wipe sampling protocol. In each of the units a "bridge room" was selected and side-by-side vacuum and wipe samples were taken in the room. The purpose of collecting these data was to build a "bridge" between the sampling method for the CAP Study, the vacuum method, and the wipe sampling method employed in the HUD demonstration and numerous other studies of lead in household dust.

With regard to window channel samples, the pilot study design called for sampling from two split windows in the "bridge" room in each unit. One window was to have both vacuum and wipe samples taken, and the other was to have either two vacuum samples or two wipe samples taken. However, sampling window channels turned out to be a difficult task; only four split window channels were actually sampled,

Table 6 Unit-to-Unit Correlations Among Sample Types After Correction for Renovation and Abatement Effects: Lead Concentration (μg/g)

		Vacuum						Soil		
		Air duct	Bed/rug uph	Entryway	Floor	Window stool	Window channel	Boundary	Entryway	Foundation
Vacuum	Air duct	0.99ᵃ	0.45	0.76	0.59	0.24	0.80	0.92	0.89
		0.09ᵃ	0.70	0.45	0.60	0.85	0.41	0.26	0.30
		(2)ᵃ	(2)	(2)	(2)	(2)	(2)	(2)	(2)
	Bed/rug/ uph		0.43	0.75	0.59	0.26	0.80	0.92	0.88
			0.72	0.46	0.60	0.83	0.41	0.26	0.32
			(2)	(2)	(2)	(2)	(2)	(2)	(2)
	Entryway			0.35	-0.04	0.26	0.20	0.54	0.33
				0.65	0.96	0.74	0.80	0.46	0.67
				(3)	(3)	(3)	(3)	(3)	(3)
	Floor				0.92	0.79	0.97	0.93	0.96
					0.08	0.21	0.01	0.07	0.04
					(3)	(3)	(4)	(3)	(3)
	Window stool					0.77	0.94	0.76	0.88
						0.23	0.06	0.24	0.12
						(3)	(3)	(3)	(3)
	Window channel						0.70	0.61	0.61
							0.23	0.39	0.39
							(3)	(3)	(3)
Soil	Boundary							0.90	0.98
								0.10	0.02
								(3)	(3)
	Entryway								0.95
									0.05
									(3)
	Foundation								

ᵃ Top value is estimated correlation coefficient, middle value is observed significance level, and bottom value is degrees of freedom.

and only one window was sampled with both the vacuum and wipe sampling methods. Therefore, the vacuum and wipe sampling methods were not formally compared for the window channel data.

The paired floor lead loadings are plotted in Figure 3. In the figure, lead loadings from wipe samples are plotted vs. lead loadings from vacuum samples. A reference line that represents complete agreement between the two sampling methods is also included. With one exception, the lead loadings from wipe samples exceed the lead loadings from vacuum samples. A statistical analysis was performed to quantify this relationship.

Both the vacuum lead loadings and wipe lead loadings were assumed to follow a lognormal distribution. For this reason a log-linear model was employed to characterize the relationship between wipe and vacuum lead loadings. This model was fitted to the six pairs of floor lead-loading measurements plotted in Figure 3, and the hypothesis of a fixed multiplicative bias was tested. The estimate of the multiplicative bias of wipe over vacuum measurements was 9.76, with a 95% confidence interval of 1.88, 50.62. This result implies that, on average, the wipe lead loadings are 9.76 times larger than matching vacuum lead loadings, on floors.

The paired window stool lead loadings are plotted in Figure 4. The statistical analysis performed for floor lead loadings was repeated for the window stool lead loading data. For window stool lead loadings, the estimated multiplicative bias of wipe over vacuum measurements was 4.50, with a much tighter 95% confidence interval of 3.48, 5.82. This result implies that, on average, the window stool wipe lead loadings are 4.50 times larger than the matching vacuum lead loadings.

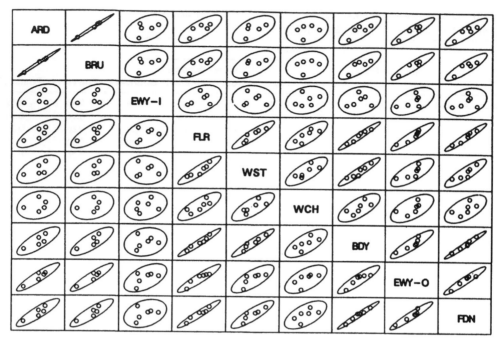

Figure 2 Scatterplot matrix of geometric unit means for different sample types after correction for renovation and abatement effects: lead concentration (µg/g).

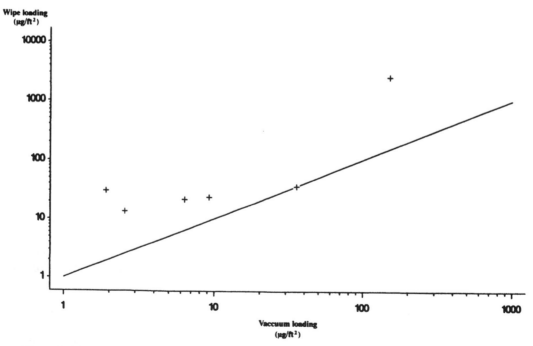

Figure 3 Vacuum vs. wipe comparison: geometric means by housing unit for floor lead loadings (µg/ft²).

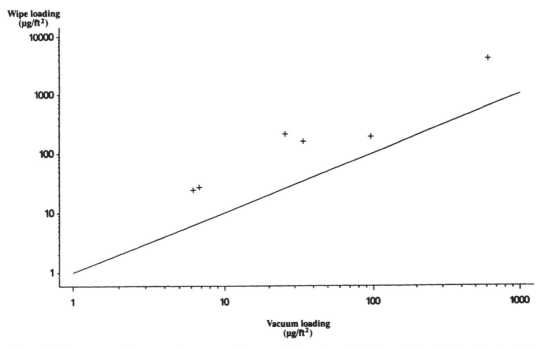

Figure 4   Vacuum vs. wipe comparison: geometric means by housing unit for floor window stool loadings (μg/ft²).

REFERENCES

1. Battelle Memorial Institute and Midwest Research Institute, Quality Assurance Project Plan for the Abatement Performance Pilot Study, report to U.S. Environmental Protection Agency, Office of Pollution Prevention and Toxics, Washington, D.C., 1991.
2. Battelle Memorial Institute, Final Report for the Comprehensive Abatement Performance Pilot Study, Vol. I, Results of Lead Data Analyses, report to U.S. Environmental Protection Agency, Office of Pollution Prevention and Toxics, Washington, D.C., 1993.

Chapter 7

Incidence of Severe Lead Poisoning in Children in Trinidad Resulting from Battery Recycling Operations

I. Chang-Yen, C. Emrit, and A. Hosein-Rahaman

CONTENTS

I. INTRODUCTION

Lead poisoning in children has recently become a very topical issue,[1] largely due to the realization that blood lead levels previously thought safe, could result in adverse long-term effects.[2] Effects attributed to chronic lead poisoning include microcytic anaemia,[3] reduced growth in stature,[4] hearing and speech impairment,[5] retarded mental development,[6–8] and reduced IQ levels.[9,10] In extreme cases, frank anaemia, nephropathy, encephalopathy, and even death may occur.[2]

Like adults, children may be exposed to lead from a variety of sources. However, children absorb a greater percentage of ingested lead, and are also more susceptible to lead poisoning,[11,12] than adults. Common sources of exposure to lead include exhaust fumes from vehicles burning leaded gasoline,[13,14] lead-based paints,[15,16] lead-glazed ceramics,[17] and even food[18] and water.[19]

In the Caribbean, cases of lead poisoning have been reported in Jamaica,[20] involving children playing with discarded lead battery plates. Likewise, lead smelting[21] and battery repair operations[22] have resulted in elevated blood lead levels of children involved in, or living close to, these activities. In Barbados a community of potters using lead glazes suffered chronic lead poisoning.[23] In Trinidad and Tobago, where leaded gasoline and lead-based paints are still used, lead is ubiquitous in the local environment, particularly at roadsides.[24,25] In addition, although regulations exist governing the manufacture and repair of lead batteries,[26] there is rarely enforcement. Similarly, while a code of practice for lead workers exists, it is hardly ever observed.[27]

The present study is an example of severe environmental pollution that resulted from lead battery recycling operations, and subsequent toxicity to an entire extended family in Trinidad.

II. INVESTIGATION OF LEAD POISONING INCIDENTS

A. INITIAL CASES

On December 26, 1991, a 2-year-old boy from Tunapuna, north-central Trinidad, was admitted to the Port of Spain General Hospital, suffering from epileptic seizures and paralysis of the left side of his body. On January 2, 1992, his 6-year-old sister was admitted to the same hospital, but to a different ward, with paralysis of the ankles. Unfortunately, the relationship between the two children was not recognized until a day later when the possibility of lead poisoning was finally raised. Further delay in confirming lead poisoning was experienced due to the inability of local regulatory laboratories to carry out blood lead analyses. Blood samples finally reached us on January 7 for lead determination. In the meanwhile the condition of the 2-year-old worsened, and following a series of cardiac arrests, he died on January 9, 1992. His death, based on blood lead analysis (218 µg/dl), was attributed to lead poisoning. The 6-year-old with a blood lead level of 235 µg/dl was chelated with calcium disodium EDTA and survived.

1-56670-113-9/95/$0.00+$.50

Table 1 Blood Lead Concentrations in Children

Family	Age of child (years)	Blood lead (µg/dl)	Poison risk class
A	0.3	57	IV
	0.3	46	IV
	2 (deceased)	218	V
	4	104	V
	6	235	V
	14	35	III
	17	71	V
	19	32	III
	(mother, 37)	42	
B	5	88	V
	7	49	IV
	16	65	IV
	(parents not available)		
C	11	47	IV
	12	50	IV
	14	78	V
	(father, 44)	10	
D	4	20	III
	9	34	III
	(mother, 32)	17	
E	2	65	IV
	7	26	III
	10	25	III
	11	25	III
	(mother, 29)	20	

B. FOLLOW-UP STUDIES

Following the above incidents, medical public health, and laboratory staff visited the home and environs of the children. Investigations revealed that a plot of land, approximately 30 × 35 m, was the original site of metal salvaging operations, including lead recovery from batteries. Although such operations had ceased over 13 years earlier, the remains of batteries that had been used, together with topsoil, as a landfill were clearly evident. A large number of fragments of battery cases protruded from the ground, with others lying loosely on the surface, easily accessible to children. The surface dust was loose, fine, and ubiquitous around the site. Five houses had been constructed on the site, and were occupied by descendants of the operators of the metal salvaging operations.

In addition to the two children hospitalized, at least 18 others had lived for most of their lives on the site. Three of these, although having relocated about a year ago, were still frequent visitors.

Blood Lead Determination

Venous blood samples were collected from children and their parents into EDTA-containing vials and stored at 5°C until analysis.

A Varian SpectrAA 300 atomic absorption spectrophotometer, a Varian GTA96 graphite tube atomizer fitted with pyrolytic-coated partition graphite tubes, and an automated sample dispenser were used for all blood analyses. The furnace temperature program used was based on that of the Centers for Disease Control.[28]

Blood lead controls (Kaulson Labs, Inc., NJ) were appropriately diluted with a matrix modifier[28] and used to establish the accuracy of the analytical procedure. Blood samples from an international blood lead proficiency testing program (Centers for Disease Control, Atlanta) were also analyzed with samples. Multiple standard additions of a 250-µg/l Pb^{2+} standard to the control samples were used to calibrate the procedure. Mean slope values were then used to calculate sample lead concentrations. Controls were inserted at regular intervals among samples, to maintain accuracy of analyses. Results of blood-lead analyses are given in Table 1.

Soil Lead Determination

Soil samples were taken from various areas of the contaminated site, that were easily accessible to children (Figure 1). Surface samples (<1 cm depth) over approximately 1 m^2 at each sampling point were collected

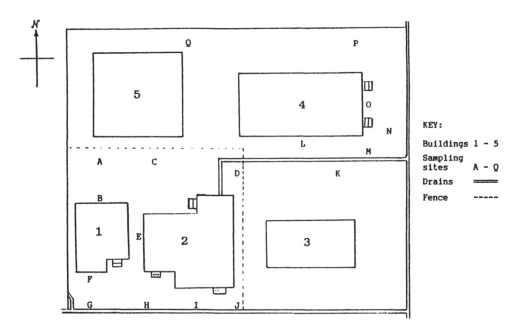

Figure 1 Lead-contaminated site in Trinidad.

Table 2 Soil Lead Levels at Lead-Contaminated Site

Sampling point	Lead level (wt. %)
A	0.20
B	1.64
C	0.56
D	0.61
E	0.53
F	2.39
G	0.59
H	1.95
I	2.73
J	1.25
K	0.18
L	1.45
M	0.76
N	1.46
O	1.18
P	0.20
Q	1.35

with a plastic scoop into resealable polyethylene bags. All samples were transferred to glass beakers in the laboratory and dried at 105°C for 24 h.

Each sample was sieved through a 600-μm stainless steel sieve (BS 410, Endecotts Ltd., London) and the larger particles discarded. The sieved portion of each sample was thoroughly mixed, and replicate aliquots (about 0.5 g each) were transferred to boiling tubes. Analytical grade nitric acid (10 ml) was added to each tube, and predigestion at room temperature was allowed for 18 h. Samples were then digested at 130°C for 2 h. Following dilution with distilled water, each sample was filtered through a Whatman #42 filter into a 100-ml volumetric flask and made up to the mark with rinsings of filter and residues. After appropriate dilution, samples were analyzed by flame atomic absorption spectrophotometry, with deuterium background correction. After correction for acid and filter blanks, the results of soil analyses are given in Table 2.

III. RESULTS AND DISCUSSION

A. BLOOD LEAD LEVELS

As Table 1 shows, of the 20 children monitored, the two who were initially hospitalized had blood lead levels exceeding 200 µg/dl. These levels are more than eight times the 1985 intervention level, or 25 times the level now adopted for young children.[2] It was no surprise that the 2-year-old died, and while the 6-year-old survived, she has been assessed as having a mental age of 3 and still suffers paralysis of the limbs.

Blood lead levels of the others showed that four other children suffered class V poisoning, requiring immediate medical and environmental management.[2] Seven children suffered class IV poisoning, requiring prompt medical and environmental intervention, while the other seven with class III poisoning required further investigation. Ongoing studies have shown that blood lead levels of children not normally exposed to sources of lead are less than 10 µg/dl. The generally elevated blood lead levels of all children tested, ranging in ages from 4 months to 19 years, are consistent with exposure to a common source of lead. Even the children of family C, who were nonresident, but still frequent visitors to the site, had blood lead levels as high as the others.

Of the four parents tested, three had blood lead levels lower than those of their children. The fourth parent, who was breast feeding her 4-month-old twins at the time, had a blood lead level as high as three of her children. The reason for this is unclear, but may be linked to her maternal activities.

B. SOIL LEAD LEVELS

Table 2 shows the soil lead levels of sampling points A to Q in Figure 1. Lead concentrations in the <600-µm fractions, containing particles likely to be ingested by children, ranged from 0.20 to 2.73%. All sampling points were readily accessible to the children, who spent much of their spare time playing on the site. Hand-to-mouth activities or pica were reported to be common, such activities being reported to be a major route of entry of lead into children.[12,29] The high lead level of the 17-year-old in family A (71 µg/dl) thus appears to be the cumulative effect of continuous exposure to his lead-contaminated surrounding.

Blood lead levels in children have been correlated with soil lead levels around secondary lead smelters,[21] battery repair shops,[22] and dusts in urban environments.[30] The peak soil lead levels recorded in this study are well in excess of those reported in the other studies and were considered responsible for the elevated blood lead levels in the affected children.

IV. FOLLOW-UP ACTION

Based on the results of the above investigations, specific remedial action was taken by local municipal and health authorities. Medical attention was provided for children suffering from classes IV and V poisoning. Chelation therapy was administered to most of them, resulting in significant reduction in their blood lead levels. The families were relocated to allow environmental remediation of the site. Two of the houses were dismantled, and the upper 20 to 35 cm of soil, containing visible lead battery wastes, were excavated. After mixing with cement clinker dust, the wastes were removed to a sanitary waste disposal site. Uncontaminated gravel was placed on the site, but has yet to be spread over the excavated area.

However, much more still needs to be done. Continued therapy and monitoring, as well as mental assessment of the lead-poisoned children are still required. Likewise, environmental monitoring of the site and those adjacent to it, and runoff into drains leading to agricultural holdings downstream, are urgently required. Unfortunately, lack of funding, and a legal contest over ownership of the site, have halted all further monitoring for the time being.

A recent survey also confirmed that a large number of lead battery repair and recycling, and lead smelting, operations exist country-wide, and almost all are unregulated. We anticipate that in the absence of an effective national policy to control and monitor such activities, lead poisoning in local children will continue to be experienced.

ACKNOWLEDGMENTS

The authors are grateful to the Tunapuna Regional Corporation for its prompt assistance in our investigation.

REFERENCES

1. Emerson, T., Waldman, S., Marshall, R., Hall, C., and Waldrop, T., Lead and your kids, *Newsweek*, 34, 1992.
2. Preventing Lead Poisoning in Young Children, USDHHS 1, Center for Disease Control, U.S. Department of Health, Education, and Welfare, Atlanta, 1991.
3. Boeckx, R.L., Lead poisoning in children, *Anal. Chem.*, 275A, 1986.
4. Shukla, R., Bornschein, R.L., Dietrich, K.N., Buncher, C.R., Berger, O.G., Hammond, P.B., and Succop, P.A., Fetal and infant lead exposure: effects on growth in stature, *Pediatrics*, 84, 604, 1989.
5. Schwartz, J. and Otto, D., Blood lead, hearing thresholds, and neurobehavioral development in children and youths, *Arch. Environ. Health*, 42, 153, 1987.
6. Bellinger, B., Leviton, A., Waternaux, C., Needleman, H., and Rabinowitz, M., Longitudinal analyses of prenatal and postnatal lead exposure and early cognitive development, *N. Engl. J. Med.*, 316, 1037, 1987.
7. Dietrich, K.N., Kraft, K.M., Bornschein, R.L., Hammond, P.B., Berger, O., Succop, P.A., and Bier, M., Low level lead exposure on neurobehavioral development in early infancy, *Pediatrics*, 80, 721, 1987.
8. McMichael, A.J., Baghurst, P.A., Wigg, N.R., Vimpani, G.V., Robertson, E.F., and Roberts, R.J., *N. Engl. J. Med.*, 319, 468, 1988.
9. Needleman, H.L., Gunnoe, G., Leviton, A., Reed, R., Peresie, H., Maher, C., and Barret, P., Deficits in physiologic and classroom performance in children with elevated dentine lead levels, *N. Engl. J. Med.*, 300, 689, 1979.
10. Needleman, H.L. and Gatsonis, C.A., Low level lead exposure and the IQ of children, *JAMA*, 263, 673, 1979.
11. Lin-Fu, J.S., Vulnerability of children to lead exposure and toxicity, *N. Engl. J. Med.*, 289, 1289, 1973.
12. Lin-Fu, J.S., Undue absorption of lead among children — a new look at an old problem, *N. Engl. J. Med.*, 286, 702, 1972.
13. Caprio, R.J., Margulis, H.L., and Joselow, M.M., Lead absorption in children and its relationship to urban traffic densities, *Arch. Environ. Health*, 28, 195, 1974.
14. Schutz, A., Ranstam, J., Skerfving, S., and Tejning, S., Blood-lead levels in school children in relation to industrial emission and automobile exhausts, *Ambio*, 13, 115, 1984.
15. Chisolm, J.J., Jr., Mellits, E.D., and Ouaskey, S.A., The relationship between the level of lead absorption in children and the age, type, and condition of housing, *Environ. Res.*, 38, 31, 1985.
16. Farfel, M.R. and Chisolm, J.J., Jr., Health and environmental outcomes of traditional and modified practices for abatement of residential lead-based paint, *Am. J. Public Health*, 80, 1240, 1990.
17. Klein, M., Namer, R., Harpur, E., and Corbin, R., Earthenware containers as a source of fatal lead poisoning, *N. Engl. J. Med.*, 283, 669, 1970.
18. Survey of lead in food. Second Supplementary Report. Working Party on the Monitoring of Foodstuffs for Heavy Metals, Tenth Report, Ministry of Agriculture, Fisheries, and Food. H.M.S.O., 1982.
19. Moore, M.R., Meredith, P.A., Campbell, B.C., Goldberg, A., and Pocock, S.J., Contribution of lead in drinking water to blood level, *Lancet*, ii, 661, 1979.
20. Burke, L.M., Chronic lead poisoning, *West Indian Med. J.*, 6, 105, 1957.
21. Matte, T.D. and Burr, G.A., NIOSH Health hazard evaluation report. Technical assistance to the Jamaican Ministry of Health, Kingston, Jamaica, HETA 87–371–2000, 1989.
22. Matte, T.D., Figueroa, J.P., Burr, G., Flesch, J.P., Keenlyside, R.A., and Baker, E.C., Lead exposure among lead acid-battery workers in Jamaica, *Am. J. Ind. Med.*, 16, 167, 1989.
23. Koplan, J.P., Wells, A.V., Diggory, H.J.P., Baker, E.K., and Liddle, J., Lead absorption in a community of potters in Barbados, *Int. J. Epidemiol.*, 6, 225, 1977.
24. Bernard, G., A study of lead pollution resulting from automotive exhaust emission. M.Phil. thesis, University of the West Indies, St. Augustine, 1979.
25. Chang-Yen, I., Pooransingh, N., and Miles, M., Lead levels in roadside dust: a commuter problem. 2nd Caribbean Conf. on Transportation and Traffic Safety, May 14–16, Port of Spain, Trinidad.
26. The Electric Accumulator (Manufacture and Repair) Order 1974, Trinidad and Tobago.
27. Chapman Boyd, B., Lead workers in Trinidad and Tobago — problems of health care, *Caribbean Med. J.*, 29, 37, 1978.
28. Determination of lead in blood by graphite furnace atomic absorption spectrometry with deuterium background correction, Method 1080A, Centers for Disease Control, Center for Environmental Health and Injury Control, Atlanta, 1990.
29. Jacobziner, H., Lead poisoning in childhood: epidemiology, manifestations and prevention, *Clin. Pediatr.*, 5, 277, 1966.
30. Committee on Environmental Hazards/Committee on Accident and Poison Prevention, Statement on childhood lead poisoning, *Pediatrics*, 79, 457, 1987.

Part II
Program and Policy Issues

Chapter 8

The U.S. Environmental Protection Agency's Broad Strategy to Address Lead Poisoning

J. S. Carra

CONTENTS

I. LEAD AS AN ENVIRONMENTAL PROBLEM

In February of 1991 the U.S. Environmental Protection Agency (EPA) officially launched a concerted attack against lead poisoning. More often than not, EPA is forced to confront environmental and public health issues, with limited information about the potential for a toxic substance to produce a harmful effect. The case of lead is, however, far different from the usual situation where we make policy decisions in the face of limited scientific data. Some of the health consequences of exposure to lead have been known since ancient times. Because lead has been widely used for centuries, today it is ubiquitous in the environment.

The major concerns about lead exposure revolve, of course, around children. While preschool-age youngsters are not the only group that can experience adverse health effects from such exposure, they are the population group most vulnerable.

According to an October 1991 Centers for Disease Control (CDC) report, "Lead poisoning is one of the most common and preventable pediatric health problems today." The CDC estimates that three million children in the U.S. have blood lead levels above 15 µg/dl of blood, a level higher than the current level of concern. After reviewing new data showing significant adverse effects of lead exposure in children at blood lead levels previously thought to be safe, in 1991 the CDC lowered its intervention level from 25 to 10 µg/dl.

At the same time that the intervention level has been continuously lowered, substantial progress has been made in reducing childhood blood lead levels. For example, deaths from lead poisoning are rare today. Nevertheless, as the CDC has stated, large numbers of children continue to have blood lead levels high enough to cause concern.

There are three primary sources of lead exposure in the U.S. Lead-based paint still found in older homes continues to be the major source of high-dose lead poisoning in children. Lead in urban soil and dust is believed to be another significant source of exposure. Soil and dust have become contaminated by deteriorating lead-based paint, by past use of leaded gasoline, and by industrial point sources. Lead in drinking water is the third primary area of concern. This is the result of the use of lead in solder, brass fittings, and service lines and generally contributes low to moderate exposures to relatively large populations. Although lead poisoning knows no racial, ethnic, or economic boundaries, poor and minority children are at greatest risk of being affected.

In recognition of lead poisoning as a significant public health problem, about 4 years ago EPA began to develop a broad strategy for EPA, and indeed federal government-wide, involvement in dealing with lead exposures.

II. EPA'S LEAD STRATEGY IS BROAD BASED

After several months of developmental work, in February of 1991 EPA unveiled its strategy for attacking the human health and environmental consequences of lead contamination. The EPA strategy is a multimedia approach involving the coordinated efforts of a variety of program offices. EPA also coordinates with other agencies, such as the Department of Housing and Urban Development (HUD), the Centers for Disease

Control (CDC), and the Occupational Safety and Health Administration (OSHA), which offer resources for dealing with the lead problem.

The strategy is managed by an agency-wide workgroup that functions to share information and data and to provide a unified, consistent approach. The Office of Pollution Prevention and Toxics (OPPT) chairs this workgroup and plays the leadership role in coordinating EPA lead-related initiatives. The lead strategy is unique among EPA activities in that we are looking across the board at sources of lead exposure. This multimedia approach is different from the usual tendency to look in a piecemeal approach at soil, water, or air pollution.

EPA's lead strategy has one goal: to reduce lead exposure to the fullest extent practicable, with particular emphasis on reducing the risk to children. In order to reach this goal, we have established two objectives to help us set priorities and to measure program success:

1. First, we want to significantly reduce the incidence of blood lead levels above 10 µg/dl in children, while taking into account the associated costs and benefits. Children are the most sensitive population, and this is consistent with the CDC direction on lowering the level of concern.
2. Second, we want to significantly reduce, through voluntary and regulatory efforts, unacceptable lead exposures that are anticipated to pose risks to children, the general public, or the environment. Again, these sources are primarily lead in paint, in urban soil and dust, and in drinking water, but other sources may contribute to unacceptable exposures in some populations.

Activities that EPA is undertaking to implement the lead strategy fall into certain categories:

1. Addressing the proper management and abatement of lead hazards from *past uses*
2. Vigorously *enforcing* our *current standards*, such as the National Ambient Air Quality Standards for lead
3. Conducting investigations as to whether additional regulations are needed to address *current uses* of lead that present the greatest risks
4. Establishing a system to monitor *new uses* of lead, so that we avoid uses that may subject people to additional significant risks from lead
5. *Public education* to let parents know what *they* can do to reduce risks from lead to their children
6. *Research* to develop better methods to identify, assess, and cost effectively abate the risks from lead

The lead strategy enjoys high visibility and high priority at EPA, both in headquarters and in the field. Lead poisoning as a public health issue has also become a popular topic outside the EPA, as is demonstrated by the many conferences, workshops, publications, etc., to which EPA staff are asked to contribute. Also, it is of considerable interest on Capitol Hill, where a variety of legislative proposals pertaining to lead exposures are pending. In fact, at the request of Congress, EPA is currently working on a comprehensive lead strategy that will truly be government-wide. This was completed in 1993.

The priority that lead enjoys in EPA is illustrated by the resources that have been devoted to addressing lead issues. Many of EPA office directors serve on a lead committee. At the next lower level, a lead cluster group comprises 80 staff members representing 15 offices and 8 regions. This cluster group is further divided into six activity areas:

1. Research
2. Urban lead issues
3. Regulatory implementation and enforcement
4. Public education
5. Risk assessment
6. Evaluation and planning

As needs change and we identify new areas of concern, EPA's structure for addressing lead issues may be revised. It is a safe assumption, nevertheless, that lead will continue to be the object of study and action, in the years ahead. Already EPA has accomplished several things and has many more activities underway in its lead program. A summary of these activities follows.

III. REGULATORY AND NONREGULATORY ACTIONS TO REDUCE LEAD EXPOSURES

OPPT has primary responsibility for several EPA lead activities. At the same time, a number of other offices are cooperating with initiatives of their own. The following pages summarize EPA's lead activities, with those which are regulatory in nature presented first.

A. CURRENT EPA REGULATORY ACTIVITIES

A TSCA Significant New Use Rule (SNUR) currently under consideration would reduce the potential for risks from new lead-containing products. This would allow EPA to review proposed new uses of lead, before products enter the marketplace, and would include a list of ongoing uses that would be exempt from review.

Still another activity relates to implementation of the National Primary Drinking Water Regulation, which was promulgated in the spring of 1991. Under this rule large and medium size water systems have completed a first round of tap water monitoring for lead. A variety of actions may be required, depending on what the data show. These actions include corrosion control treatment, source water treatment, public education, and lead service-line replacement.

Efforts to ensure that all areas of the country meet the National Ambient Air Quality Standards for lead comprise another piece of the EPA's lead program. Existing state implementation plans called for under the Clean Air Act are being enforced and revised for lead nonattainment areas of the country. EPA is currently conducting monitoring and carrying out enforcement actions, designed to identify these areas and bring them into attainment.

B. NONREGULATORY ACTIVITIES

While EPA is traditionally viewed as a regulatory agency, there are many nonregulatory initiatives we can pursue to address significant environmental problems. These complement the stronger regulatory tools with which we are most often associated. Many of EPA's nonregulatory lead initiatives are designed to address risks from paint, dust, and soil and will answer questions of (1) what to abate, (2) how to abate, (3) who will abate, and (4) what individuals can do.

On the question of what to abate, there are five specific activities in progress.

1. Field test kits for commercial and home use are being developed and evaluated to test for lead-based paint.
2. We are refining methods for detecting and measuring lead in paint, soil, and dust, both in the laboratory and in the field.
3. We have established a National Lead Laboratory Accreditation Program. This includes development of protocols and standard reference materials.
4. We are expanding an existing model to define acceptable levels of lead in paint, soil, and dust, and from this we will develop interim health-based standards.
5. A community-based guide to assist local governments in creating multimedia primary prevention programs is near completion.
6. We have published a guideline on the proper conduct of remodeling and renovation activities.

Certain initiatives are also directed toward understanding and explaining *how* to abate lead problems:

1. There is a study underway on repair and maintenance activities. This will assist in the development and evaluation of lower-cost abatement techniques.
2. Another study of low-cost abatement focuses on in-place management of lead-based paint and public education.
3. A complete abatement performance study analyzed the long-term efficacy of lead-based paint abatement methods used by HUD in its Private Housing Demonstration Study.
4. An abatement debris study will be included in a Report to Congress on the applicability of the Resource Conservation and Recovery Act to lead-based paint debris.
5. We are testing a new protocol for evaluating lead-based paint encapsulants.
6. We are completing a major field evaluation of XRF and home test kit devices.

On October 28, 1992, the Housing and Community Development Act of 1992 (P.L. 102–550), which includes Title X—"The Residential Lead-Based Paint Hazard Reduction Act of 1992," was signed into law. Title X provides for a comprehensive national approach to dealing with lead-based paint in the nation's housing stock. It changes the program's philosophy from total abatement to a program of abatement and in-place management of priority hazards, and allows EPA and HUD to focus on hazardous conditions rather than the mere presence of lead-based paint.

Title X, which amended TSCA by adding Title IV, mandates that EPA promulgate a number of regulations. These include:

1. TSCA Section 402(a), Lead-Based Paint Activities Training and Certification Regulations; TSCA Section 404(d), Model State Program
   - Requires EPA to promulgate regulations ensuring that individuals engaged in lead-based paint activities are trained, training programs are accredited, and contractors are certified, and to set standards for performing abatement activities.
   - Requires EPA to promulgate model State programs for compliance with training and accreditation regulations, including application process and compliance monitoring.
2. TSCA Section 403, Identification of Dangerous Levels of Lead
   - Requires EPA to promulgate regulations to identify lead-based paint hazards in paint, dust, and soil.
3. Section 1018, Disclosure of Information Concerning Lead upon Transfer of Residential Property
   - Requires EPA and HUD to jointly promulgate regulations mandating that:
     - (i) Purchasers/lessees receive EPA's lead pamphlet;
     - (ii) Sellers/lessors disclose all known lead hazards to purchasers/lessees;
     - (iii) Purchasers have a 10-day period for inspection for lead-based paint hazards;
     - (iv) Sales contracts contain a lead warning statement.
4. TSCA Section 406(b), Renovation Information Rule
   - Requires EPA to promulgate rule requiring renovators and remodelers to furnish customers with copy of EPA brochure prior to beginning work.

Other activities will improve the skills of those who work in the abatement field:

1. Model training courses for workers, inspectors, and supervisors are in various stages of development, and some are completed.
2. We have funded five university-based consortia to establish regional lead training centers. This program began operation in July 1992 and will facilitate the development of professionals qualified to identify and reduce risks from lead in paint, soil, dust, and water.
3. EPA and HUD have given grants to labor-management trust funds to assist nonprofit groups conducting lead-based paint abatement worker training.
4. A congressionally mandated lead worker accreditation program is being developed. This model plan will help guide state certification and licensing of contractors and workers. It will also provide an overall framework for the training, testing, and licensing of lead inspectors, contractors, and workers.

Another nonregulatory area relates to assisting the general public. In our efforts to educate people about lead poisoning and how it can be avoided, we will be taking advantage of a variety of types of activities. These include preparing publications, using both television and print advertisements to make the public aware, and operating government clearinghouses and hotlines to disseminate information. Certainly the area of public education is a central part of our overall strategy. We have to get the word out on lead if any of our other efforts are to pay off. The EPA-managed National Hotline is operational at 800-424-5323.

In addition to the foregoing, several other nonregulatory initiatives are underway. EPA has completed its Three-City Study in Boston, Baltimore, and Cincinnati to evaluate the effect of removing lead-contaminated soil and dust on children's blood lead levels. This project will help us to understand the movement of lead through the urban environment.

Also on the urban lead front, our waste management program is planning to issue a directive on soil cleanups, including guidance on a numerical soil cleanup level. An interagency lead-based paint task force is developing risk assessment tools for lead in soil. A multitude of questions still remain on the extent of contaminated soil, the effect of soil on blood lead levels, the effectiveness of various abatement techniques, and related matters. EPA will continue to work through these issues.

At the same time that all these activities are underway, we will be continuously reviewing and evaluating our lead program. A special group has been formed to examine progress over the past year and to recommend updates and improvements to the lead strategy, as well as activities for the future. In a related project, the Comparative Risk Study being conducted by the Office of Policy, Planning, and Evaluation is progressing. This analyzes the severity and magnitude of risks from a variety of pathways.

Research is a major component of the lead program. In 1991 our Lead Research Subcommittee produced the first cross-media research plan for lead (as opposed to a separate plan for each EPA media office). In 1992 our Office of Research and Development shifted to an issue-based budget planning process — away from a media-based plan — and has prepared plans for 37 issues. Lead is included as one of the 37 research issues being addressed. A lead research plan covers the 1993–1997 fiscal years.

It is obvious from this extensive list of projects that lead is a hot topic at EPA. Certainly it is the focus for one of the most comprehensive approaches to problem-solving we have seen in the environmental field.

IV. CONCLUSION

Lead presents unique concerns regarding its effects on human health and the environment. As a toxic substance it is also unique in that we know a great deal about it and can feel confident about investing resources in controlling risks from exposures. So often we have to act with limited information and on faith that the benefits of our actions are worth the costs entailed.

Because young children are the most seriously affected by lead poisoning, public attention and support have been particularly focused on addressing lead contamination of our environment. While children are not the only beneficiaries of EPA's lead activities, they are the ones to whom EPA's broad-based strategy is primarily directed. This strategy aims to control risks across environmental media. Such an approach is a departure from the more ordinary one of dealing with pollution medium by medium.

Virtually all of EPA is involved, to one degree or another, in conducting lead initiatives. Further, other agencies that can contribute to a comprehensive attack on this problem are working closely with EPA. EPA and other agencies are constantly seeking to refine our overall approach and to identify the best methods of combatting lead poisoning. We feel we have the support of Congress and of the public in our decision to invest heavily in a comprehensive lead strategy. It is a reasonable prediction that interest in lead will continue to be high and that a variety of lead-related activities will occupy our efforts, for some time to come.

Chapter 9

The U.S. Department of Housing and Urban Development's Lead Strategy and Lead Based Paint Program

R. J. Morony and B. T. Cook

CONTENTS

I. INTRODUCTION

The Lead-Based Paint Poisoning Prevention Act, as amended by Section 566 of the Housing and Community Development Act of 1987, required HUD to prepare and transmit to the Congress a comprehensive and workable plan, for the prompt and effective inspection and abatement of privately owned single-family and multifamily housing, including housing assisted under section 8 of the United States Housing Act of 1937. In December 1990 the U.S. Department of Housing and Urban Development (HUD) released its lead strategy, the "Comprehensive and Workable Plan for the Abatement of Lead-Based Paint in Privately Owned Housing." This strategy describes the extent of lead hazards in the nation's housing stock, and the components of a national plan to reduce exposure through the abatement of lead-based paint. This plan is one of many activities that include research, demonstration, and policy actions that were undertaken in response to the 1987 amendments to the Lead-Based Paint Poisoning Prevention Act. Other activities include (1) a national survey to better estimate the extent of lead hazards in our nation's housing stock, (2) a major multicity demonstration to identify the most cost-efficient methods for lead hazard abatement, and (3) the development of interim technical guidelines for the testing and abatement of lead hazards in public housing. In response to Congress' 1992 appropriations, a new office, the Office of Lead-Based Paint Abatement and Poisoning Prevention (OLBPAPP), was created within the Office of the Secretary.

II. THE COMPREHENSIVE AND WORKABLE PLAN FOR THE ABATEMENT OF LEAD-BASED PAINT IN PRIVATELY OWNED HOUSING

The comprehensive and workable plan is one of a series of research, demonstration, and policy actions initiated by HUD in response to the 1987 amendments. The lack of public awareness, coupled with the high cost of testing and abatement, has produced relatively little public or private action to address this public health concern. In response, HUD proposed a comprehensive plan to mitigate the problems that have inhibited efforts to address lead-based paint hazards. Activities planned by the department include

1. *Updating HUD lead-based paint regulations.* The department is conducting a review of existing program regulations and will propose modifications to make all HUD programs consistent with regard to recommendations to address lead-based paint.
2. *Expanded information and education effort.* There is a lack of public awareness of the seriousness of lead exposure and ways to avoid it. Efforts at HUD will accelerate the transmittal of information to the public.
3. *Research and demonstration activities.* HUD realizes the need for expanded health, epidemiologic, and environmental research, and various demonstration activities are necessary to support the effective elimination of lead poisoning. This includes research in cost-effective testing and abatement of lead in paint, dust, and soil, and providing a better understanding of the contribution of lead paint to blood lead levels.

4. *Capacity building and local program development.* In the U.S., state and local governments have primary responsibility for regulating housing conditions. Presently, most devote few resources to the problem of lead-based paint. The federal government must assist state and local governments to develop the capacity to assume a leadership role in regulating and managing large-scale and effective programs of lead-based paint hazard elimination.

5. *Fiscal assistance for lead-based paint abatement.* Options to provide additional financial support for single-family and multifamily residential abatement is needed. Low- and moderate-income homeowners, and/or landlords with priority hazards, and in housing occupied by families with children, would be eligible for abatement assistance.

III. HUD LEAD-BASED PAINT ACTIVITIES

The following is a partial list of the major HUD lead-based paint accomplishments and efforts that are currently underway.

1. *HUD-EPA MOU.* HUD and EPA executed a Memorandum of Understanding (MOU) on lead-based paint (LBP) in April 1989. This MOU has become the basis for the Federal Interagency cooperation on LBP, including the Federal Interagency Task Force. In addition, the MOU provides for an agreement between the two agency's on EPA technical support to HUD's congressionally mandated data collection efforts.

2. *Abatement grants.* HUD issued a notice of fund availability (NOFA) in the summer of 1992 for the HUD grant program to state and local governments for the abatement of low- and moderate-income housing. Approximately $46.4 million was awarded in fiscal year 1993, and approximately $93.4 million was awarded in fiscal year 1994. $139.4 million will be awarded in fiscal year 1995.

3. *Risk assessment.* OLBPAPP worked closely with the Office of Public and Indian Housing (PIH) to develop an appropriate protocol for determining the level of lead-based paint hazards in public housing, and setting priorities for action.

4. *Interim containment plan.* HUD is developing an in-place management protocol that will reduce the health threat in public housing units, containing LBP, that are not scheduled for abatement in the near future. It was published as part of the Public Housing's Risk Assessment.

5. *HUD task force.* A HUD task force on HUD lead-based paint regulations has been established and is responsible for the development of a common set of LBP regulations that will govern all HUD housing programs. All draft regulations will be completed in 1995.

6. *Development of final lead-based paint abatement guidelines.* HUD has entered into an interagency agreement with the National Institute of Standards and Technology (NIST) to carry out a critical review of the HUD document "Lead-Based Paint: Interim Guidelines for Hazard Identification and Abatement in Public and Indian Housing" and to recommend changes, if any, to plan and implement research needed to improve the technical basis for the guidelines; to prepare draft standardization through the voluntary consensus process; and to provide leadership and technical support in the organization and operation of the ASTM Subcommittee on Standards for Abatement of Hazards from Lead-Based Paint in Buildings.

7. *National clearinghouse and hotline.* HUD worked closely with EPA and CDC in the development of a national lead poisoning clearinghouse and toll-free "hotline." The hotline is managed by EPA: 800-424-5323

8. *Alliance grant.* HUD negotiated with the Alliance to End Childhood Lead Poisoning and EPA to establish a strategic framework for addressing lead-based paint and dust hazards in all categories of housing. The project will be conducted jointly with the EPA-funded project on hazard definition, priority setting, and appropriate responses using the Alliance consensus process.

Chapter 10

CDC's Perspective on Preventing Lead Poisoning in Young Children

S. Binder

CONTENTS

I. HISTORY OF PUBLIC HEALTH INVOLVEMENT IN CHILDHOOD LEAD POISONING PREVENTION

A. EARLY HISTORY OF CHILDHOOD LEAD POISONING

The modern era of childhood lead poisoning began in the 1890s, when Australian investigators reported on a cluster of cases of childhood lead poisoning for which no source of lead could be found. Twelve years later the source was identified as lead-based paint on walls and veranda railings, in the homes of these children. Subsequently, many cases of lead poisoning in children were identified in the U.S. and other nations. Lead paint in houses and on furniture and toys, and burning of battery casings for fuel in homes, were among the most important sources of lead in the early 1900s.[1]

Case finding efforts to identify lead-poisoned children started in Baltimore in the 1930s. Public health departments in other communities, however, did not get interested in childhood lead poisoning and its prevention until the 1950s, when caseworkers in a few large cities attempted to find lead-poisoned children. In 1966 Chicago began the first mass screening program, followed shortly by New York and other cities.[1]

B. HISTORY OF THE CENTERS FOR DISEASE CONTROL INVOLVEMENT IN CHILDHOOD LEAD POISONING PREVENTION

The Lead Paint Poisoning Prevention Act, passed in 1971, initiated a national effort to identify children with lead poisoning and to ensure abatement of the sources of lead in their environments. For most years of this program, federal funds appropriated under this Act were administered by the Centers for Disease Control (CDC). More than $89 million were distributed, and over a quarter of a million children were identified with lead poisoning and received referrals for environmental and medical intervention.[2]

In 1981 the Omnibus Budget Reconciliation Act created the Maternal and Child Health (MCH) Services Block Grant Program and consolidated many categorical programs, including that for childhood lead poisoning prevention, into the Block Grant. The MCH Block Grant Program continues to be a major source of funding for childhood lead poisoning prevention activities.[2]

The Lead Contamination Control Act of 1988 authorized CDC to once again administer a childhood lead poisoning prevention grant program. Under this law $4 million were appropriated in fiscal year 1990, $7.8 million were appropriated in 1991, and $21.3 million were appropriated in 1992.

C. THE CURRENT CDC PROGRAM IN CHILDHOOD LEAD POISONING PREVENTION

Currently, through the grant program authorized by the Lead Contamination Control Act, CDC funds childhood lead poisoning prevention programs in 31 states and cities (Figure 1). In addition, CDC provides technical assistance to programs, including on-site visits to CDC grantees. Special software (System for Tracking Elevated Lead Levels and Remediation [STELLAR]) has been developed to assist childhood lead poisoning prevention programs in case and data management.

CDC also provides assistance to laboratories, particularly those of childhood lead poisoning prevention program grantees. This assistance includes provision of reference materials and assistance in quality control and assurance. In addition, CDC, in conjunction with the University of Wisconsin and the Health Resources and Services Administration, conducts a proficiency testing program for analysis of lead and erythrocyte protoporphyrin levels in blood.

CDC efforts in surveillance and epidemiology complement grant program activities. CDC is developing national surveillance for lead levels in children, which will allow better targeting of resources, as well as evaluation of our progress in eliminating this disease. Studies and evaluations conducted by CDC focus on critical issues for the cost-effective expenditure of funds: for example, the effectiveness of interventions in reducing children's blood lead levels.

CDC also has placed great emphasis on the coordination with other child health programs serving similar populations. For example, CDC has conducted a series of meetings in regional offices, to bring together representatives of WIC, EPSDT, community health centers, etc., to identify ways in which these programs could coordinate better. CDC also works with other federal agencies involved in such related areas as environmental protection, housing, worker safety, and food safety.

Among CDC's most critical accomplishments has been the development and dissemination of two major policy statements: "Preventing Lead Poisoning in Young Children" and the "Strategic Plan for the Elimination of Childhood Lead Poisoning." These are described below.

II. THE CDC STATEMENT ON PREVENTING LEAD POISONING IN YOUNG CHILDREN

The CDC, and the Public Health Service before it, have been publishing statements on childhood lead poisoning, since 1970. The 1991 statement, "Preventing Lead Poisoning in Young Children,"[3] provides guidelines on childhood lead poisoning prevention for a number diverse groups. Public health programs that screen children for lead poisoning look to this document for guidance on screening regimens and public health actions. Pediatricians and other health-care practitioners look to this document for information on screening and guidance on the medical treatment of poisoned children. Government agencies, elected officials, and private citizens seek guidance about what constitutes a harmful level of lead in blood and what blood lead levels should trigger environmental and other interventions.

Since 1970, when the first lead statement was published, our understanding of childhood lead poisoning has changed substantially. As investigators have used more sensitive measures and better study designs, the generally recognized level for lead toxicity has progressively shifted downward. Before the mid-1960s, a level above 60 µg/dl was considered toxic.[4] By 1978 the defined level of toxicity had declined 50%, to 30 µg/dl. In the 1985 CDC lead statement, lead poisoning was defined as a blood lead level 25 µg/dl associated with an erythrocyte protoporphyrin level of 35 µ/dl. The 1991 statement identifies 10 µ/dl as the blood lead level of concern. Figure 2 shows how the federal definition of an elevated blood lead level has changed over the years.[3]

Other major differences between the 1985 and 1991 statements are as follows:

- Because the erythrocyte protoporphyrin test, which was identified as the screening test of choice in 1985, is not sensitive enough to identify children with elevated blood lead levels below about 25 µg/dL, the screening test of choice is now blood lead measurement.
- The 1985 statement emphasized screening children at high risk for lead poisoning. However, since virtually all children are at risk for lead poisoning, a phase in of universal screening is recommended in the 1991 statement, except in communities where large numbers or percentages of children have been screened and found not to have lead poisoning. The full implementation of this will require the ability to measure blood lead levels on capillary samples and the availability of cheaper and easier-to-use methods of blood-lead measurement.

- The previous lead statements emphasized secondary prevention — finding children with lead poisoning and taking steps to shorten the duration of their disease. The 1991 statement emphasizes primary prevention — efforts to prevent lead poisoning before it occurs. This will require communitywide environmental interventions, as well as educational and nutritional campaigns.

III. STRATEGIC PLAN FOR THE ELIMINATION OF CHILDHOOD LEAD POISONING

Lead poisoning is the most common and societally devastating environmental disease in children in this country, and it is entirely preventable. Therefore, in the 1980s the Department of Health and Human Services (HHS) began to call for the elimination of childhood lead poisoning. At the request of HHS, CDC developed the "Strategic Plan for the Elimination of Childhood Lead Poisoning," which was released by HHS in 1991.[2] This plan describes actions that can be taken by all levels of government and the private sector, in the first 5 years of a 20-year effort to eliminate childhood lead poisoning as a public health

Figure 1 States and cities receiving CDC grant funds for childhood lead poisoning prevention activities, 1992.

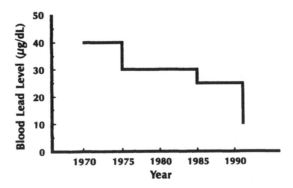

Figure 2 Changes in the blood lead level considered to be of concern by the Public Health Service and Centers for Disease Control.

problem. This plan was developed in consultation with other federal agencies, state and local childhood lead poisoning prevention programs and housing agencies, and private sector consultants.

The program agenda of this plan calls for the following:

- Increased childhood lead poisoning prevention activities
- Increased abatement of leaded paint and paint-contaminated dust in housing
- Reductions in other sources and pathways of lead exposure
- National surveillance for blood lead levels

The research agenda calls for the applied research needed to carry out the program agenda in the most cost-effective manner.

A. PROGRAM AGENDA ITEM 1: INCREASED CHILDHOOD LEAD POISONING PREVENTION ACTIVITIES

Childhood lead poisoning prevention activities include the screening of children for elevated blood lead levels, referral of poisoned children for medical and environmental interventions, and education about childhood lead poisoning and its prevention. In fiscal year 1990 around 1 million children were screened by childhood lead poisoning prevention programs receiving at least partial funding by CDC, and around 5000 children (.5%) were identified with blood lead levels ≥ 25 µg/dl. However, expansion of childhood lead poisoning prevention activities are needed to increase the number of children screened, particularly in communities with the highest levels of blood lead in children, and to assure proper follow-up of poisoned children. In addition, increased use of intensive screening methods, such as door-to-door screening, may result in identification of children who are at particularly high risk of lead poisoning and who are not receiving the benefits of other child health programs.[5]

Further efforts are also needed to improve training of public health professionals providing childhood lead poisoning prevention services; to develop easier-to-use and cheaper laboratory instruments for measuring blood lead levels, and increase screening by the private sector.

B. PROGRAM AGENDA ITEM 2: INCREASED ABATEMENT OF LEADED PAINT AND PAINT-CONTAMINATED DUST IN HOUSING

Lead-based paint and paint-contaminated house dust are still the major cause of high-dose lead poisoning in U.S. children. Lead-based paint abatement is an integral part of the treatment of childhood lead poisoning and a crucial step in the prevention of new cases.

The Department of Housing and Urban Development has estimated that 74% of housing built before 1980 contains some lead-based paint, but not all of these residences pose an imminent hazard. Priorities for abatement and other risk-reduction activities should be based largely on public health concerns. Three priority groups of housing for abatement are homes of children identified with lead poisoning, homes at high risk of housing children with lead poisoning (but in which poisoned children have not yet been identified), and homes with lead-based paint that are being renovated or remodeled for other reasons. Day-care centers and other buildings frequented by young children are also a high priority. Abatement programs must work in tandem with childhood lead poisoning prevention programs, to ensure the most efficient use of resources.

C. PROGRAM AGENDA ITEM 3: REDUCTIONS IN OTHER SOURCES AND PATHWAYS OF LEAD EXPOSURE

Lead-based paint and paint-contaminated dust account for most cases of lead poisoning in the U.S. Other sources of lead will also have to be addressed, however, to eliminate this disease. Federal agencies are continuing to regulate environmental lead in water, air, and housewares.

D. PROGRAM AGENDA ITEM 4: NATIONAL SURVEILLANCE FOR ELEVATED LEAD LEVELS

The only national data available for estimating the number of children who may have elevated blood lead levels are derived from national surveys of nutritional and health status that, in the past, have been conducted about once a decade. These data are extremely valuable for providing unbiased estimates of the blood lead levels of children and workers in the U.S. However, they cannot be used to monitor short-term trends over several months or a few years. They cannot be used to characterize geographic distributions of poisoning in the community or to target interventions where they are most needed. A national surveillance program for elevated blood lead levels in children and workers is essential for better targeting of interventions, for tracking our progress in eliminating childhood lead poisoning, and for evaluating lead exposure in abatement workers and workers in other lead-contaminated environments.

CDC is beginning the effort to develop national surveillance for blood lead levels in children. Funds to support state-based surveillance activities have been made available in the past few years. The American Academy of Pediatrics, the American Medical Association, and the Council of State and Territorial Epidemiologist have endorsed the development of such surveillance.

IV. DIRECTIONS FOR THE FUTURE

While great progress has been made in reducing blood lead levels throughout the U.S., a great deal more must be made if we are to eliminate childhood lead poisoning as a public health problem.

We must make the shift to primary prevention; that is, preventing lead poisoning before it happens. Currently, the focus of prevention efforts is mainly on identifying and treating individual children. This must change so that environmental sources of lead that may pose a hazard to children are remediated before exposure occurs. As communities begin primary prevention programs, they will have to evaluate the cost-effectiveness of the interventions conducted. The results of such evaluations will be important for ensuring that future primary prevention programs are as cost-effective as possible.

Linkages among different parts of government and between government and the private sector will have to improve. Public agencies will have to work with pediatric health-care providers to identify communities with childhood lead-poisoning prevention problems and unusual sources of lead and to ensure environmental follow-up of poisoned children. Public housing and economic-development agencies will have to integrate lead paint abatement into housing rehabilitation policies and programs. Health-care providers will need to increase the amount of blood lead screening of children. Public and private organizations must continue to develop economical and widely available blood lead tests to make such screening possible. Owners of private dwellings must bear a portion of the financial burden for abatement.

Childhood lead-poisoning prevention activities need to be integrated into the actions of other child health, environmental, housing, and other programs. For example, programs funding rehabilitation of older housing should encourage lead removal at the same time, since lead abatement adds only a relatively small amount to the cost of ongoing modernization activities. Job training programs can help persons with little training develop the skills needed for lead-based paint abatement; these persons will then be likely to vacate jobs that do not require training, resulting in employment opportunities for persons who are not currently working.

Further studies and analysis are needed to define those environmental circumstances that are not hazards to children. Human use of lead has resulted in such widespread contamination of housing, soil, and dust that resources will not be available to clean it all up in the near future. Therefore, priorities for cleanup need to be established. For example, a national soil lead standard has not been set, and the factors that should determine when lead-contaminated soil requires remediation have not been well evaluated. Cleanup standards being used in some places are so stringent that many communities would have to remediate the majority of their soils to meet them, with unclear public health benefits. The costs of soil lead remediation using very strict standards could far exceed the costs of abating lead-based paint, which is a greater potential hazard to children under most circumstances. Better understanding of the soil conditions and concentrations that are potentially hazardous to children would result in resources being used where they can have the greatest public health impact.

Efforts must continue to evaluate the effectiveness of interventions aimed at reducing or preventing lead exposure. Although several retrospective evaluations have been conducted on the impact of lead-based paint hazard reduction activities,[6-9] well-designed prospective studies currently being conducted will provide extremely useful data. The Environmental Protection Agency is conducting prospective studies that will be critical to our understanding the impact of removing lead-contaminated soil.

CDC remains committed to its goal of having childhood lead poisoning be a disease of the past. We look forward to the day when childhood lead poisoning is no longer a public health problem.

REFERENCES

1. Lin-Fu, J., Historical perspective on the effects of lead, in *Dietary and Environmental Lead: Human Health Effects*, Mahaffey, K.R., Ed., Elsevier, Amsterdam, 1985, chap. 2.
2. Strategic plan for the elimination of childhood lead poisoning: report to Congress, U.S. Department of Health and Human Services, Washington, D.C., 1991.
3. Preventing lead poisoning in young children: a statement by the Centers for Disease Control, Department of Health and Human Services, Centers for Disease Control, Atlanta, 1991.
4. Chisolm, J.J. and Harrison, H.E., The exposure of children to lead, *Pediatrics*, 18, 934, 1956.
5. Centers for Disease Control, Lead Contamination Control Act of 1988, *MMWR*, 41, 288, 1992.
6. Rosen, J. F., Markowitz, M. E., Bijur, P. E., et al., Sequential measurements of bone lead content by L-X-ray fluorescence in Ca_2 EDTA treated lead-toxic children, *Environ. Health Perspect.*, 98, 271, 1991.
7. Amitai, Y., Brown, M. J., Graef, J. W., and Cosgrove, E., Residential deleading: effects on the blood lead levels of lead-poisoned children, *Pediatrics*, 88, 893, 1991.
8. Copley, G., unpublished data, 1990.
9. States, C., Matte, T., Copley, C. G., Flanders, D., and Binder, S., Retrospective study of the impact of lead-based paint hazard remediation on children's blood-lead levels in St. Louis, Missouri, *J. Epidemiol.*, 139, 1016, 1994.

Chapter 11

National Implementation Plan for the Prevention of Childhood Lead Poisoning from Residential Exposure to Lead-Based Paint

B. T. Cook

CONTENTS

I. INTRODUCTION

In the fiscal year 1992 Appropriations Committee report, the House of Representatives directed the Environmental Protection Agency (EPA) to work with the Department of Housing and Urban Development (HUD) and Centers for Disease Control (CDC) to develop an implementation strategy for reducing childhood exposure to lead-based paint. Individual agency-specific strategies released individually by EPA, CDC, and HUD in late 1990 and early 1991 form the basis of the report to Congress. These agency-specific strategies are

1. "Comprehensive and Workable Plan for the Abatement of Lead-based Paint in Privately Owned Housing: A Report to Congress," HUD, December 1990.
2. "Strategy for Reducing Lead Exposures," EPA, February 1991.
3. "Strategic Plan for the Elimination of Childhood Lead Poisoning," CDC, February 1991.

These three agencies, as well as over a dozen other federal agencies, have been coordinating federal activities since 1989 to reduce childhood lead poisoning from exposure to lead-based paint, through a federal interagency lead-based task force.

This paper summarizes the efforts of the federal interagency lead-based paint task force, summarizes the goals of EPA's strategy, and describes how the federal agencies activities are integrated into a national implementation plan for eliminating childhood lead poisoning from exposure to lead-based paint.

II. THE FEDERAL INTERAGENCY LEAD-BASED PAINT TASK FORCE

In 1989 EPA and HUD signed a memorandum of understanding that, among other agreements, established a federal interagency task force to address the hazards posed to children from exposure to lead-based paint in housing. Since then the task force has increased in size to include the CDC and 14 other federal agencies. Other federal agency members include the Occupational Safety and Health Administration, the National Institute for Occupational Safety and Health, the Consumer Product Safety Commission, the National Institute of Environmental Health Sciences, the Agency for Toxic Substance and Disease Registry, the Department of State, the Farmers Home Administration of the Department of Agriculture, the Department of Defense, the Department of State, the Department of Veterans Affairs, the Resolution Trust Corporation, the Food and Drug Administration, the Council of Environmental Quality, and the National Institute of Standards and Technology.

The task force operates to exchange information on lead-based paint efforts among the member agencies, to coordinate lead-based paint activities to avoid duplication of effort, and to implement joint projects to maximize resources. Activities that the task force is currently engaged in include

1. The evaluation of analytical methodology, including improved laboratory protocols, improved detection technology, and the development of standard reference materials for determinations of lead in paint, soil, and dust
2. The preparation of recommendations for the development and implementation of a national voluntary laboratory accreditation program to assure that laboratories engaged in testing are proficient
3. The establishment of a federal clearinghouse and hotline for the dissemination of information on lead health hazards and federal programs to the general public, health professionals, abatement professionals, and state and local governments
4. The investigation of exposures to lead in paint and soil, and their relationship to dust lead
5. The accreditation of personnel involved in the identification of lead hazards and the abatement of these hazards, including abatement project supervisors, lead abatement inspectors, and abatement workers
6. The development of the national implementation plan on federal activities to reduce exposure to lead-based paint

III. SUMMARY OF EPA STRATEGY TO REDUCE LEAD EXPOSURES

In February 1991 EPA released its "Strategy for Reducing Lead Exposures." This plan addresses the significant health and environmental problems our society is facing as a result of lead pollution. EPA, recognizing that lead is a multimedia pollutant, has developed an approach for reducing lead exposures in air, water, paint, soil, and dust. The *goal* of the agency's strategy is to *reduce lead exposures to the fullest extent practicable, with particular emphasis to children*. To achieve this goal the agency has established two *objectives* to set program priorities and to measure success. These objectives are to

- Significantly reduce the incidence of blood lead levels above 10 μg/dl in children, while taking into account the associated costs and benefits
- Significantly reduce, through voluntary and regulatory actions, unacceptable lead exposures that are anticipated to pose risks to children, the general population, and the environment

These objectives will be achieved by implementing a set of actions. These actions are

- *Develop methods to identify geographic "hot spots."* Identifying high-exposure areas is critical to encouraging and directing abatement. Methods to locate geographical areas with high exposures to lead and potential for lead poisoning in children are needed.
- *Develop and transfer abatement technology.* Cost-effective methods and tools to reduce lead exposure will be developed and disseminated.
- *Implement lead pollution prevention programs and encourage recycling.* Identifying and encouraging technologies to recycle lead-based paint abatement wastes will reduce future exposures associated with past uses of lead-based paint.
- *Minimize human and environmental exposures through traditional control mechanisms.* Research activities that lead to health-based standards for exposure to lead will be developed as well as enforcement of existing standards in various media.
- *Develop and implement a public information and education program.* Informing and educating the public about sources of lead exposure, how to reduce or avoid exposure, and approaches to preventing additional exposures, are essential to the success of the agency's lead strategy.
- *Coordinate research programs.* The agency has reviewed ongoing and future research needs, coordinated with program offices within the agency and other federal agencies, and prioritized research efforts so that the agency's research program is supporting the most critical needs.

In addition, EPA regional offices are preparing individual lead strategies to serve as the region's framework for implementing the EPA's national strategy, while considering the needs of each region.

IV. INTEGRATED NATIONAL IMPLEMENTATION PLAN

The National Implementation Plan describes federal activities in five subject areas.

A. LEAD INDICATOR MONITORING

The national capacity for elevated blood lead levels (EBL) surveillance is being expanded and a survey conducted to determine which states have regulations or laws in place that require laboratories to report blood lead levels. These activities will help to identify geographic "hot spots" — areas with concentrations of EBL cases and lead-based paint (LBP)-contaminated housing — and monitor the extent and nature of these problems. In addition, research efforts are under way to define the extent and nature of the problems and to develop low-cost means of identifying high-priority areas for testing and abating LBP in housing units. Blood lead screening programs target children for blood lead tests and identify those requiring follow-up with medical treatment, as well as remediation of exposure sources. Grants are being awarded to states to extend these programs. Research is being conducted to evaluate the efficiency and effectiveness of different mechanisms for blood lead screening, and improved means for blood sampling are being developed. Efforts are being initiated to expand laboratory capacities for blood assay and to standardize and enhance their analytic performance.

B. PUBLIC AWARENESS ENCHANCEMENT
An informed public, aware of the problem and motivated to take appropriate action, is essential for a successful national effort to eliminate childhood lead poisoning from LBP-contaminated housing. A comprehensive plan for disseminating information to the general public and special target audiences is being developed. Part of this strategy involves establishing a federal hotline and information clearinghouse. Grants have been awarded to state and local health agencies to provide training to health-care providers. Educational materials are also being developed and disseminated.

C. TESTING AND ABATEMENT CAPACITY DEVELOPMENT
Activities in the categories described above will increase the number of housing units identified as requiring testing and abatement. Abatement can be defined as a comprehensive process of eliminating exposure to lead paint and lead dust, which must include testing, replacement, enclosure, encapsulation, or removal of the lead-based paint hazard, as well as measures for worker protection, containment of dust and debris, cleanup and disposal of waste, and clearance testing. Consequently, activities have been initiated to increase the supply of qualified inspection and abatement personnel and to improve laboratory capabilities for analyzing environmental samples. A network of five university-based centers for training inspectors and abatement personnel has been initiated, and curricula are being prepared. Additional training opportunities are being fostered by training grants through labor unions. To improve environmental laboratory capacity and performance, a laboratory accreditation program is being developed. Efforts are under way to improve environmental laboratory analysis and field measurement methods and to provide laboratories with standard reference materials.

D. ABATEMENT TECHNOLOGY EVALUATION
Both implementation and research activities are being conducted in this area. A total of $25 million is being made available by HUD to Public and Indian Housing Authorities for risk assessments and to test for the presence of LBP and other lead hazards. To build capacity to increase the number of abatements, $137.7 million, in 1993, is being provided to support abatement in low- and moderate-income owner-occupied units and in low-income privately owned rental units. Research in this area focuses on evaluating the efficacy of abatement and waste disposal methods, identifying effective low-cost abatement options, and assessing the effect of abatement on childhood blood-lead levels. Measurement methods research is investigating ways to enhance the X-ray fluorescence (XRF) technology employed by professional inspectors, developing the next generation of technologies, standardizing and improving field-sampling methods, and developing test kits suitable for initial screening for LBP by homeowners, apartment dwellers, and contractors. Another line of research is studying the environmental pathways for LBP, as well as its uptake, absorption, and biokinetics.

E. PRIMARY PREVENTION INITIATIVES DEVELOPMENT
A shift in perspective is necessary by all parties — federal, state, and local agencies — involved in lead poisoning prevention, to make the transition from a secondary prevention approach (identifying and responding to cases of lead poisoning) to a primary prevention approach that eliminates elevated blood lead levels by controlling lead exposures before they occur. Primary prevention initiatives are essential to reducing the incidence of, and ultimately eliminating, childhood lead poisoning. A model program for the prevention of childhood lead poisoning would include the following major components: (1) screening and surveillance; (2) risk assessment and integrated prevention planning; (3) outreach and education; (4) infrastructure development; and (5) hazard reduction. One effort in this area is the primary prevention strategy

handbook that will assist local and state agencies in developing a community-based primary prevention program.

When completed, the national implementation plan will contain descriptions of over 80 federal projects and activities that contribute to the reduction of exposures to lead from lead-based paint and childhood lead poisoning.

Chapter 12

A Pound of Prevention, an Ounce of Cure: Paradigm Shifts in Childhood Lead Poisoning Programs

K. W. James Rochow

CONTENTS

I. SUMMARY

The purpose of this paper is to complement the other presentations at the American Chemical Society's symposium on *Lead Poisoning in Children: Exposure, Abatement, and Program Issues*[1] by focusing on local childhood lead poisoning prevention programs.[2] The efficacy of state and municipal lead poisoning prevention programs is critical because under the system of administrative regulation in the U.S. programs are actually implemented and enforced at the local level. Three-stage theory based on the evolution of environmental regulation provides a framework for analyzing program evolution.

Eliminating childhood lead poisoning — "the No. 1 environmental health hazard facing American children" — provides a classic illustration of the necessity of a preventive program approach. Childhood lead poisoning is characterized by subclinical symptoms and often irreversible effects and at the same time, is almost completely preventable through the elimination of existing sources, most notably lead-based paint and dust.

Historically, the operation of existing programs has been characterized by medical case management and "tracking" of individual cases without effective hazard control (second stage); enforcement is typified by lawsuits with "jackpot" monetary damages (first stage). It is necessary that childhood lead poisoning programs shift to a new paradigm of prevention consistent with the third stage of regulation.

This new prevention paradigm should incorporate in a series of practicable and progressive steps: needs and resource assessments; priority-based, prevention-oriented, and coordinated program operation; continued program planning and evaluation; and engagement of private-sector resources. The prevention paradigm is not only consistent with, but may go beyond third-stage environmental requirements in such areas as assessment and community outreach. Conversely, the paradigm is potentially weaker in public accountability and enforcement.

II. THREE-STAGE EVOLUTION OF ENVIRONMENTAL REGULATION

The movement away from cure toward prevention is a critical element in the evolution of environmental regulation, which can be typified as a three-stage process, with an evolving fourth stage of postcommand and control regulation.

A. FIRST STAGE — "ADHOCRACY"
The first stage — the "adhocracy" — is characterized by individualized after-the-fact responses to environmental damage, in the form of lawsuits or complaints. Monetary compensation (damages) for harm done; the need to prove actual harm; and "balancing of the equities" between the perceived social utility of the conduct complained of and the resultant damage to individuals are signatures of the first stage.[3]

B. SECOND STAGE — ADMINISTRATIVE REGULATION
The second stage involves the movement toward administrative regulation. Characteristic features of the second stage are the establishment of environmental agencies; the promulgation of generally applicable regulations; the definition of "harm" as a violation of legal standards; and, most importantly, the establishment of permit systems. Permit systems represent a leap forward into the third stage, because they institutionalize prevention through preoperational review and permission based on legal definitions of harm.[4]

C. THIRD STAGE — PREVENTION AND PLANNING
The third-stage represents comprehensive environmental programs directed at prevention rather than belated cure. Typical features of the third stage include postoperational monitoring as well as preoperational review (every permit becomes, in effect, an experimental permit that monitors operational effects); the routine incorporation of planning and impact assessments into decisions; and the geographic (areawide/ecosystemic) and disciplinary (social/cultural/economic) extension of impact evaluation.[5]

Evolving Fourth Stage — Postcommand and Control
Within the past decade there has been increasing discussion, and some attempted implementation, of alternatives to classic command and control regulation. Market-based incentives and voluntary compliance programs are two examples of attempts at postcommand and control regulatory programs. "Postcommand and control" is a more accurate description of these types of incentive-based programs than is "alternative regulation." Almost invariably, "alternative regulation" requires legal mandates and regulatory standards for the creation and definition of market standards for noncompliance with the particular scheme.[6] The evolution of what is termed "alternative regulation" is consistent with three-stage theory, because innovative schemes of environmental protection depend upon the existence of a mature command and control regulatory structure.

III. CHILDHOOD LEAD POISONING AS A SPECIAL CASE OF REGULATION

A. CHARACTERISTICS OF CHILDHOOD LEAD POISONING
Need for Prevention
Against this *schema* of the evolution of environmental regulation into prevention, attempts to combat childhood lead poisoning pose special problems because of the nature and severity of the problem, and the historic inadequacy of the response. The concept of prevention is the key to eliminating childhood lead poisoning. While it is the most prevalent childhood disease in the U.S., childhood lead poisoning is an environmental disease that is completely preventable.[7]

Childhood lead poisoning, moreover, presents a classic (and literal) case of irreparable harm.[8] At low doses, lead is of greatest concern to children because of its neurotoxic effects, which can include IQ reductions, reading and learning disabilities, decreased attention span, hyperactivity, and aggressive behavior. The effects of childhood exposures can be irreversible and result in higher school failure rates and reduced performance later in life. The subclinical symptoms and irreversible effects of childhood lead poisoning cry out for a preventive third-stage approach. Strategic plans developed by HUD, HHS, and EPA within the past few years each stress the need to diagnose and correct problems before children are poisoned.[9]

Locus in the Housing Environment
The fact that childhood lead poisoning is both a health and environmental problem whose solution depends largely on controlling lead-based paint and dust (and sometimes soil[10]) in the housing environment, complicates solutions.[11] Property owners are not regarded as fit subjects for permits, even when they cause and maintain nuisances, in contradistinction to industrial operators who more obviously pose the potential for continuing polluting discharges.[12] The high positive value, even piety, placed on homeownership in our society confers an aura of intrinsic "innocence" that warrants modification of traditional regulatory approaches.[13] Engagement of market forces, including the home finance industry, licensed inspectors and risk assessors, and trained abatement contractors, thus becomes critical for achieving lead hazard control in the housing environment.[14]

Problems of Coordination
The definition of the problem of childhood lead poisoning has itself been compartmentalized: environmentalists have viewed it as a housing problem, housing advocates define it as a health problem, and so forth. The multifaceted nature of childhood lead poisoning requires a high level of interdisciplinary cooperation among diverse fields in order to eliminate it.

B. THE MEDICAL CASE MANAGEMENT MODEL IN CURRENT PROGRAMS
Consistent with the structure of legal authority and administrative practice in the U.S., childhood lead poisoning prevention programs are run by state and, especially, local governments.[15] Historically, lead programs have been based on a medical case management model that relies on identifying poisoned children through blood lead screening.[16] Under the case management approach, if the initial blood lead screening is a confirmed positive, the child undergoes treatment and periodic follow-up testing.[17] The case management model also includes environmental follow-up, primarily consisting of a visit to a child's home to identify (and, it is hoped, control) sources of lead.

The basic problem with the virtually exclusive use of the medical management model is that it tends, in practice, to bog down into a circular medical tracking program. As a result, diagnoses of lead poisoning in children are repeatedly confirmed, but prevention through source control and lead hazard elimination is not necessarily accomplished. Part of the reason for the lack of prevention orientation is that the medical case management model inherently focuses on treating individual symptoms. Another prime factor contributing to the overriding emphasis on reaction, rather than proactive response, has been the lack of coordination between the agencies that should address childhood lead poisoning.

Lack of Interagency Coordination in Current Programs
Consistent with the medical case management model, most childhood lead poisoning programs have been lodged in health departments. Systematic coordination between health and housing and environmental departments has been conspicuously lacking in many cases. Blood lead screening efforts and housing inspections, for example, are typically the responsibility of different agencies that do not effectively communicate with each other. As a result, programs lack assured follow-up to reported cases of childhood lead poisoning (much less a preemptive response before poisoning occurs) in the form of routine inspections of the housing environment.

C. THREE-STAGE THEORY AS APPLIED TO CURRENT PROGRAMS
Stuck in the Second Stage
In current operation, "childhood prevention program" is often a misnomer.[18] The existence of an administrative structure, but one that does not typically incorporate prevention, defines current lead programs as stuck in the early second stage of regulation. The fact that childhood lead poisoning cuts across traditional disciplinary barriers and defined areas of concern exacerbates the difficulties of shifting from the current mix of first- and second-stage elements to a new third-stage paradigm of prevention. Childhood lead poisoning prevention programs will begin moving into the advanced second and third stages only when

they effectively coordinate housing, health, and environmental policy, in the interest of prevention. To take one example of how to make the transition into prevention, local programs should systematically collect and analyze data relating to the age and condition of housing stock. Those data relating to the likely presence of lead-based paint hazards — not to the presence of an already lead-poisoned child — would enable lead-hazard inspections and control in homes on a priority basis.

First-Stage Lawsuits
Reflecting the current system's lack of prevention orientation, the main mechanism for legal redress of childhood lead poisoning belongs to the first stage. Private lawsuits (mostly sounding in tort) emphasizing monetary damages for harm incurred to the child victim, rather than prospective relief such as enforcement of source control measures, are characteristic. Even in jurisdictions with more comprehensive programs, such as Massachusetts, where actions are brought to enforce source control requirements, a substantial number of personal injury lawsuits for lead poisoning continue to be brought.

IV. SHIFTING TO A PREVENTION PARADIGM

A. NEED FOR A PARADIGM SHIFT
The medical case management model standing alone is demonstrably inadequate for incorporating prevention into ongoing lead programs. To the contrary, in its present form it has tended to inhibit the movement of local programs from reaction to genuine prevention. No one of good faith denies the need to care for and, to the maximum extent possible, cure the existing victims of childhood lead poisoning. The medical case management model in isolated operation, however, has the ironic effect of perpetuating childhood lead poisoning, by virtue of its emphasis on circular "tracking" of individual cases of already poisoned children. Consequently, an accelerated evolution into a prevention paradigm is essential to eliminating this disease.[19]

B. GENERAL REQUIREMENTS FOR PREVENTION PROGRAMS
Development of childhood lead poisoning prevention programs can be analyzed in terms of program elements and functional tasks.[20] For the purpose of analyzing the development of local prevention programs from the perspective of three-stage theory, four basic principles undergird both of those modes of analysis. First, programs need to be tailored to local needs and circumstances. Second, programs should take an active, community-based approach to build constituencies and to educate people about prevention. Third, programs should be based on hazard priorities and opportunity points for fuller abatement (such as unit turnover or housing vacancy), rather than on blanket requirements of immediate complete deleading. Fourth, programs need to use all available resources, private as well as public, to build markets and capacity for lead hazard reduction.[21]

C. MAKING THE SHIFT TO THE PREVENTION PARADIGM
Preventive Features of the Medical Case Management Model
Analyzing elements and tasks necessary to develop childhood lead poisoning prevention programs in terms of the three stages of regulation can clarify the key steps necessary to make the shift to the prevention paradigm. Certain preventive features, actual or latent, in the medical case management model itself serve as a departure for analysis. "Anticipatory guidance" is used in pediatric medicine to refer to education, in the clinical setting, concerning ways of preventing disease in the patient's day-to-day existence. In the context of childhood lead poisoning, anticipatory guidance would include informing child patients and their parents of the nature and sources of childhood lead poisoning and suggesting simple prevention measures, ranging from special household cleaning procedures to improved nutrition and hygiene (hand washing).[22]

The screening program activity analogous to anticipatory guidance is the home visits that screening program outreach workers may make to follow up indications of childhood lead poisoning, by taking samples of children's blood for lead concentration analysis. The home visit provides an opportunity to impart the same kind of basic information to children and parents that is given as part of anticipatory guidance.

During the home visit the outreach worker may also conduct an environmental investigation, which includes a visual inspection for peeling and chipping lead-based paint and for other obvious sources of lead in the household. Environmental investigations fit into the medical case management model even under strict definitions, because source identification and control is an essential part of an effective treatment regime. If such sources are identified, the outreach worker may make a referral to the housing department for lead cleanup and abatement. Treating expenses incurred in lead abatement as deductible medical expenses, for income tax purposes, is another way public policy can fit abatement directly into the medical model.

Screening as a Link to Prevention

Blood lead screening is the core program element of the medical model. Childhood lead prevention programs are currently organized around its provision, reporting, and follow-up. Screening is the activity that triggers case management, when it results in confirmed reports of elevated levels of lead in children's blood. Screening itself does not constitute primary prevention, because its *raison d'etre* is to identify already lead poisoned children.

Although screening is not primary prevention per se, screening has a variety of implicit relationships to primary prevention. For one thing, screening can be part of a clinical setting that includes anticipatory guidance. Screening, in addition, can prevent future generations of children from contracting childhood lead poisoning, by providing the means for identifying environments that have demonstrably contributed to childhood lead poisoning.[23] Screening can also create a constituency for developing prevention programs, by delineating the extent of the problem in a given community. Most importantly, screening generates data on actual cases which, when combined and collated with data from other sources such as census and housing stock data, can target high-risk neighborhoods and localities for concentrated preventive action.

Moving to Prevention in the Absence of a Screening Program

Although most current childhood lead poisoning prevention programs are essentially screening programs, there may be localities that do not have any childhood lead poisoning prevention programs, even blood lead screening programs. These cases test the proposition that it may be possible to move to prevention directly without an intervening screening program. This type of program would target environments — housing environments in general and lead-based paint dust in particular — for lead hazard control activities, based on such data as the age and condition of housing. Even in such cases, however, screening of children 6 years of age and under should constitute a required part of monitoring the effectiveness of lead hazard control.

V. CONCLUSION: CONSISTENCY OF THE PREVENTION PARADIGM WITH THIRD-STAGE REGULATION

The prevention paradigm is designed to achieve prevention in practicable, progressive steps. In operation the paradigm is basically consistent with the development of third-stage regulation.[24] The potential weakness of the paradigm lies in public accountability and enforcement. Because the principal sources of exposure to childhood lead poisoning are found in the home environment (particularly lead-based paint and dust), efforts to mandate lead hazard control must address homeowners and owners of rental property. The classic permit system that arose out of the second stage of regulation does not fit the circumstances of such putatively "innocent" property owners: certainly not the positive public image conjured up by private homeownership in our society.

In lieu of property owner permits, the paradigm calls for a system of licensed private inspectors and risk assessors and ongoing operational and postoperational monitoring, including provisions for self-monitoring. Although this privatized system has elements of the evolving fourth stage of regulation, it lacks clear financial incentives (such as emissions trading units). Recognition of the benefits of a lead-safe house, combined with increasing awareness of the link between preventive maintenance and lead hazard control, provides the opportunity for incorporating financial incentives and resources for prevention into the real estate financing system.

At the same time the paradigm must take into account the fact that legal requirements and the threat of liability will drive the creation of opportunities and markets for lead hazard control. In addition, to insure that lead hazard control is done properly, the paradigm needs to incorporate public oversight in such forms as random inspections, monitoring of abatement results on a selected basis, and quasi-permits in the form of lead hazard control plans that the agency could use to translate inspection results and risk assessment recommendations into enforceable obligations.

In other respects the paradigm advances third-stage regulation beyond the requirements of such characteristic features as environmental impact assessments. As an initial step, effective lead poisoning prevention requires that programs concurrently conduct needs and resource assessments. Needs assessments are designed to customize the program by identifying the characteristics of the childhood lead poisoning problem in particular communities (including lack of information about the problem). Resource assessments are designed to identify, engage, and coordinate all potential community resources and actors that might contribute to solving the childhood lead poisoning problem. Such assessments could begin at the basic level of analyzing current lead-related programs, and their connections, if any, with each other. The assessments should expand to analyze how potentially lead-related efforts, such as home improvement and energy conservation, as well as social service programs, such as Head Start and the Supplemental Food Program for Women, Infants, and Children (WIC), could be synthesized into the prevention program.

The needs and resource assessments are both a way of analyzing the problem and conjunctively identifying and organizing programs and resources to deal with it. This recommended process suggests that integrating health, resource, and program issues into problem identification and impact assessment can provide the basis for reciprocally defining how to deal with a problem in a preventive fashion while fine-tuning the definition of the problem itself. Even functions that are not inherently preventive — notably, the case management model — can be incorporated into the assessment through reporting of screening and environmental investigation data.

Education and community outreach is also an essential element of a prevention program. While it is important that the program educate both general and specialized audiences concerning the nature of childhood lead poisoning and program requirements, it is even more important that the program build a supportive constituency by actively involving the community in its ongoing operation, through intensive outreach effects. Where funds are severely limited, demonstration programs can be set up in particular communities. The active public participation inherent in community outreach and community-based demonstration programs contrasts with the relatively passive model of public hearings on proposed projects and permits that informs even the third stage of environmental regulation.

FOOTNOTES

1. August 25–26, 1992, Washington, D.C.
2. This paper is based, in part, on the conclusions of the Primary Prevention Strategies project that the Alliance To End Childhood Lead Poisoning is currently undertaking. The purpose of this EPA-funded project is to develop a handbook setting forth a model or "blueprint" to help local programs move to effective prevention of childhood lead poisoning. The author wishes to thank the Environmental Protection Agency for its continuing support of this project. The ritual auctorial disclaimer obtains: The opinions expressed in this paper are solely those of the author and not necessarily those of EPA nor the Alliance To End Childhood Lead Poisoning.
3. Rochow, K. W. J., The continuing vitality of the common law in the regulatory state, *Ritsumeikan Law Rev.*, 4, 81, 1989.
4. Rochow, K. W. J., The far side of paradox: state regulation of the environmental effects of coal mining, *W.V. Law Rev.*, 81, 559, 1979 (reprinted in Storm, F. A., Ed., *Land Use and Environment — 1980*, Clark Boardman, New York, 1980, 275.
5. *Ibid.*
6. Attempts to devise market-based pollution control programs — such as the trading of discharge rights — in Eastern Europe, in the absence of a matured and effective system of environmental controls, will test the proposition that so-called alternative regulation depends upon the development and maturation of command and control regulation.
7. Lead poisoning has been declared, by both the Environmental Protection Agency and the Department of Health and Human Services, as "the No. 1 environmental health hazard facing American children." See Sullivan, L. W., Remarks for the 1st annual conference on childhood lead poisoning, in Preventing Childhood Lead Poisoning (Final Report of the First Comprehensive National Conference), Washington, D.C., 1991, A-1. Based on the federal government's most recent estimates, almost 10% of U.S. children under age 6 are lead poisoned (defined as blood lead levels equal to or above 10 μg/dl). See Brody, D. J. et al. Blood lead levels in the U.S. population: phase 1 of the third national health and nutrition examination survey (NHANES III, 1988 to 1991). *Journal of the American Medical Association*, 277 et seq., 1994. Lead-based paint and dust in the home is a major cause of childhood lead poisoning and the source of intensive exposures. Not surprisingly, the problem is most severe in dilapidated older housing; the worse the condition of the home, the greater the risk of lead exposures to children. While almost one in ten of all American children are lead poisoned, the prevalence rate among low-income black children is almost 30% nationwide.
8. The evolution of "irreparable harm" as an equitable criterion for the issuance of a preliminary injunction (a form of relief often requested in environmental and public health cases) itself traces the three-stage pattern. Originally, the term had something of a literal meaning, requiring a showing of a serious level of actual environmental harm. "Irreparable harm" has come to mean the threat of environmental harm or the violation of law, or both. See Rochow, K. W. J., The

continuing vitality of the common law in the regulatory state, *Ritsumeikan Law Rev.,* 4, 81, 1989; *Commonwealth v. Barnes & Tucker,* 1 Pa. Commonwealth Ct. 552 (1970).

9. Strategy for Reducing Lead Exposures, Environmental Protection Agency, Washington, D.C., 1991.
10. Contamination of the soil in the immediate vicinity of the house is often caused by flaking from exterior lead-based paint.
11. See n. 7, *supra.*
12. Cf. *National Wood Preservers et al. v. Commonwealth,* 489 Pa. 221, 414 A.2d 37 (1980) with *Philadelphia Chewing Gum Corp. v. Commonwealth,* 35 Pa. Commonwealth Ct. 443, 387 A.2d 142 (1978) (appealed *sub nom. National Wood Preservers et al. v. Commonwealth)* (Landowner liability under the Pennsylvania Clean Streams Law).
13. See Comprehensive Environmental Response, Compensation, and Liability Act (commonly known as "Superfund"), 42 U.S.C.A. §§ 9601 (35), 9607(b)(3). Based on the author's experience with the Primary Prevention Strategies Project advisory committees, landlords are viewed as much more appropriate subjects for coercive regulation than are homeowners.
14. See Section V, pp. 93–94, *infra.*
15. In contrast to many types of environmental regulation, there is no federal statute establishing national uniform minimum standards and oversight of state and local lead programs. Some state statutes, however, themselves define something resembling a comprehensive childhood lead poisoning program that municipalities must carry out in whole or part. See 16 Mass. Ann. Stat., ch. 111, § 190 *et seq.* (1993).
16. Definitions of case management differ. Generally, "case management" is a service concept that incorporates two functions: (1) providing individualized counseling to clients, and (2) linking clients to community services and support networks. Rothman, J., A model of case management: toward empirically based practice, *Social Work,* 520, 1991. The emphasis in the second part of the definition on community support could provide a springboard to tying case management to community-based lead poisoning prevention programs. See Moore, S. T., A social work practice model of case management: the case management grid, *Social Work,* 444, 1990.
17. In severe cases of childhood lead poisoning, treatment can consist of chelation — the use of chemical agents to purge the body of lead through excretion. In less-severe cases, treatment can consist of nutritional counseling, parental education, and close medical monitoring. In all cases treatment should include environmental monitoring and source control and abatement.
18. Therapeutic criticism of existing childhood lead poisoning programs should not be taken as deprecating in any way the undoubted dedication and commitment of existing program staffs. As discussed in the text *infra,* provision and efficient use of both public and private resources, as well as program restructuring, are needed to change the orientation of existing programs to prevention.
19. "Paradigm" as used in this paper refers to the dynamic complex of prevailing opinion, policy formulation, program operation, and funding priorities that constitute a sort of collective mindset at a particular moment in time. The idea is borrowed loosely from T. Kuhn's seminal book, *The Structure of Scientific Revolutions,* 2nd ed., University of Chicago Press, Chicago, 1970.
20. The seven basic elements that serve as the "building blocks" of effective childhood lead poisoning prevention programs are (1) assess the problem in each community; (2) educate the public about prevention; (3) coordinate public sector lead-related programs; (4) build constituencies through community outreach; (5) increase capacity for lead hazard reduction; (6) prioritize lead hazard reduction; (7) leverage private sector resources.

    The six tasks that represent the functional corollary of the prevention program elements are (1) assess the problem; (2) develop resources technology; (3) organize coordinating council; (4) design and implement pilot programs; (5) develop skills and capacity; (6) train community groups.

    The analytic categories of program elements and tasks are compatible. The former is "programocentric" and has the advantage of providing practical guidance for the step-by-step development of prevention programs. The latter is task oriented and has the advantage of illuminating the kind of capacity-building in the private, as well as the public, sector necessary to achieve prevention.
21. This analysis of how programs might most effectively shift to prevention is generally based on the conclusions of the Primary Prevention Strategies Project. See n. 2, *supra.*
22. Basic cleaning and maintenance recommendations would include paying special attention to keeping friction and abradable surfaces — such as window stools and sashes — free of lead dust through regular cleaning; keeping alert to surfaces with deteriorated paint; and, if practicable under the circumstances, more advanced cleaning techniques such as the use of special cleaning agents and filter vacuum cleaners. Anticipatory guidance could also include nutritional counseling.
23. This often-made argument is somewhat circular because it depends on the existence of lead hazard control programs that follow through on screening results, which is not typically the case as the text discusses *passim.*
24. This section of the paper will not discuss all the elements of a prevention program, see n. 20, *supra.* Rather, it will highlight several key elements that incorporate features of third-stage regulation. The paradigm is outlined in the PPS Handbook, which consists of three volumes: *Childhood Lead Poisoning: Blueprint for Prevention* (Vol. I) (1993); *Childhood Lead Poisoning: Developing Prevention Programs and Mobilizing Resources* (Vol. II) (1994); and *Childhood Lead Poisoning: Resources for Prevention* (Vol. III) (1994). (All volumes published by the Alliance To End Childhood Lead Poisoning, Washington, D.C.)

Chapter 13

Coordinated National Strategy on Childhood Lead Poisoning: A National Action Plan

A. M. Guthrie and D. Ryan

CONTENTS

I. INTRODUCTION

Based on the federal government's estimates, at least 10 to 15% of U.S. children under age 6 have neurotoxic levels of lead in their blood — as many as 3,000,000 children. Yet, until quite recently, this epidemic had remained overlooked by health professionals, policy makers, and the general public. After two decades of inaction and inattention, important changes have begun to occur, beginning with the federal government's acknowledgement of lead poisoning's epidemic proportions in the Department of Health and Human Services's (HHS) 1988 Report to Congress, "The Nature and Extent of Childhood Lead Poisoning in the U.S."

Since December 1990, HHS, the Environmental Protection Agency (EPA), and the Department of Housing and Urban Development (HUD) have each released strategic plans or reports on lead poisoning prevention.* These plans have proven to be important in raising public and congressional awareness, and they include many useful elements. However, they reflect the jurisdictional limitations of their agencies and thus do not add up to a national policy. In many respects, critical gaps are obscured rather than highlighted. Priorities have not been clearly established, and confusion persists over federal agencies' responsibilities in several areas critical to long-term progress in preventing childhood lead poisoning, such as training requirements, cleanup standards, and quality control for contractors and laboratories.

As a national, nonprofit, public interest organization working with people from a broad range of disciplines from around the country, the Alliance To End Childhood Lead Poisoning is in a unique position to update and build on the framework provided by the federal agency plans to produce one integrated document. (We are aware that EPA, at the direction of Congress, has just completed a new plan that is currently undergoing review within the administration and is not yet public.)

We believe that our effort is important for several reasons. Some issues must be updated to reflect technological advances or policy changes that have resulted from the recent unprecedented level of professional and public attention paid to this issue. Other issues need to be carefully addressed to recognize the need for priority setting caused by the scarce resources generally available to many governmental and private entities. In addition, the cross-disciplinary nature of lead poisoning, and the technical and political hurdles associated with permanent progress, require a conscientious effort to build consensus around critical objectives.

The goal of the project is to develop a document that recommends clear priorities and reasonable objectives for next steps in preventing childhood lead poisoning, and suggests appropriate roles and responsibilities for different levels of government and key private sector participants. As such, the plan will

---

* U.S. Department of Housing and Urban Devlopment, "Comprehensive and Workable Plan for the Abatement of Lead-Based Paint in Privately-Owned Housing," 1991; U.S. Department of Health and Human Services, "Strategic Plan for the Elimination of Childhood Lead Poisoning," 1991; and U.S. Environmental Protection Agency, "Strategy for Reducing Lead Exposure," 1991.

1-56670-113-9/95/$0.00+$.50

reach across disciplines to address a broad range of relevant topics, including screening and surveillance, hazard assessment, abatement and hazard reduction, worker protection, research and technology, public education, insurance and liability, and resources and financing. It will be action oriented, seeking to promote the policies, programs, and resources essential for implementation.

The Alliance is convinced that a national action plan is essential to sharpen the discussion and illuminate the critical policy issues requiring attention. Congress is demonstrating visible interest in lead poisoning, with enactment of major legislative initiatives possible in the next few months. Even if Congress were to enact all of the pending legislation, there would still be many necessary tasks unassigned and unclaimed. In addition, coordination is needed at many levels — federal, state, and local — across both the public and private sectors to prevent critical pieces from falling between the cracks. In short, what is needed is a national implementation plan, a national plan for action.

This plan will

- Identify the progress to date and critical obstacles to further progress;
- Set clear priorities and objectives for the next few years;
- Pinpoint key policy, program, resource, and coordination steps;
- Sort out the confusion in federal agency jurisdictions left by the three federal agency plans; and
- Identify appropriate roles and responsibilities for different levels of government and the private sector.

II. PLAN DEVELOPMENT AND DISSEMINATION

Our first national conference in 1991 assembled over 800 participants from various fields and disciplines, serving to identify many obstacles to progress and to raise core issues for the plan. The purpose of this project is to go the next step: to produce an insightful action plan that will enlighten legislative efforts and federal policy in the coming years.

The initial phase of the project involves a research effort designed to learn from the considerable efforts, to date, of experts and practitioners in this field. Specifically, we have been interviewing, and are continuing to interview, a wide range of individuals on their views about progress, obstacles, and priorities in childhood lead poisoning. This effort includes outreach to federal agency staff and officials, legislative staff, state and local officials, low-income housing providers, public health and housing agencies, labor unions, small businesses, environmental organizations, and real estate and other affected private sector interests. We are trying to interview individuals with diverse views, to ensure that the plan reflects enlightened thinking and balances broad viewpoints. We do plan to include an appendix listing the identity and affiliation of all persons interviewed in the course of preparing the plan. Should an individual or organization prefer not to be listed, we will honor such a request.

After we have completed the interview stage, we will begin developing a draft plan that will be widely circulated in an effort to gain input and suggestions. All who are interviewed will have an opportunity to review a working draft and to offer technical, policy, or editorial comments and suggestions for improvement. We will carefully consider all relevant comments and produce the final plan.

Because this plan will have direct relevance to policy implementation, the process — especially with respect to consensus building and information dissemination — will continue after the plan's completion. The Alliance will seek the formal endorsement of the plan by as many interested groups as possible. However, the plan will be an Alliance document, and the Alliance will be solely responsible for the contents. It is not intended to be, nor will it be, represented as a consensus document. The final National Action Plan was released in 1993 and distributed to policy makers and agency program staff.

III. INTERIM FINDINGS

We are currently in phase one of the project: completing interviews and weighing the multiple viewpoints and suggestions that have been made. Consequently, we cannot yet share our final recommendations. However, we have noticed some interesting trends and thought-provoking comments in our interviews to date, which we would like to share. But, we must caution that this is still an ongoing effort, and the comments that follow can best be described as anecdotal. It would be premature to suggest that these ideas will reflect the final plan or that they represent the views of the Alliance or its board of directors.

A. PROGRESS TO DATE

Virtually without exception, the individuals we have interviewed have indicated that increased public awareness of lead poisoning has made the biggest difference in advancing efforts to prevent lead poisoning. They credited extensive media attention and the Administration's acknowledgment of the problem for the rise in awareness. Some interviewers indicated that increased screening of children has been a major source of renewed awareness, since action, to some extent, depends on a ''body count.'' Closely related is increased demand by parents to have *their* children screened.

Those who have been working in the field for a number of years indicated that the most significant historical event in preventing lead poisoning has been the phase out of leaded gasoline. The population-wide decreases in blood lead levels that have resulted from this change have cleared the way for us to address other sources of lead in the environment.

Some individuals brought an action-based viewpoint to this question, indicating that the most significant recent progress has been a revised mindset of advocates that recognizes the enormity of the problem and the limited resources available to all sectors and seeks to promote a prioritized, risk-based approach to the problem, specifically with respect to lead paint abatement and the introduction of the concept of ''hazards.'' We were told that this is an approach that makes the problem manageable and wins credibility, especially in the current fiscal environment.

B. OBSTACLES

When asked to identify the major obstacles to current efforts, people repeatedly mentioned two things. First, the multidisciplinary nature of lead poisoning was frequently mentioned as a source of confusion, as it compounds the general disarray surrounding the problem and makes it easy for no one to take responsibility. From a purely programmatic standpoint, it raises the familiar bureaucratic challenge of reliance on interagency coordination to address a problem. This was described as something that will always be a complicating factor, even when workable programs are in place.

The other frequent response was, to no one's surprise, the lack of resources for agencies at all levels. Funds for abatement, research, and screening have been mentioned most often as urgent needs.

Another obstacle that has been mentioned is the fact that ''lead-free'' or ''lead-safe'' homes are not widely recognized as having more value in the real estate market, either formally (through appraisal standards or mortgage underwriting guidelines) or informally (through increased market value or appeal). When combined with the cost of abatement, many property owners have only disincentives to test for lead paint. Since the answers are likely to be negatives (known lead paint hazards or costs of abatement), they'd rather not know. This poses a particular problem in developing policies to promote abatement of properties with very low market values — sometimes even less than the cost of needed abatement.

What people have *not* said has also been interesting. No one has mentioned skepticism about lead's toxicity as a problem, even at low levels. No one has yet named specific ''opponents'' or industry interests as an obstacle, although some have expressed frustration at the expenditure of time and effort needed to debunk specific efforts to misuse the science or throw out red herrings to stall policy.

C. GAPS/RECOMMENDATIONS

We quickly merged our question about biggest gaps and priority recommendations because the same answers were frequently provided for both questions.

Without a doubt, the most frequently mentioned need has been that for research on the short- and long-term effects of alternative environmental interventions. Most expressed the need explicitly with respect to ways of dealing with lead paint in homes: i.e., what strategies are protective for what price? No one has yet expressed the need more broadly, addressing other media beside paint (although other media were mentioned in other questions). A related research suggestion was the need for better and cheaper field tools to measure lead levels in dust, paint, blood, etc.

An interesting criticism was that we are spending way too much money on medical research examining the effects of low-level lead toxicity. Instead, it was argued, we should be using these resources to examine alternative tools for prevention; that is, environmental intervention.

Another high-priority recommendation was the need for clear assignment of roles and responsibilities, across all levels of government and including parents and the private sector. No one suggested that the entire responsibility rests with a particular component of society. Implementing this recommendation is made all the more complex by the multidisciplinary nature of the problem, which poses problems even in developing legislative mandates, since the committee jurisdictions in Congress (and in state legislatures) make it difficult to enact legislation establishing a coherent and coordinated prevention program.

Low-income housing has clearly been identified as a particular challenge. Legislation specifically targeted to lead-paint abatement in low-income housing was recommended as critical, since abatement is very expensive and there is no other way to ensure action in this segment.

Protection of workers in the abatement and renovation fields was also frequently identified as a major need. Most expressed a desire for the Occupational Safety and Health Administration (OSHA) to develop the needed standard, but also expressed skepticism that, as it is now stands, OSHA would be able to do so. Effective action will be difficult because of the number of small businesses in the abatement, home renovation, and construction trades. We have noted that many people view worker protection as inseparable from other prevention efforts, since strategies to protect workers would ultimately be protective for occupants of the property and for workers' children, too.

IV. CONCLUSION

Although these observations just scratch the surface of the information shared with us during the initial interviews, they do convey a startling amount of consensus, as well as some clear indications of the challenges that lie ahead. We hope that the process of articulating the next steps in an honest and forthright manner will help to energize all involved parties. We recognize that there is some danger to a process like this, in that opponents can seize on the unknowns in an attempt to stall progress. For instance, some will argue that we should not proceed with wide-scale lead-based paint abatement until we have definitive answers about the most cost-effective means of eliminating or controlling lead paint hazards. In fact, we have already heard such arguments made in opposition to pending federal legislation.

However, we would like to describe an analogy that was raised in one of our interviews. In medicine, health-care providers are routinely faced with patients about whose conditions the definitive answers are unknown. In the absence of answers, we do not deny treatment, but instead do the best we can under the circumstances. In the case of lead poisoning, we know the cause, we know how children are exposed, and we know how to prevent it. Thus, there is no basis on which to deny action: either primary or secondary prevention. We can, of course, be smarter and more efficient in what we do, and that is why we should constantly be evaluating our successes and failures, so as to enlighten our ongoing decision making.

The Alliance hopes to take advantage of the considerable momentum, interest, and expertise in this issue to produce an ambitious, but realistic, plan that enjoys the support of public and private organizations involved in lead poisoning and that can enlighten the important policy decisions that need to be made in the coming months.

Chapter 14

Encapsulation of Lead-Based Paint

B. A. Leczynski, J. G. Schwemberger, and R. J. Cramer

CONTENTS

I. BACKGROUND

Encapsulants are liquid coatings that are applied to a lead-based painted surface to make the lead paint inaccessible by sealing the painted surface. These encapsulant coatings are normally applied by brush or roller or by spraying using conventional surface coating equipment. After the product cures, the coating is intended to provide a durable, easy-to-maintain surface that does not flake, chip, or peel, effectively preventing further exposure to lead paint chips and dust. Encapsulation of lead-based paint by applying a coating to cover and seal a painted surface offers a safe, low-cost, and potentially effective means of reducing the exposure of lead in paint.

There are a number of companies that produce and commercially market encapsulation products. These products include liquid-applied coatings, composite reinforced resins, and inorganic cementitious coatings. Although these products are commercially available and have been used in a variety of abatement situations, there is a lack of understanding and general concurrence on the performance requirements of these materials. These requirements include a list of product attributes; associated testing protocols; and criteria with which one could evaluate the product safety, durability, and effectiveness. Acceptable performance standards and consistent data are either not available or have not yet been centralized in a form that can readily provide answers on how well encapsulants perform in reducing lead exposure.

In response to this concern and to the 1987 and 1988 amendments to the Lead-Based Poisoning Prevention Act of 1971 (LPPPA),[1,2] the U.S. Department of Housing and Urban Development (HUD) requested the American Society for Testing and Materials (ASTM) to establish a subcommittee to formally document lead abatement procedures, including the use and application of acceptable encapsulants. At the state level, 1988 amendments to the 1971 Massachusetts Childhood Lead Poisoning Prevention Act mandated the Director of the Massachusetts Childhood Lead Poisoning Prevention Program to investigate, field test, and approve new methods of abating lead hazards.[3] In response to this mandate, the Commonwealth of Massachusetts has embarked on a program to establish performance standards for encapsulant coatings, with the anticipation of incorporating encapsulation as an acceptable form of abatement into their lead legislation. In addition, the U.S. Environmental Protection Agency (EPA), as part of their national support of the lead-based paint program, has initiated cooperative research efforts to obtain additional data to better understand and support the development of performance standards and protocols for application and use of these coatings in lead abatement.

This paper will offer a review of the approach presently being used to define encapsulation product performance and the protocols for their use in lead abatement.

II. ASTM SUBCOMMITTEE ON THE ABATEMENT OF LEAD HAZARDS IN BUILDINGS AND RELATED STRUCTURES

This subcommittee was organized with the goal of developing standard practices for abatement of hazards from lead-based paint in buildings. The standard practices reviewed to date include test methods for lead in paint, dust, air particulates, and soil; removal of lead-based paint; encapsulation of lead-based paint; accreditation of laboratories, field testing companies, and contractors; as well as defining the terminology associated with lead abatement practices. Task groups were established to identify the issues and to develop draft documents in these areas.

The task group on encapsulation technology identified the following areas of concern that require development of standard guides: encapsulation product performance, product qualification and product approval, use and application standard practice, and a postabatement inspection and maintenance program. These documents are being prepared with the intent of providing voluntary standards to be used with confidence by state and local authorities, abatement contractors, and even private homeowners, as consistent guides for abatement of lead-based paint with encapsulation.

III. ENCAPSULATION PRODUCT PERFORMANCE

Defining the performance requirements of lead-based paint encapsulants is not a trivial task. There are a wide variety of materials and product compositions that could be considered for use as encapsulants in lead-based paint abatement for interior and exterior surfaces. There is also a host of substrates to consider, such as wood, metal, plaster, etc., and some areas experience a significant amount of wear and tear. Some unique approaches of categorizing these products and determining the performance requirements needed for coating specific substrates in different environments were devised.

Since there are a wide variety of product compositions that differ extensively in their chemistry, predicting performance acceptability, based on product composition, was not a feasible approach. Rather, minimum performance requirements would be defined for four general groups of products. Products could be categorized as liquid-applied encapsulant coatings and reinforced liquid-applied products. Two other options identified as techniques for reducing lead exposure through encapsulation are flexible coverings, which can be applied with appropriate adhesives, and rigid enclosures, which are mechanically fastened and sealed at the joints.

Categorizing products in this manner reduced the need of using the chemical composition as a basis for assessing the acceptability of the materials' performance. In other words, any product could be used as long as the product meets the defined minimum performance specification.

A product classification scheme based on substrate, product type, and product grade was also devised. Substrates were categorized as wood, metal, cementitious, plaster, and gypsum board, with the note that all other types of substrates could fall under these categories, based on like physical properties. There are two product types to consider, depending on whether the product is suitable for interior (type I) or exterior (type II) application. Of course, the option still exists for products that have the potential for both interior and exterior use. Two grades of products were also defined. Grade A includes products that meet the requirements for use on abraded exposed areas and grade B for use in nonabraded areas. Abraded exposed areas are defined as substrate locations subjected to abrasion and ongoing wear. An additional class of products that depicts the materials' ability to be used on nonflat surfaces was also included.

Minimum performance parameters and test methods to determine product performance have been proposed for evaluating these materials. Some of the parameters are specific to durability and the mechanical properties of the encapsulant, such as impact resistance, etc., while other aspects of performance are targeted at insuring the product does not in itself pose a toxicity risk. The proposed test methods are well-documented standard methods that may be modified to be suitable for evaluating encapsulation products.

This scheme was developed with the expectation of having a means of judging a materials' acceptability for specific applications, based on the performance data. As examples, a wood type I grade A product would meet the requirements for application on interior wood substrates that are abraded exposed areas, such as floor boards, wood stair cases, and interior door jams. A cement type II grade A product would be suitable for the abatement of the exterior cement side of the house. If the technology permits, a type I grade AB product could be used in all abatement situations.

IV. ENCAPSULATION PRODUCT FORMULATION HEALTH RISK

It is also important to determine that the chemistry does not itself pose a health risk to workers during application, and to the residents of the property where encapsulants were applied.

The Office of Pollution Prevention and Toxics (OPPT), as part of its Existing Chemicals Program at the U.S. Environmental Protection Agency (EPA), performs risk assessments on chemical concerns, to identify and review potential hazards and exposure scenarios. This process provides information to determine if regulatory or nonregulatory actions are required to mitigate the risks associated with the use of certain chemicals and technologies.

A sampling of encapsulant coating product compositions was collected and submitted for risk assessment in the Existing Chemicals Program, to determine if this process could serve as a model for evaluating the risk of products used in lead abatement.

This review process included a chemical assessment, economic assessment, structure activity review, occupational and consumer exposure assessment, and a dose-response assessment. These assessments address issues such as the exact chemical composition and chemical properties; the extent the products are manufactured or are intended to be manufactured; and the health concerns, exposure levels, and risks to exposure to the specific chemicals in the formulations.

Since the OPPT existing chemicals review process has not been completed in its entirety, no information on its findings are presently available. However, it has been determined that screening estimates of the exposure risk of the chemicals contained in encapsulation products can be calculated. The relative risks of each formulation will not be obtained by OPPT at this time, since this assessment is only being performed on a limited number of formulations. Upon completion of the process, the details of how this risk assessment process has been conducted and how it could be expanded will be made available to interested parties concerned with evaluating encapsulant coating product risk.

V. ENCAPSULATION PRODUCT QUALIFICATION AND APPROVAL PROCESS

There has also been much discussion about how to evaluate encapsulant coatings against a standard of performance, and who would have the authority to develop an approved product list. A guide is being developed that describes the elements of a product certification program.

This program is simply based upon a third-party certification process. This process proceeds with the product manufacturer submitting performance data and chemistry information to an approving or certifying body. The certifying body would initially review the submission to check if the test methods used, and the data obtained, were in accordance with the performance standard described earlier. These data would then be forwarded to a third-party laboratory or other disinterested organization, to review the performance data for quality, reproduce some or all of the tests, if necessary, and assess the product hazard risks. Results of this review are returned to the certifying body, and a decision is then made to provide the product with a type and grade designation and indicate its approval for use in lead-based paint abatement. If approval is not granted, the results are returned to the manufacturer, with an explanation of the deficiencies. The certifying body may maintain the product on an approved product list. This list would then be made available to abatement contractors, for use in specifying products for lead-based paint encapsulation.

Approving officials may be constituents of local lead-based paint programs. If the entire process from submission of the performance data through the review and approval is executed according to the standard guides, then it would appear feasible that if more than one state or local program conducted product qualifications, the approved product list could be reciprocated between states.

VI. ENCAPSULATION PRODUCT USE AND APPLICATION

It is recognized that in addition to the inherent qualities of a product itself, the performance of encapsulant coatings depends on the integrity and compatibility of the substrate. Some surfaces, such as those that are rotting, may not be suitable until repaired, washed, deglossed, or otherwise prepared. In order for encapsulants to be used effectively and appropriately in lead-based paint abatement, inspection procedures and protocols need to include the assessment of the surface for suitability for encapsulant use, the type of preparation needed, worker protection requirements, and criteria to determine whether an encapsulant has been adequately applied.

Also very important is the issue of postabatement inspection and long-term maintenance of an encapsulated surface. Proper documentation and record keeping must be done for future reference to surfaces that still contain lead-based paint, but have been encapsulated. Information needs to be provided to landlords about appropriate maintenance procedures and methods for inspecting for deterioration.

A continuing education program curriculum needs to be instituted for abatement workers, that provides instruction and training on encapsulation technology, types of approved products available and their uses, and the long-term inspection procedures and protocols mentioned above. The EPA is currently addressing these issues through their state cooperative programs.

VII. SUMMARY

Voluntary standards in the area of lead-based paint encapsulant performance, application, and use would provide significantly needed guidance to the abatement community. Much more information is needed to further improve the technology of encapsulation, but these interim provisions will certainly provide relief to organizations that need to make decisions today on abatement procedures. Of course, encapsulation would not be a solution to all abatement problems, but under appropriate circumstances where the existing paint layer and substrate are intact, coating the surface may be a viable low-cost option.

REFERENCES

1. Lead-Based Paint Hazard Elimination. Final Rule. 53 FR 20790. U.S. Department of Housing and Urban Development, Washington, D.C., 24 CFR Parts 35, 200, 510, 570, 882, 886, 941, 965, and 986, 1988.
2. Lead-Based Paint Poisoning Prevention Act. Public Law. 42 U.S.C. 4821–4846, June 6, 1988.
3. Deleading. 454 CMR 22.00, Massachusetts Division of Industrial Safety, 1988.

Chapter 15

Successful Low-Cost Risk Communication and Public Education Programs

D. L. McAllister

CONTENTS

I. INTRODUCTION

As the services of Childhood Lead Prevention Programs (CLPPP) become more accessible and acceptable to more people, the number of children that will be found to have blood lead concentrations above 9 µg/dl would place impossible pressures on existing resources, without an emphasis on primary prevention through health education. Human behavior directly causes lead poisoning and is critical to its control. Health education is that dimension of health care concerned with influencing behavior. Based on scientific principles, health education bridges the motivational gap between health information and practice, by predisposing, enabling, and reinforcing the examination of options, weighing of consequences and informed decisions to voluntarily adopt attitudes and habits that promote social and behavior changes conducive to health. These educational activities have been highlighted in the following summary:

- Programs, materials, curricula
- Presentations, campaigns, fairs
- Networks, liaisons, coalitions
- Trainings, workshops, conferences
- Clearinghouse, library, resource database
- Information and referral hotline
- Technical assistance, newsletter

II. TARGET AUDIENCE

The first and most crucial step is to identify the at-risk target populations. Once this critical step is taken, educational activities can be tailor-made for that audience. Get to know all you can about the people you wish to influence: the women of child-bearing age, parents, other caretakers, as well as the children themselves, especially those without a usual source of health care. There are other important players in lead

1-56670-113-9/95/$0.00+$.50

poisoning prevention. Be sure to specialize messages and approaches to meet the different needs of health professionals; para-professionals; policy and decision makers; public health and other related governmental agencies; community-based groups; educators; real-property owners; community, church, banking, and business leaders; housing authorities; contractors and their organizations and unions; environmental equity groups; special interest groups; and institutions that serve the at-risk target audience. Detailed suggestions for practical low-cost risk communication and outreach follow.

III. HOTLINE

Any program with an extra phone line can train operators to run an information and referral phone service that logs and answers inquiries from the public. A cheerful, clear "Good morning or afternoon, you have reached a lead poisoning information and referral service. My name is _____ , how may I help you?" can open the conversation, which should always be polite and unhurried. Make every effort to answer each question directly. Callers are in a "teachable moment" and are open to receiving a great deal of education over the phone. Have the operators follow a written guide, which assures all calls are handled in the same way by different operators. This protocol also reminds operators to ask key questions and include the educational messages you want callers to hear. Never give callers the run around for information, although referrals given for services not provided by the program may often be necessary. Log all calls in a public inquiries log, by categories of caller, inquiry referral, and whatever demographics you feel are important. You may even make a mailing list of interested callers. If you cannot dedicate one full-time operator to the hotline, alternate the assignment among workers. At the end of the day an answering machine can be used. A typical announcement is, "Hello, you have reached a lead poisoning information and referral service available workdays, staff permitting. Since all local numbers will be called back, the only message you need leave is to slowly and clearly give your name and workday phone number, after the beep." At the beginning of the next workday, the next operator clears the machine, returning all calls.

IV. CLEARINGHOUSE

A noncirculating collection of articles, books, pamphlets, curricula, programs, files, and rolodex of referral sources and contacts locally and in other regions can be easily catalogued and made available to others.

V. NEWSLETTER

A program newsletter may be issued from time to time, using your own or other mailing lists. Even with desktop software, a newsletter will take its share of time, effort, and cost on a recurring basis, but offers big returns in morale and keeping your program in the minds of its readership.

VI. TECHNICAL ASSISTANCE

Your employees and volunteers should be available to attend meetings and provide educational technical assistance to groups who can assist your program in its mission, the eradication of lead poisoning.

VII. ACTION COUNCILS

Employees and volunteers can act as lead poisoning liaisons to network with local community leaders, community-based organizations, churches, schools, health and other service providers to the general public, and especially to your target audiences. As community needs assessments and surveys are made and resource directories developed, coalition building creates "community action councils." These local councils should have representation from all the players outlined above. A council that has come together with a shared interest and a commitment to preventing childhood lead poisoning by promoting the program's agenda can share information and resources with the target community.

VIII. EDUCATIONAL MATERIALS

Your employees and volunteers can assess needs, develop drafts, and pretest educational materials in focus groups. Well-thought-out and received materials can be produced, marketed, and distributed by local businesses, as public services creating good will in their communities. You can make educational public service announcements, posters, fliers, announcements, logos, stickers, banners, brochures, exhibits, photographs, animations, audiotapes, slides and slide shows, filmstrips, films, videos, novelty items, etc., to achieve educational outcomes in specific target audiences. Local media are likely to cosponsor and even produce programs in the public interest and can often enlist the support of celebrities. After lending our expertise to a 20-min show hosted by Geraldo Rivera, we were given permission to distribute it for not-for-profit screening by target audiences. You can still send us a request on your letterhead for a free copy of "What's Poisoning Your Children," in exchange for a new VHS tape. Keep a record of all materials distributed by your program, by developing an order form and a system for generating reports on this important service. An order form is best distributed on the back of a cover letter that explains the services available from your program.

IX. SPEAKERS BUREAU

If you lack educators or have too few, train more educators by establishing four-day monthly trainings, utilizing the resources of your program and its supporting institutions. A winning combination is to mix staff lectures with educational videos, reference articles, and field trips. Include practice in public speaking, with or without videotape feedback. Don't be afraid to use a pre- and posttest or an audience feedback or evaluation form. To complete the training, the speaker in training must give a successful presentation evaluated by someone from your program. Then issue a certificate and keep track of the events completed by each graduate.

X. EDUCATIONAL EVENTS

Lead poison awareness events range from health fairs, poster contests, and full-fledged publicity campaigns to simple lectures and presentations at community meetings. Don't leave important details to chance. Be sure to encourage proper planning with some kind of event planning form. Send a confirmation letter with an easy photocopyable poster on the back. Arrange for the timely mailing or delivery of materials to the event. A reminder phone confirmation is always prudent just before the event. Always have your events overseen by a staff person from the host agency. Have the host handle the logistics and remain in charge in the room throughout the event. Ask the host to document the event with some kind of evaluation form. You can expand your outreach if you ask each one in your audiences to go back and teach one of their neighbors or co-workers. Give them extra educational materials and have them practice what they would say as part of your presentation. Target risk groups and provide in-service staff trainings to those who also serve them. Professionals can also be reached through conference workshops, especially when Continuing Medical Education credits are arranged for health-care providers. Audience feedback and suggestions for improvement should be elicited periodically from all types of audiences. Keep these forms on file and generate statistics with them. A monthly chart showing the audiences reached by these events is a great morale and productivity booster.

There are many other possibilities for low-cost risk communication and outreach; be creative and document your process so that others can learn from your experience.

The following appendix contains sample forms we use, which you may adapt to your program.

Appendix A

EVENT FORM-PLANNING (*side one*)

Date of event:_____ from:_____to_____ Type of event:_____

Number of times event to be given that day:__

Number anticipated: __parent; __student; __teacher;__landlord;__tenant;__contractor; __leg. __med. __soc. service provider; __media; ____other(_____)

Organization:_____Address:_
_____zip
plus five:_____-_____
Public Transportation:_____
Contact Person:_____
Title:_____Fax:_____
Phone:_____ Best time:_____
Referred by_____

CONTACTS NARRATIVE

HOST ASSIGNED TO STAY IN ROOM
NAME:
LOCATION:
PHONE:

LITERATURE   AMOUNT

LESSON PLAN

Event initiated by_____ on _____.
Unit chief approval signed_____ on _____.
Confirmation letter sent att:_____ on _____.
Compensatory time request submitted via PMO on _____.
Materials mailed, or delivery was requested on _____.
Phone confirmation by_____ on _____.

Appendix B

EVENT FORM-EVALUATION

Speaker_____ Date_____ Total
number of times event given today:____
Grand total attended:____parents;___students;
_____teachers; _____landlords; _____tenants;
____contractors; _leg. _med. ___soc. service
providers; __media; ____other(_____)

Observed by_____
Title_____ Phone_____ Would
you like to repeat this event with another audience
s o m e t i m e ? Y e s ___ N o ___
When_____Would you
recommend this event to others? Yes___ No___

Why?_____
Who(include phone)_____
Who else_____

Any suggestions for improvements:_____
_____
_____
_____

PLEASE RATE THE SPEAKER: Excellent-Good-Fair-Poor
    Poise,confidence,enthusiasm----4------3-----2---1
    examples:_____

    Voice (volume,tone,speed)------4------3-----2---1
    examples:_____

    Language (clear,correct)-------4------3-----2---1
    examples:_____

    Knows subject------------------4------3-----2---1
    examples:_____

    Rapport with audience----------4------3-----2---1
    examples:_____

    Multicultural sensitivity------4------3-----2----1
    examples:_____

    Listens------------------------4------3-----2----1
    examples:_____
    _____

    Invites discussion-------------4------3-----2----1
    examples:_____

    Audiovisual Aids and Handouts--4------3-----2----1
    examples:_____
    _____

Appendix C

AUDIENCE FEEDBACK FORM

EVALUACION   orador/a speaker:_____
         fecha   date:_____
EVALUATION   local place:_____

                                              SI/YES   **NO**

	SI/YES	NO
The speaker was easy to understand. El orador/a fue facil para comprender		
The speaker invited audience participation El orador/a invito a la audiencia a participar		
The speaker was willing to answer questions. El orador/a estaba dispuesto/a a responder a las preguntas.		
The presentation taught me the sources of lead Con la presentacion, aprendi mas informacion acerca del envenenamiento por plomo.		
I have more resources for free screening and information on tenant's rights. Tengo mas informacion sobre recursos para el examen gratis, tambien acerca de mis derechos de inquilino/a.		
I know enough to teach others about lead. Tengo suficiente informacion para compartir con amigos acerca del envenenamiento por plomo		

How can we improve this presentation?_____ _____

_____
_____
_____
_____(continue on other side)

Appendix D

PUBLIC INQUIRIES LOG

```
1:PARENT 2:STUDENT 3:TEACHER 4:LANDLORD
5:TENANT 6:CONTRACTOR 7:LEGAL 8:MEDICAL
9:SOCIAL SERVICE PROVIDER O:OTHER
```

A:child Pb-B (BCH Regional Offices)
B:Group Pb-B/results (DOH screeners)
C:peeling apt. (HPD Cntrl Cmplnts 960-4800)
D:peeling   pvt.   home   (DOH   sanitarians)
E:Trng/Ed/Outreach (DOH Ed. & Trng.)
F:water (DOH test 693-4637/EPA 800-426-4791)
G:American Council of Labs (202) 887-5872
H:adult Pb-B/Occ/Hobby (Env. Epi. 788-4380)
I:media (Office of External Affairs 788-5290)
J:private abatement/contractor (HUD 264-0903)
O:Other-specify referral in "other" column

date	codes	mat.	referred by	other(specify)

Appendix E

LETTER OF CONFIRMATION (*side one*)

CITY OF NEW YORK, DEPARTMENT OF HEALTH
ENVIRONMENTAL HEALTH SERVICES
LEAD POISONING PREVENTION PROGRAM, BOX 58
TRAINING AND EDUCATION UNIT
311 Broadway, Second Floor
New York, New York 10013
tel: (212) BAN–LEAD

Dear colleague,

We are pleased to confirm that _____ of the Education and Training Unit will be providing the Lead Poisoning Prevention Event: _____ to be held _____ (day), _____ (date) from _____ to _____ in room _____ at _____ .

The unit resources include lectures, various media, and strategies requiring audience participation. We hope to reach many others through this audience, expected to number _____ , as they pass their learning and our materials to those who have not been reached by this message back in their environment.

Please provide _____ , and have a staff person in attendance to facilitate the event and evaluate the speaker. A flier advertising the event can be photocopied on colored paper, from the art work provided on the other side of this confirmation letter. Thank you for your interest and support toward eradicating this serious health problem.

Sincerely,

Appendix F

CONFIRMATION LETTER (*side two*)

TRAINING and EDUCATION UNIT
311 Broadway - Second Floor - BOX 58
New York, New York 10013

PHONE (212) BAN-LEAD

NEW YORK CITY DEPARTMENT OF HEALTH

CHILDHOOD LEAD POISIONING PREVENTION PROGRAM

H A B R A   U N   E V E N T O :

E V E N T :

P R E V E N C I O N

P R E V E N T I N G

E N V E N E N A M I E N T O

L E A D   P O I S O N I N G

P O R   P L O M O

DATE/FECHA: _____

TIME/HORA: _____

PLACE/LOCAL: _____

_____

ROOM/SALON: _____

Appendix G

COVER LETTER

CITY OF NEW YORK, DEPARTMENT OF HEALTH
ENVIRONMENTAL HEALTH SERVICES
LEAD POISONING PREVENTION PROGRAM
TRAINING AND EDUCATION UNIT — BOX 58
311 BROADWAY — Second Floor
New York, New York 10013
Phone (212) BAN–LEAD

Dear colleague,   date: _____

     Enclosed please find samples of some current Lead Poisoning Prevention educational materials. An "original" order form for materials available from the Education and Training Unit is printed on the other side of this letter. Please retain it for future use, by filling out photocopies of it as needed.

     Our "Speakers Bureau" offers health fair participation and workshops to the general public, service providers, and small children. We also offer a four-day monthly speaker's training. Other trainings and technical assistance are considered on a case-by-case basis. Our "Leadline," 212/BAN–LEAD, provides the public with information and referral. We also have a noncirculating reference library including a display of the sources of lead, and a lead information database and resource clearinghouse. A 20-minute video "What's Poisoning Your Children?" is available in exchange for a new VHS tape.

                   Sincerely,

Appendix H

MATERIALS REQUISITION

```
City of New York, Department of Health, Environmental Health
        Lead Poisoning Prevention Program — Box 58
   311 Broadway — Second Floor, New York, New York 10013
```

RETAIN THIS BLANK ORIGINAL FOR THE FUTURE — USE PHOTOCOPIES

FAX: (212) 941-1582

Organization:_____
Contact person:_____
Title:_____
Address:_____
City:_____
State:_____zip+5:_____-_____
Phone:_____Best time:____
You serve:__parents;__students;
__teachers;__landlords;__tenants
__contractors;__legal,__medical,
__social, service prov.;__media
Referred by:_____

ALLOW FOUR WEEKS FOR DELIVERY

	Request	Stock	OFFICE USE
Annual Report Submission FY1992_____		LD100	
20-min video (send new VHS tape)_____		WPIX1	
Lead Tests Results Card_____		LD000	
Dear Mom and Dad (Eng)_____		LD001	
Dear Mom and Dad (Sp)_____		LD002	
Eat foods Rich in Iron (Eng/Sp)_____		BN001	
CDC '92 Statement_____		CDC01	
Help for Parents (Eng/Sp)_____		LD010	
Help for Parents (Eng/Ch)_____		LD014	
Help for Parents (Fr/Cr)_____		LD015	
If Your Child Has Lead Poisoning (Eng)_		LD006	
If Your Child Has Lead Poisoning (Sp)__		LD009	
Lead Poisoning in Children (Eng)_____		NYS01	
Lead Poisoning Prevention (Eng)_____		LD011	
Lead Poisoning Prevention (Sp)_____		LD012	
Lead Poisoning Prevention (Fr)_____		LD013	
Nutrition and Lead (Eng)_____		LD003	
Nutrition and Lead (Sp)_____		LD004	
Know the Facts (Eng)_____		LD006	
Owners/Tenants (Eng)_____		HPD01	
Owners/Tenants (Sp)_____		HPD02	
Owners/Tenants (Ru)_____		HPD03	
Owners/Tenants (Ch)_____		HPD04	
Owners/Tenants (Fr)_____		HPD05	
Owners/Tenants (Cr)_____		HPD06	
Protecting Your Kids_____		DEP01	
Stop Lead Poisoning (Eng)_____		NSC01	
Stop Lead Poisoning (Sp)_____		NSC02	
Q&A About Lead_____		DEP02	

Received_____; Sent_____ by_____

Appendix I

EDUCATIONAL MATERIALS EVALUATION (*side one*)

TITLE OF MATERIAL_____
EVALUATED ON_____BY_____
TITLE _____
ORGANIZATION_____
ADDRESS_____
_____

PHONE_____ Best time to call= _____ FAX_____
INSTRUCTIONS: Use the following scale for evaluating educational material as it relates to your target audiences. Each descriptive statement is followed by (1), (2), (3), (4), (NA). Indicate your assessment by circling the appropriate number in the scale:
(1)Definitely Yes (2)Yes (3)No (4)Definitely No (NA)Not Applicable

OBJECTIVES

Identified outcomes may be obtained through use of the material.

1 2 3 4 NA

The material is representative of good educational objectives and help further those objectives.   1 2 3 4 NA

ABILITY RANGE

The material provides for the range of abilities and aptitudes of the target audience.   1 2 3 4 NA

CONTENT

The content is factually accurate.   1 2 3 4 NA

The content is clear.   1 2 3 4 NA

The content is contemporary.   1 2 3 4 NA

The material is controversial.   1 2 3 4 NA

The material presents alternative views.   1 2 3 4 NA

The material is balanced and not partisan to any product, company or political cause.   1 2 3 4 NA

The nature and scope of the material content is adequate to meet educational objectives.   1 2 3 4 NA

The material introduces experiences that would not otherwise be available to the target audience.   1 2 3 4 NA

The material suggests other resources.   1 2 3 4 NA

Appendix I

EDUCATIONAL MATERIALS EVALUATION (*side two*)

UTILIZATION CHARACTERISTICS

The content is logically organized into parts and can easily be discussed or processed in stages.   1 2 3 4 **NA**

The time required to utilize the material is appropriate for use with the target audience.   1 2 3 4 **NA**

The quantity and diversity of information given is appropriate to the target audience.   1 2 3 4 **NA**

The degree of difficulty, and complexity of the technical information is appropriate to the target audience.   1 2 3 4 **NA**

The material is appropriate for the interest level of the target audience.   1 2 3 4 **NA**

The language of the presentation matches the literacy of the target audience.   1 2 3 4 **NA**

The material is attractive to the target audience.   1 2 3 4 **NA**

The material can be identified with by multicultural target audiences.   1 2 3 4 **NA**

The material motivates to the target audience.   1 2 3 4 **NA**

PRESENTATION

The style of the presentation is likely to lead the target audience to accomplishing the basic educational goals.   1 2 3 4 **NA**

The intended use is clearly understood.   1 2 3 4 **NA**

The production quality of the material is acceptable.   1 2 3 4 **NA**

Overall evaluation of the usefulness of this material? _____
_____

This evaluation is based on usage with a target audience numbering _____ consisting of _____

In the next year it is anticipated that this material will be used _____ times with target audiences totaling _____ consistingof_____

PLEASE MAIL TO NEW YORK CITY HEALTH DEPARTMENT, ENVIRONMENTAL HEALTH SERVICES LEAD POISONING PREVENTION PROGRAM BOX 58 311 BROADWAY - Second Floor New York, New York 10013

Part III
Chemical Measurement Methods

Chapter 16

Phosphate Addition to the Delves Cup Method

F. Ruszala, D. Worsley, and J. Hogan

CONTENTS

I. INTRODUCTION

A blood sample placed on filter paper can be analyzed for lead by the Delves Cup Method (DCM). DCM, forerunner of modern graphite furnace atomic absorption spectroscopy (GFAAS), decomposes the sample by placing it within the flame, which releases the sample cloud in a quartz or ceramic tube.[1,2] Continuous operation of the spectrometer measures the absorbance signals. The results of the DCM are complicated by a double peak consisting of a smoke peak followed by a lead peak. With low to moderate amounts of lead (1 μg/dl to 20 μg/dl), the smoke peak frequently combines with the lead peak to make quantitation difficult, whether done manually or with electronic integration.

DCM, however, does have a number of advantages, and acetylene flame atomic absorption spectrometers (FAAS) operated by trained personnel are still quite common in clinical laboratories. The DCM is an inexpensive, fast, and uncomplicated method for the analysis of blood lead levels. The DCM is simple to manage; an operator can be trained quickly, and a skilled operator can process up to 500 samples per 8-h shift. The person assigned to perform the sample preparation and the data entry does not need to be highly skilled. The highest skill level required is that of the analyst involvement in data interpretation and being challenged with the problem of coalescing smoke and lead peaks. Unfortunately, the current public health campaign against lead poisoning in children corresponds to a substantial decrease in the use of the DCM as it is replaced by the more accurate and sensitive GFAAS method. It has been demonstrated that phosphate solutions of [H_3PO_4, $NH_4H_2PO_4$, and $(NH_4)_2HPO_4$] can be used as matrix modifiers for lead analysis of biological samples.[3–10] The phosphate solution is added to the sample either prior to the beginning of the FAAS process or after the sample has been dried in the graphite furnace tube. Some instrument manufacturers recommend the use of phosphate for FAAS lead determinations.[11]

II. EXPERIMENTAL

A Perkin Elmer Model 460 atomic absorption spectrometer, with a lead element tube, at 283 nm was coupled to a Fisher Recordall Series 5000 stripchart recorder at the 10 mV output of the spectrometer. Simultaneously, a Perkin Elmer Model LCI 100 integrator was connected to the 1 V output of the spectrometer.

The integrating recorder was used to obtain peak times, peak heights, and peak areas. The stripchart recorder was used to produce signals for the operator to monitor.

The $Pb(NO_3)_2$ and $(NH_4)_2HPO_4$ used for this study were ACS reagent grade, and distilled water was used for dilution to achieve the final concentrations.

Human blood samples containing 1, 3, 5, 7, 10, 15, 20, 25, 29, and 51 μg/dl of lead were placed on lead-free filter paper and air dried at 120°C. The filter paper was obtained from Schleicher and Schell, Keene, NH. Filter paper disks for the Delves Cup Method were prepared from the paper samples, using a 3/16-in. hand puncher from M. C. Mieth Inc., Port Orange, FL.

Phosphate spiking solutions were delivered using an Eppendorf 10-μl fixed-volume pipet. After spiking, samples were air dried at 120°C. A series of $(NH_4)_2HPO_4$ solutions ranging from 0.001 M to saturation were prepared at 25°C. These were delivered to pairs of samples with equal blood lead concentrations, and run in the spectrometer to study the effect of phosphate on the resolution of the lead peak. A second series

122

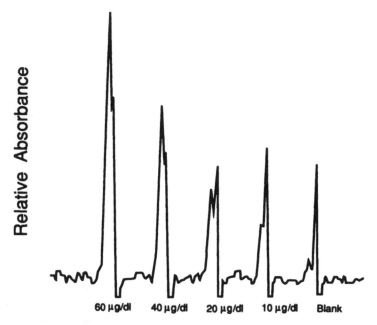

Figure 1 Blank (1 µg/dl Pb) dried blood samples with the addition of nitrate standards (read right to left) blank, 10 µg/dl Pb, 20 µg/dl, 40 µg/dl, and 60 µg/dl Pb. A consistent initial smoke peak is noted, followed by increasing lead recoveries. Peak overlap compromises integration capability.

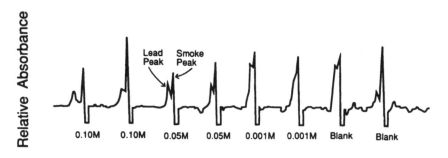

Figure 2 Pairs of 5 µg/dl dried blood samples with increasing concentrations, 0.001, 0.05, and 0.10 M(NH$_4$)$_2$HPO$_4$ (read right to left). Note separation of the lead and smoke peaks at higher concentrations.

using 0.05, 0.10, 0.15, 0.20, 0.25, and 0.30 M (NH$_4$)$_2$HPO$_4$ spiking solutions were delivered to seven samples with the same lead concentration, to further refine the optimum concentration of phosphate required. A third series, using 0.20 M (NH$_4$)$_2$HPO$_4$ delivered to 10 samples with the same lead concentration, was prepared and tested to obtain statistical data.

III. RESULTS AND DISCUSSION

Figure 1 shows a series of lead standards, ranging from 10 to 60 µg/dl, where phosphate was not used. As can be seen, the smoke peak (the sharper peak on the right of each pair) can coalesce with the lead peak (the broader of the two), making quantitative interpretation difficult.

The addition of phosphate solution to the samples has a dramatic impact on the separation of the two peaks, as can be seen with the 0.05 M (NH$_4$)$_2$HPO$_4$ in Figure 2. The difference between a series of blood

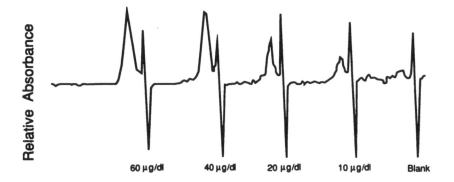

Blood Lead Standard Spiked With 10 µl of 0.20M (NH$_4$)$_2$HPO$_4$

Figure 3 Lead standards in concentrations of 10, 20, and 40 µg/dl, 60 µg/dl prepared as in Figure 1, with the addition of phosphate (read right to left). Note the marked improvement in peak separation due to phosphate addition.

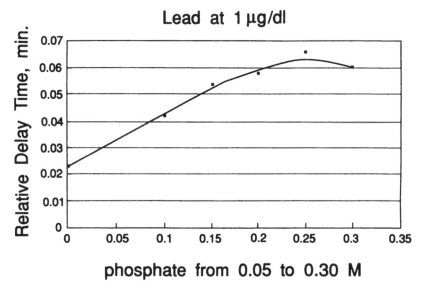

Figure 4 At low lead concentration (1 µg/dl), increasing phosphate concentrations are plotted against lead peak retention relative to the smoke peak. An increased delay facilitates automated integration.

lead standards where a 10-µl spike of 0.20 M (NH$_4$)$_2$HPO$_4$ has been added is dramatic, as demonstrated in Figure 3.

The variation of time between the lead and smoke peaks, as a function of increasing phosphate concentration for a fixed blood-lead concentration, is shown in Figures 4 and 5. The most effective amount is approximately 0.20 M (NH$_4$)$_2$HPO$_4$, above which no increase in time delay is obtained, and below which the peaks are closer to one another. At concentrations of phosphate exceeding 1 M, the lead peak begins to broaden and decrease in height.

Ten samples per each blood lead concentration were analyzed to compare the statistical difference between samples with and without phosphate. Figures 6–9 show the relative peak height and the coefficient of variation vs. increasing blood lead concentration, resulting from this comparison.

Although the curve is linear when no phosphate spiking is utilized, 60% or more of the samples neither split nor were able to be integrated, demonstrating erratic behavior as shown in Figure 6. Figure 7 shows the coefficient of variation vs. lead concentration, which also confirms poor performance.

When a phosphate spiking solution is utilized, not only is the curve linear, but 98% of the samples produced split peaks that were successfully integrated (Figure 8). The coefficient of variation vs. lead

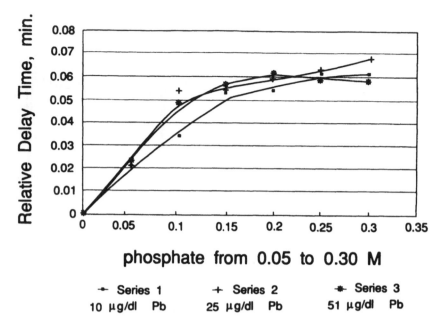

Figure 5 Higher concentrations of lead are shown vs. increasing phosphate concentration. Above 0.2 *M*, the delay of the lead peak is essentially constant.

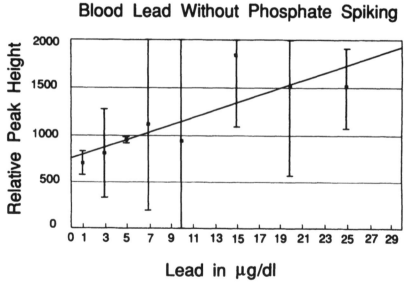

Figure 6 Peak height vs. increasing lead concentrations without phosphate using the Delves Cup Method. A gross linear relationship is noted.

concentration shows an orderly behavior pattern (Figure 9). At blood lead concentrations of 5 to 10 µg/dl, samples give coefficient of variation of about 8.5%. This is significant in that it demonstrates that the DCM can achieve adequate sensitivity and accuracy at the current CDC action level of 10 µg/dl.

IV. CONCLUSIONS

The addition of 0.2 *M* $(NH_4)_2HPO_4$ to blood samples improves the resolution of the Delves Cup Method by effectively splitting the smoke and lead peaks. Additionally, phosphate improves sensitivity and repro-ducibility and allows the use of electronic integration, thus freeing the analyst from subjective interpretation

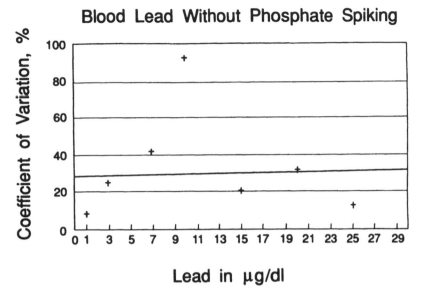

Figure 7 Coefficient of variation vs. lead concentration using the standard Delves Cup Method. Note that the variance increases and is inconsistent.

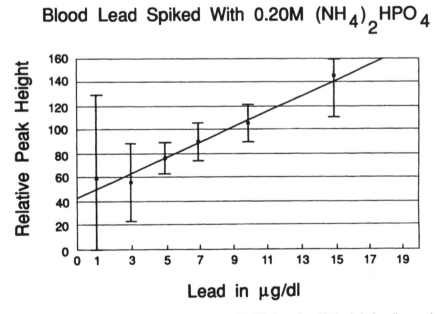

Figure 8 Peak height vs. lead concentration using the modified Delves Cup Method. A clear linear relationship is apparent.

of poorly formed data. This improved modified Delves Cup Method should stimulate a reevaluation of the value of this technique in clinical blood lead determinations. It should allow clinical laboratories with existing FAAS instrumentation and expertise to provide quality analyses in meeting the increased national demand for blood lead analyses in children.

126

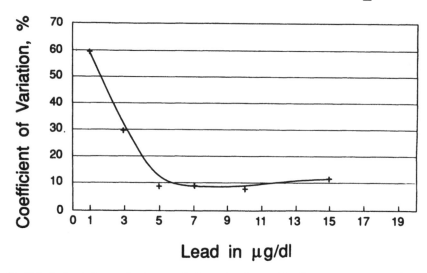

Figure 9 Coefficient of variation vs. lead concentration using the modified Delves cup method. Consistently lower variance at higher lead concentrations is demonstrated, indicating much improved reliability that exceeds standard Delves cup methodology.

REFERENCES

1. Delves, H.T., A micro-sampling method for the rapid determination of lead in blood by atomic-absorption, *Analyst*, 95, 431, 1970.
2. Fernandez, F. and Kahn, H.L., The determination of lead in blood by atomic absorption spectrometry with the delves cup technique, *At. Absorption Newslett.*, 10, 1, 1971.
3. Kaiser, M.L., Koirtyohann, S.R., and Hinderberger, E.J., Reduction of matrix interferences in furnace AA with the L'vov platform, *Spectrochim. Acta*, 36B, 773, 1981.
4. May, T.W. and Brumbaugh, W.G., Matrix modifier and L'vov platform combination for the elimination of matrix interferences in the analysis of fish tissues for Pb by graphite furnace AAS, *Anal. Chem.*, 54, 1032, 1982.
5. Koirtyohann, S.R., Kaiser, M.L., and Hinderberger, E.J., Food analysis for Pb using AA and L'vov platform, *J.A.O.A.C.*, 65, 999, 1982.
6. Rains, T.C., Rush, T.A., and Butler, T.A., Innovations in AAS with electrochemical atomization for determining Pb in foods, *J.A.O.A.C.*, 65, 994, 1982.
7. Giri, S.K., Shields, C.K., Littlejohn, D., and Ottaway, J.M., Determination of Pb in whole blood by electrochemical AAS using graphite probe atomization, *Analysts*, 108, 244, 1983.
8. Manning, D.C. and Slavin, W., The determination of trace elements in natural waters using the stabilized temperature platform furnace, *Appl. Spectrosc.*, 37, 1, 1983.
9. Pruszkowski, E., Carnrick, G.R., and Slavin, W., The blood Pb determination with the platform furnace, *At. Spectrosc.*, 4, 59, 1983.
10. Slavin, W., Carnrick, G.R., Manning, D.C., and Pruszkowska, E., Recent experience with the stabilized temperature platform furnace and Zeeman background correction, *At. Spectrosc.* 4, 69, 1983.
11. Techniques in Graphite Furnace Atomic Absorption Spectrophotometry, No. 0993–8150, Perkin Elmer Corp., April 1985, 134.

Chapter 17

New Developments in Lead-Paint Film Analysis with Field Portable X-Ray Fluorescence Analyzer

S. Piorek, J. R. Pasmore, B. D. Lass, J. Koskinen, and H. Sipila

CONTENTS

ABSTRACT

Lead-based paint is recognized as the principal source of lead in dust and soil. From here it can easily enter a child's organism, as the result of hand-to-mouth activity typical for young children at play. The preferred method for testing paint for lead is field portable X-ray fluorescence (XRF). A typical portable XRF lead analyzer provides direct readout of lead mass per unit area, and is very easy to operate. The battery-operated analyzer is equipped with a radioactive source to excite lead X-rays, and a gas proportional detector to register the lead radiation intensity. Associated electronics processes a detector signal and converts it into lead mass per unit area. The analyzer described has measurement precision better than 0.05 mg/cm² of lead at 10 s per sample measuring time. The calibration curve shows very little bias caused by substrate variations. A newly introduced analyzer features a high-resolution semiconductor probe that allows for accurate, substrate-independent lead analysis using K-lines of lead excited with the isotope Cd-109.

I. INTRODUCTION

Chronic exposure to even low levels of lead may result in serious problems in young children, including impairment of the central nervous system, behavioral disorders, and mental retardation. Paint, dust, and soil and, to a lesser extent, food, water, and air are all potential sources of exposure to lead in housing units.[1] However, lead-based paint is recognized as the principal source of lead in dust and soil. From here it can easily enter the child's system, as the result of hand-to-mouth activity typical for young children at play.

These concerns were addressed with federal regulatory efforts as early as 1971, with the enactment of the Lead-Based Paint Poisoning Prevention Act (LBPPPA). As amended, the Act specified requirements for testing and abatement of lead-based paint (LBP) in housing and designated the Department of Housing and Urban Development (HUD) as the primary agency in the federal effort to eliminate the hazard of LBP. Specifically, under Section 302 of the LBPPPA, public housing agencies (PHA) and Indian housing agencies (IHA) are required, by 1994, to randomly inspect all their housing projects for lead-based paint.[1]

Any LBP hazards equal to or greater than 1.0 mg of lead per square centimeter must be abated, as well as any applied to surface paint that contains more than 0.5% lead by dry weight.

II. ANALYSIS OF LEAD IN PAINT

While a number of methods can be used for analysis of lead in wall paint, the circumstances of testing, and the sheer number of tests to be performed in limited time, strongly favor field portable methods of lead analysis. At present two basic methods are recommended as acceptable to detect lead in wall paint.[2] The first is a field portable X-ray fluorescence (FPXRF), which is nondestructive, rapid, relatively inexpensive, and rugged. The other method is laboratory analysis of paint chips collected on site, using recognized techniques such as atomic absorption spectrometry (AAS) or inductively coupled plasma-atomic emission spectrometry (ICP-AES). A third method, spot testing using sodium sulfide or sodium rhodizonate, can be used as a qualitative tool.

The preferred method for testing paint for lead is FPXRF. In addition to the features mentioned above, the typical analyzer provides direct readout of lead mass per unit area and is very easy to operate. The battery-operated analyzers are equipped with a radioactive source to excite lead K-series X-rays, and a detector, to register the lead radiation intensity. Associated electronics process the detector signal and convert it into lead mass per unit area. The radioactive source used in commercially available analyzers is Co-57, which emits radiation of about 120 keV. This is sufficient to excite the K-series radiation of lead (70- to 88-keV range). The use of K-series radiation of lead allows for analysis of lead in a layer that may be buried under many layers of nonleaded paint. However, the relatively high energy emitted by the Co-57 source poses some radiation hazards to the operator and public. Operators using equipment containing a Co-57 radioisotope must complete a radiation safety course approved by the Nuclear Regulatory Commission.

All currently available lead-paint analyzers experience difficulties with measurement of lead in paint on different substrates. An analyzer calibrated for lead on wood panels will not yield correct readings for lead on sheetrock, steel, concrete, etc. These inaccuracies result in either false negatives or false positives. Both problems have significant practical consequences. One possible exception to this problem is the so-called spectrum analyzers. These use the whole energy spectrum, rather then just the lead K-line intensity alone, to determine the lead concentration. This type of lead analyzer, however, requires a higher level of operator training.

III. UNIVERSAL, FIELD PORTABLE X-RAY ANALYZER MEASURES LEAD IN PAINT

In many practical situations, a coat of paint is only a few layers thick. Therefore, lead can be measured using its L-series lines, which are less penetrative. Excitation of the lead L-series radiation requires a much less energetic source, such as Cd-109 or Cm-244. These two isotopes are much safer in use and storage than is Co-57.

An instrument that employs this approach, the X-MET™ 880, was successfully used for several years in various lead testing projects. Multiple layers of paint are handled by scraping and taking consecutive readings on exposed layers of paint. Substrate interferences such as buried nails, copper wires, etc., are automatically corrected for by using information provided by the spectrum. The 4-lb probe is equipped with a state-of-the-art, high-efficiency, and high-resolution proportional counter. This allows for excellent precision of measurement (better than 0.05 mg/cm^2) with as short as 5 to 10 s of counting time.

By using the L-series excitation, the influence of buried pipes and conduits on the lead readings is practically eliminated. The narrow "footprint" of the probe, 1.5 in. wide, allows its use on many building components, such as window wells.

Figure 1 shows the X-MET, fitted with its surface probe, being used to measure lead in wall paint. Figure 2 represents a typical calibration curve for lead in paint on a wood substrate. It was obtained with a Cm-244 radioisotope, which has the advantage of a long half-life (17.4 years). The instrument has software capabilities that allow for background and spectral overlap corrections so that the same calibration curve can be used on the other substrates, such as glass, steel, sheetrock, etc., with only about 0.1 mg/cm^2 bias due to the substrate difference. This is demonstrated in Table 1. The precision of a single measurement is better than 0.05 mg/cm^2 lead for lead concentrations in the range of 0 to 3 mg/cm^2 at 10 s of measurement time. Equivalent results can be obtained with a Cd-109 radioactive source that, however, has a shorter half-life of 1.3 years.

The X-MET 880 can be classified as either a "direct reader" or "spectrum analyzer."[2] While the result of every measurement is presented to the operator as a direct lead readout in mg/cm^2 on the display, it is always extracted from the whole X-ray spectrum of the sample, which, if needed, can be stored and examined by the operator.

A recent practical example of its application for lead in paint analysis has been provided by Southwestern Laboratories from Houston, TX[3] (see Figure 3). Last year, for one summer week, Southwestern Labs

Figure 1 X-MET Portable X-ray Analyzer.

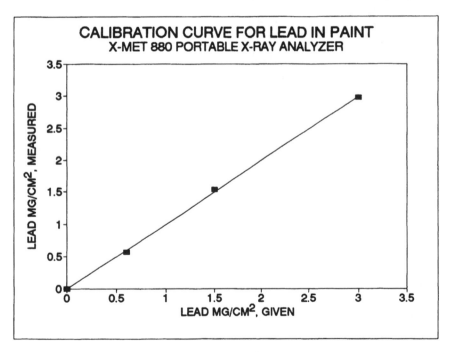

Figure 2 Calibration curve for lead in paint.

performed in excess of 2000 measurements of all painted surfaces in the Texas State Capitol Building in Austin. The measurements were performed with an X-MET Portable X-ray Analyzer equipped with a surface probe and a Cd-109 source. The system was calibrated for lead, with standards of varying paint thickness and lead concentrations, measured on wood, sheetrock, and steel substrate. Exposure time was 5 s per area tested, providing precision of a single measurement better than ±0.06 mg/cm^2 lead for lead concentrations in the range from 0 to 10 mg/cm^2. Levels of lead from 0 to 8.0 mg/cm^2 were detected throughout the structure.

25

Table 1 Results of Lead Measurements on Three Paint Standards as Obtained with X-MET 880 Field Portable X-Ray Analyzer Equipped with a High Resolution Gas Proportional Detector Contained in a Hand-Held Surface Probe

Substrate	0.6 mg/cm² STD	1.5 mg/cm² STD	3.0 mg/cm² STD
Wood	0.570±0.006	1.537±0.021	2.977±0.040
Sheetrock	0.577±0.006	1.576±0.022	3.091±0.040
Concrete	0.572±0.006	1.566±0.022	3.051±0.040
Steel plate	0.559±0.006	1.459±0.020	2.848±0.037
1/4 in. glass	0.578±0.006	1.520±0.021	3.007±0.039

Notes: Measurement time 10 s per sample.

Results reported in milligram of lead per square centimeter.

Each result reported with one standard deviation counting statistics error.

Figure 3 Measurement of lead on painted columns with X-MET.

IV. SUBSTRATE SENSITIVITY PROBLEM OF CONVENTIONAL LEAD ANALYZERS

The lead analyzers employing a Co-57 source and a poor-resolution detector (such as conventional proportional counter or a scintillation counter) suffer a major disadvantage originating in signal intensity variations of an X-ray peak representing the portion of the source radiation backscattered from the sample. The variation of this peak is a manifestation of an elementary phenomenon of gamma and X-ray radiation interaction with matter.

In simple terms, there exists a strong inverse correlation between the average, effective atomic number of the sample and the intensity of the incoherently (or Compton) scattered radiation. This means that the intensity of a Compton peak from a sample of wood will be much higher than that from a steel plate or a glass. As a matter of fact, this correlation is so significant that in certain cases it is utilized for a quite accurate measurement of the average effective atomic number of the substrate and to appropriately correct lead measurements. For conventional lead analyzers, however, varying intensity of a Compton peak is an interference, rather than an advantage, in case of lead in paint analysis. This is due to the fact that the desirable K-alpha peak of lead excited by a Co-57 source is always superimposed over this varying intensity of the low-energy tail of the Compton peak. This is illustrated in Figure 4.

Since the resolution of either proportional or scintillation counters does not allow for direct extraction of the net lead peak intensity from an X-ray spectrum, indirect methods are employed. These methods

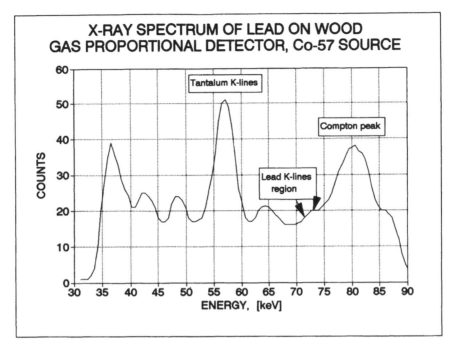

Figure 4 X-ray spectrum of lead in paint on wood obtained with a Co-57 source and a conventional proportional detector.

require two consecutive measurements of the sample, without and with a special X-ray filter between the sample and detector. The difference between the two measurements is used as a measure of the net lead signal. This approach is limited, as it does not correct for variances in the energy distribution of the Compton scattered radiation, which, in turn, changes the transmission characteristics of the filter used. This phenomenon is the main reason why all conventional lead analyzers are substrate sensitive.

V. NEW, HIGH-RESOLUTION, SUBSTRATE-INSENSITIVE PROBE

The situation described above can be readily corrected by replacing the Co-57 source with a Cd-109 radioisotope and by using a high-resolution semiconductor detector to collect the spectrum. Figure 5 illustrates a spectrum of lead paint excited on wood, with a Cd-109 source, and recorded with a liquid nitrogen–cooled Si(Li) detector-based probe. It is seen that the K-alpha and K-beta peaks of lead are clearly resolved from each other and are positioned now on the right side of the Compton peak. Their net intensities can be easily determined by interpolation of the background line under each peak. It can be appreciated that as long as it is possible to interpolate the background line, it does not matter for the net lead peak intensity whether the Compton peak varies or not.

The net peak intensity calculations based on the Si(Li) detector spectra pose no problem for contemporary microprocessor-based analyzers. By using a high-resolution probe and a Cd-109 excitation source, measurements of lead in wall paint can be made truly substrate independent. Tests performed to date confirm this conclusion.

Recently a new probe, featuring a high-resolution Si(Li) detector, was tested for lead in paint analysis. The excitation of K-series radiation of lead was accomplished with a Cd-109 source rather than the more dangerous and short lived Co-57. The Cd-109 isotope emits a small fraction of its primary radiation at 88.2 keV which very efficiently excites lead K-series radiation. Therefore, a moderate activity of 10 to 20 mCi of the Cd-109 isotope can be used with satisfactory results. The Cd-109 excitation also significantly improves the signal to background ratio for lead K-lines, when compared with Co-57 excitation. Initial tests indicate that a statistical precision of measurement better than 0.1 mg/cm^2 can be achieved in a 10-measurement time.

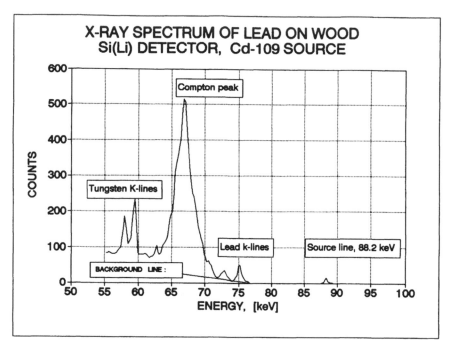

Figure 5 X-ray spectrum of lead in paint on wood obtained with a Cd-109 source and a high-resolution Si(Li) detector.

VI. BROADER ENVIRONMENTAL ASPECT OF LEAD IN PAINT PROBLEM

Numerous reports show that to control lead contamination of a household environment, it is not sufficient to just measure lead in painted surfaces. As the leaded paint weathers or flakes over time, lead-containing particles are transported to the floor dust, air, and surrounding soil.

Both the gas proportional detector probe and the new semiconductor detector probe can be used for measurement of lead (and other inorganic contaminants) in bulk soil or dust at 40- to 60-ppm levels.[4] Figure 6 shows a typical calibration for lead in soil, while Table 2 illustrates the typical performance of the analyzer for selected metals in soil.

The accepted analytical technique for lead measurements in household dust is taking wipes from the floor and window sills. One square foot of the surface is wiped until no more visible residue is collected on the wipe. Next, the lead is extracted from wipes and analyzed using flame atomic absorption spectrometry.

An alternate dust collection method is to use a small, customized vacuum cleaner to pick up the dust and deposit it onto a thin-membrane substrate. The filtered deposit is, from the point of view of X-ray fluorescence, a thin layer.[5] As such, it can be easily analyzed using the same probe that measures lead on painted surfaces. This method has the added advantage of providing the result "on the spot" for total lead. By filtering the dust collected from 1 ft^2 (929 cm^2) onto a 37-mm-diameter membrane of 10.8 cm^2 effective area, an 86-fold preconcentration of the sample is realized. The practical implication of this is increased sensitivity of measurement with XRF.

The maximum permissible levels of lead in dust on floors following abatement is 200 μg/ft^2; on window sills, 500, and in window wells, 800 μg/ft^2. These numbers translate to 0.22, 0.55, and 0.88 μg lead per square centimeter, respectively. This is less than the detection limit for a typical portable lead X-ray analyzer. However, when preconcentrated on membrane filters, 200 μg of lead per square foot represents as much as 18.6 μg of lead per square centimeter. This is about ten times above the lead detection limit of either analyzer.

Similar deposits on membrane filters can be obtained from personal monitors of breathing zone air, used by workers occupationally exposed to polluted air. Lead, and other elements in airborne dust, can be measured with both probes, after being filtered from air onto a substrate such as Millipore™ cellulose membrane. Analysis down to approximately 1 μg/cm^2 is possible.[6,7]

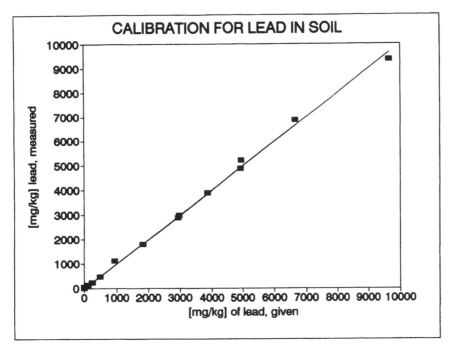

Figure 6 Typical calibration curve for lead in soil.

Table 2 Typical Performance of the Si(Li)-Based Portable X-Ray Analyzer for Metals in Soil

Element	Conc. range (mg/kg)	RMS error (mg/kg)	Precision (mg/kg)	Det. limit (mg/kg)
Cr	0–6200	<160	50	<100
Cu	0–9000	80	12	40
Zn	0–9000	<120	25	<80
As	0–10000	60	30	<80
Pb	0–10000	<80	10	<40

Notes: RMS error is one standard deviation of the data points from the empirically fitted calibration curve.
Detection limit is equal to three standard deviations of the single measurement as obtained on the sample containing no element of interest [but possibly the interfering element(s)].
All data determined for a 200-s measurement time and a 20-mCi Cd-109 excitation source.

Table 3 lists capabilities of the portable analyzer for metals filtered from air, on membrane substrate. It is seen that the detection limit for lead, offered by the X-MET FPXRF Analyzer, is at least an order of magnitude lower than any of the permissible lead levels for household surface contamination, as long as samples of dust are collected on filters.

VII. CONCLUSIONS

Lead decontamination of the household environment has to be a multifaceted approach. Simple monitoring of lead on painted surfaces is not enough. A comprehensive approach of monitoring lead in the home environment must include analysis of areas that would collect the dust and particles of peeled and weathered paint. In this manner all sources of lead poisoning can be monitored and controlled.

The present choices of lead analyzers focus primarily on paint analysis. They cannot perform analysis of lead in dust, dirt, or filters. The X-MET analyzer recognizes and addresses the total need of the lead abatement market. It is a field portable XRF analyzer that can perform analysis of various sample types, at levels exceeding the regulatory criteria. As a full-spectrum analyzer, it can compensate for the varying substrates encountered in analysis of lead in paint. Additionally, the collection of a full spectrum allows the instrument to analyze different sample types, such as filters, dust, and dirt. Detection limits and precision of analysis are easily obtained with short counting times of 5 to 10 s.

Table 3 Performance of Field Portable X-Ray Analyzer in Monitoring Ambient and Work Place Air Particulates (Gas-Filled, Proportional Detector)[a]

Element	Detection Limit µg/cm²	Detection Limit µg/m³	PEL[b] in air µg/m³
Cr	1.8	18	1000
Mn	2.4	24	50
Fe	2.4	24	10^4 (as Fe_2O_3)
Ni	1.5	15	1000
Cu	1.5	15	100–1000
Zn	1.5	15	$(1–5) \times 10^3$
As	2.4	24	10
Se	2.4	24	200
Pb	1.5	15	50
Cd	3.3	33	100–3000
Sn	3.3	33	2000

[a] Detection limits for a Si(Li) detector-based probe are on average 1.5 to 2 times better.

[b] Permissible exposure limit in work place air, PEL (as per 29 CFR 1910.1000, Jan. 1, 1977).

Notes: The data reported for Cd and Sn were obtained with 30-mCi Am-241 source, for a 600-s measurement time per sample. The data for the other elements listed in a table were obtained with a 10-mCi Cd-109 source and 200-s measurement time.

Detection limits are reported as three standard deviations due to counting statistics, as measured on a blank membrane filter.

Detection limits reported in µg/m³, assume that 1 m³ of air is pumped by personal air pump through 10-cm² area filter during an 8-h period.

The X-MET analyzer also offers other advantages to lead analysis. Replacement of the Co-57 source, with Cd-109, provides for a source with inherent advantages. First, the source is safer. Second, the Cd-109 source and X-MET features solve the problems of varying background signals caused by varying substrates. The X-MET also permits analysis of lead with either K or L X-rays, with the Cd-109 source. Both analytical approaches have their pluses and minuses, depending upon the job at hand. The X-MET, however, lets the user configure the instrument, to meet the needs of the monitoring job.

REFERENCES

1. Williams, E.E., Estes, E.D., and Gutknecht, W.F., Analytical Performance Criteria for Lead Test Kits and Other Analytical Methods, Research Triangle Institute Report to Office of Toxic Substances, U.S. Environmental Protection Agency. EPA Contract No. 68–02–4550, RTI Project No. 91U–4699–066, 1991.
2. Lead-Based Paint: Interim Guidelines for Hazard Identification and Abatement in Public and Indian Housing, Office of Public and Indian Housing, Department of Housing and Development, Washington, D.C., September 1990.
3. Reilly, S.R., private communication, Southwestern Laboratories, Houston, TX, 1991.
4. Piorek, S. and Pasmore, J.R., A Si(Li) Based High Resolution Portable X-Ray Analyzer for Field Screening of Hazardous Waste, presented at 2nd Int. Symp. on Field Screening Methods for Hazardous Wastes and Toxic Chemicals, February 12–14, 1991, Las Vegas, NV, Conference Proceedings, pp. 737–740.
5. Makov, V.M., Losev, N.F., and Pavlinski, G.V., Zavod. Lab., Vo. 34, No. 12, 1968, pp. 1459–1460.
6. Piorek, S., Application of a Field Portable X-Ray Analyzer to On-Site Analysis of Workplace Air Contaminants, paper presented at 1991 Pittsburgh Conference, March 1991, Chicago, IL.
7. Rhodes, J.R., Stout, J.A., Schindler, J.S., and Piorek, S., Portable X-ray Survey Meters for In Situ Trace Element Monitoring of Air Particulates, Toxic Materials in the Atmosphere, ASTM STP 786, American Society for Testing and Materials, Philadelphia, 1982, pp. 70–82.

Chapter 18

The Analysis of Lead-Based Paint Layers: A Qualitative Comparison of Methods

*R. J. Narconis, Jr., V. Divljakovic, S. L. Barnes, and A. M. Krebs**

CONTENTS

I. BACKGROUND

Exposure to lead comes from a variety of sources, such as paint, factory and automobile emissions, ground and surface water, pipe, and certain kinds of solder. The resulting adverse health effects on young children include impairment of the central nervous system, mental retardation, and behavioral disorders. Although children under the age of 7 and unborn fetuses are at the greatest risk, adults may suffer harmful effects as well.[1] A recent report by the Agency for Toxic Substances and Disease Registry (ATSDR) estimates that some 42 million homes contain lead-based paint (LBP) and house approximately 12 million children.[2]

Lead was a major ingredient in many types of house paint prior to, and through, World War II. In the early 1950s, lead compounds were still used in some pigments and as drying agents. Federal regulations began with enactment of the Lead-Based Paint Poisoning Prevention Act (LBPPPA) in 1971, and in 1973 the Consumer Product Safety Commission (CPSC) established a maximum lead content in paint of 0.5% by weight in a dry film of newly applied paint. In 1978 CPSC lowered the allowable level in paint to 0.06%.[3]

The action level for LBP established in the LBPPPA amendments in the 1987 Housing Act is a lead content of 1.0 mg/cm², as measured by an XRF analyzer. Under section 302 of this act, as amended, Public Housing Authorities (PHAs) are required, by 1994, to inspect randomly all their housing projects for lead-based paint. The stated preferred method for measuring the quantity of lead in a painted surface is the portable XRF.[4]

II. INTRODUCTION

Although wet chemical analysis (ICP or AAS) and XRF both provide total lead content, the boundaries of individual layers are, for the most part, not defined. We have attempted to compare analytical results from samples subjected to both techniques, and to include a third: Scanning Electron Microscopy/Energy Dispersive X-Ray Analysis (SEM/EDXA), which may aid in measurement of layer thickness, uniformity, and chemical composition.

In addition, layer identification could allow us to examine the possible effects of the depth of lead layers within the sample on XRF analytical results. Since there is some concern about the reliability of XRF results at these levels, the guidelines recommend back-up chemical testing utilizing AAS or ICP.[3] When using chemical testing, the action level is either 0.5% by weight or 1.0 mg/cm² of LBP surface area.

---

*Industrial Testing Laboratories, St. Louis, Missouri.

III. MATERIALS AND METHODS

A. XRF

A Scitec Model MAP-3 Spectrum Analyzer using a 40 mci CO^{57} source was used for all portable XRF (X-ray fluorescence spectrometer) testing. The 1.5-in.^{2} areas were marked with a template, and the rubber boot surrounding the collection window on the front of the scanner was placed against the sample surface within the marked area. Subsequent readings for lead content were obtained, and the data was recorded. Similar testing was done on prepared "standards" and on field samples.

B. SEM/EDXA

Following XRF analysis, the paint layers within the 1.5-in.^{2} marked area were mechanically removed while taking great care to include all the paint within the area and to keep layer juxtaposition intact. Some samples were clamped in a LKB ultramicrotome, and using a glass knife, cross-sectioned slices were made in order to smooth the surface to be analyzed. Others were quite smooth upon fracture at ambient temperature or after immersing in liquid nitrogen.

A JEOL 5300 Scanning Electron Microscope (SEM) equipped with a NORAN Model 5300 Energy Dispersive X-Ray Analyzer (EDXA) was used to image the paint-layer cross-sections and to perform elemental analysis, backscattered imaging, and elemental dot-mapping.

C. ICP

Samples were analyzed by inductively coupled argon plasma techniques (ICP). The instrument used was a Thermo-Jarrell Ash Plasma 300. The 1.5-in.^{2} samples collected from the same location at which XRF data was obtained were weighed and transferred to a disposable culture tube. Three milliliters of redistilled nitric acid and 1 ml of redistilled perchloric acid were then added to the tube, using a graduated pipette. The tubes were then placed in a hot-water bath (190–200°F) and heated until the reaction subsided. The tubes were transferred to a block heater set at 300°F. The samples remained in the heater until complete digestion and white perchloric fumes were evident. After allowing the samples to cool, 0.5 ml of redistilled nitric acid and 2 ml of hydrochloric acid were pipetted into the tubes. The samples were then reheated in a hot water bath, to redissolve salts. After allowing the samples to cool, each sample was transferred quantitatively into 50-ml volumetric flasks. Finally, each sample was mixed and filtered using a 10-cc syringe with a 0.2-μm acrodisc attached. Filtered into clean disposable tubes, the samples were analyzed for lead on ICP at 220.35 nm. Results were originally expressed in micrograms per milliliter and, for comparison to XRF units, were converted to milligrams per square centimeter, using the original sampling area of 1.5 in^2 for area in this conversion. Since ICP is the only analytical method of the three performed that is destructive, it was done last.

D. STANDARDS

Three in-house standards were prepared by applying an oil-based primer/sealer to 1/2-in. × 1 ft^2 plain white pine boards. Two of these boards were subsequently painted with different concentrations of lead from a mixture prepared by combining lead oxide (Fisher Brand), linseed oil, and turpentine. The third board was not painted with a lead mixture, but with white lead-less enamel instead. ICP analysis was performed and percent lead concentrations were

	Pb, % Conc.
Standard Sample #1	8.03
Standard Sample #2	2.59
Standard Sample #3	<0.05

IV. RESULTS

After initial evaluation by portable spectrum XRF, lead paint chips taken from the same area were prepared and imaged in the SEM. As was anticipated, cross-sectional views of standard sample #1, using the "normal" secondary electron (SED) mode, showed very little layer differentiation (Figure 1).

However, when the backscattered mode (which differentiates elements by "brightening" higher atomic numbered elements)[5] was employed, the lead, primer, and wood layers became quite prominent (Figure 2). Additionally, elemental dot-mapping (which was used in this case to specifically outline higher concentrations of lead)[6] was performed on the same field as Figures 1 and 2 and readily defines the location and thickness of the lead layer only (Figure 3).

When an overall EDXA elemental analysis was done across all paint layers, most of the expected constituents were detected (Figure 4), and when top and primer layers were evaluated by the same technique, the prominent lead peak was evident in the first (top) paint layer, but not in the second (primer) layer (Figures 5 and 6).

Similar analysis was performed on standard sample #2, again using SEM (SED), backscattered, and dot-mapping (Figures 7, 8, and 9, respectively), as well as EDXA of cross-section and first and second layers (Figures 10, 11, and 12, respectively). The reduced thickness of the lead layer is evident in this series.

Standard sample #3 revealed layer differentiations in backscattered mode (Figures 13 and 14) because of the dissimilar atomic number between the higher titanium concentration in the primer vs. the high silicon in the enamel coat.[6] A cross-sectional EDXA analysis of sample #3 appears in Figure 15. Since there was no detectable lead in this sample, lead dot-mapping was not done.

In our comparisons between XRF and ICP data, we saw considerable variations in results. Generally, while there was some correlation between converted ICP data and XRF-L shell values, XRF-K shell values were inconsistent and often much higher than either ICP or L data. This applied to both standard samples and field samples (Figures 16, 18, 20, 21, and 24).

SEM/EDXA information was very useful when related to adjacent layers. Since this technique only identifies elements that its detector can "see," most elements lighter than sodium are not detected. Elements that *are* detected are normalized to 100%. As a result, element percent concentrations are often much higher when done by EDXA than by other means, which would include "organics" in their concentration calculations. For example, a sample that contained 50% lacquer, 20% lead, and 30% titanium, could be reported as 40% lead and 60% titanium by EDXA. Another possible drawback for EDXA is its relatively high detection limit. Various factors, including sample composition, surface roughness, and accelerating voltage, also play a role. We have used a detection limit of approximately 1%, on our graphs as a guideline.

EDXA lead analysis by layer proved to be most useful in identifying the location of the highest concentrations (Figures 17 and 19), and even multiple layers were easily differentiated by SEM (Figure 22) and EDXA (Figure 23). Since the system normalizes all elements detected to 100%, there is not necessarily a direct correlation between ICP/XRF data and EDXA data.

V. DISCUSSION

Existing portable XRF equipment used in lead detection can be divided into two distinct groups. The direct-read instruments are based on the subtraction of the counts of characteristic X-ray photons obtained on the substrate, from the counts obtained when the surface and the paint are excited simultaneously. In other words, a minimum of two readings must be taken, and paint layers are usually removed in order to read the substrate.

The second group of portable XRFs are spectrum analyzers. As a general rule, these instruments acquire the entire spectrum, utilizing a multichannel analyzer. Various algorithms are employed to allow compensation for the various matrices found in the field. There is no need for scraping of paint, and theoretically, only one reading is needed. In our tests we used the spectrum analyzer type of equipment.

Potential sources for error in this study would include, for example, effects of completeness of paint scraping within the targeted area. Scraping too little (that is, insufficient area *or* depth) or scraping too much (outside marked boundaries) could affect lead measurements per unit area. EDXA values can be affected by surface geometry, layer thickness, and interference from elements with very similar characteristics.

Spectrum-type XRFs could present serious problems in determination of lead concentration in paint buried deep in the subsurface, or in situations where lead-containing material (pipes for example) are near or behind the surface being sampled.

Operation of the portable spectrum XRF equipment of the type we used can be optimized by use of both internal (algorithm) "compensation" for substrates and calibration against actual samples of various substrate standards carried to the field and used as references. The operator should know how to use the equipment and, to some extent, interpret field results. He may like to know, for example, if lead layers are deep or shallow within the sample, to better understand the significance of K and L readings.

Both the XRF manufacturer[7] and, recently, the EPA have issued detailed SOPs for the use of this equipment.[4] We believe the rigorous application of these procedures would yield very good results. But we also suspect the actual use of this equipment by field personnel might yield results that more closely

138

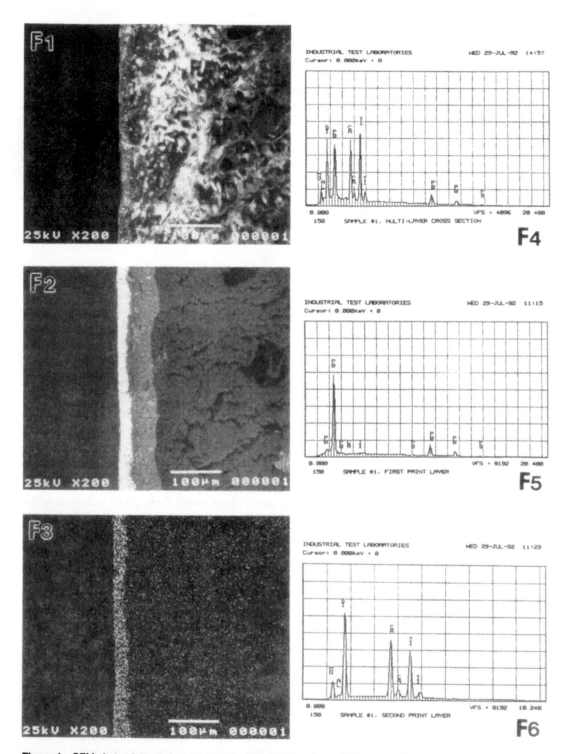

Figure 1 SEM photomicrograph cross-section, standard sample #1 (SED mode). Bar = 100 µg. **Figure 2** Same area as Figure 1, backscattered mode (BSE). **Figure 3** Same area as Figures 1 and 2, elemental dot-mapping for lead. **Figure 4** EDXA graphic printout of cross-section through paint layers in Figures 1–3. **Figure 5** EDXA printout of top layer in Figures 1–3. **Figure 6** EDXA printout of second layer in Figures 1–3.

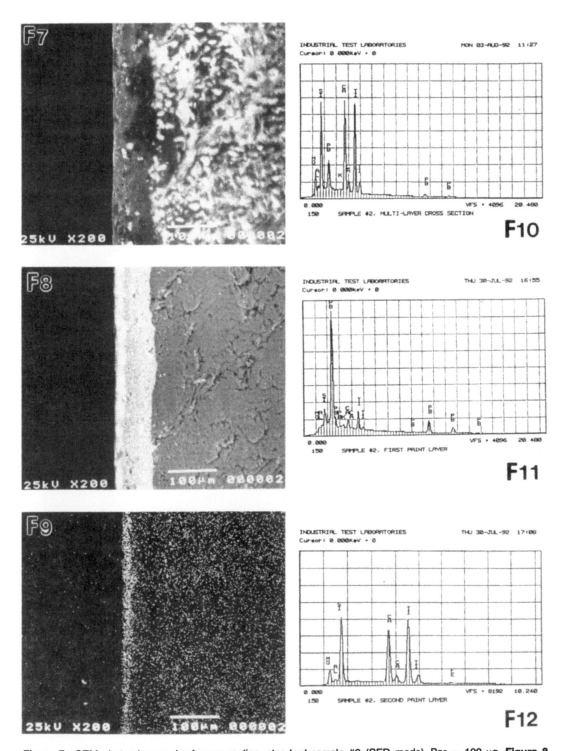

Figure 7 SEM photomicrograph of cross-section, standard sample #2 (SED mode). Bar = 100 µg. **Figure 8** Same area as Figure 7, backscattered mode (BSE). **Figure 9** Same area as Figures 7 and 8, elemental dot-mapping for lead. **Figure 10** EDXA graphic printout of cross-section through paint layers in Figures 7–9. **Figure 11** EDXA printout of top layer in Figures 7–9. **Figure 12** EDXA printout of second layer in Figures 7–9.

140

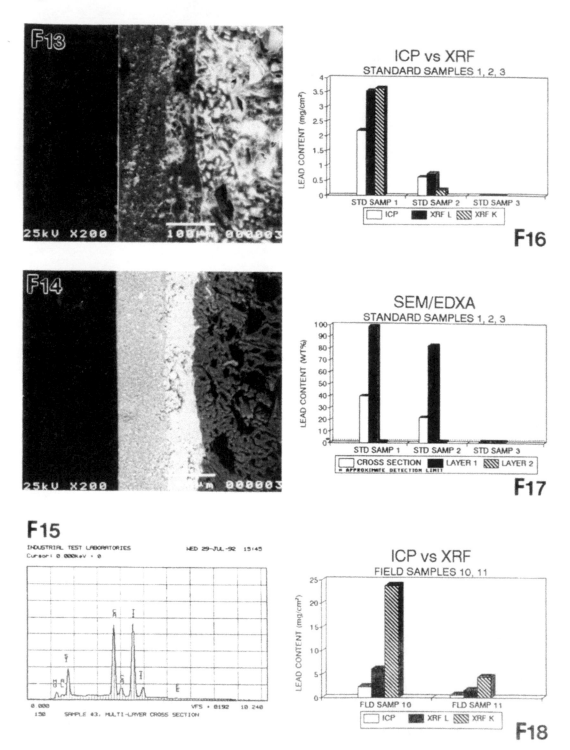

Figure 13   SEM photomicrograph of cross-section, standard sample #3 (SEM mode). Bar = 100 μg. **Figure 14** Same area as Figure 13, backscattered mode (BSE). **Figure 15**   EDXA graphic printout of cross-section through paint layers in Figures 13–14. **Figure 16**   ICP data (converted to mg/cm²) vs. XRF data on standard samples #1, 2, and 3. **Figure 17**   SEM/EDXA data for cross-section and individual layers on standard samples #1, 2, and 3. **Figure 18**   ICP data (converted to mg/cm²) vs. XRF data on field samples #10 and 11.

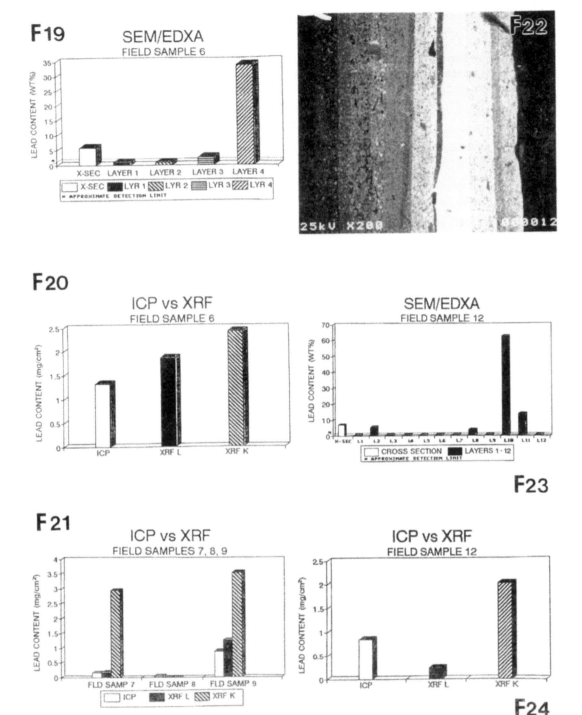

Figure 19 SEM/EDXA data for cross-section and individual layers on field sample #6. **Figure 20** ICP data (converted to mg/cm²) vs. XRF data on field sample #6. **Figure 21** ICP data (converted to mg/cm²) vs. XRF data on field samples #7, 8, and 9. **Figure 22** SEM photomicrograph in backscattered mode (BSE) of field sample #12 showing multiple layers. Bar = 100 µg. **Figure 23** SEM/EDXA data for cross-section and individual layers on field sample #12. **Figure 24** ICP data (converted to mg/cm²) vs. XRF data on field sample #12.

resemble those reported here. In practical application, we would suggest, if possible, simpler operation and generation of information that provides the user with exact concentrations of lead, regardless of the relative *position* of the lead-containing paint layer.

Continued collection of data from our future daily analytical work should prove beneficial in the refinement of the techniques mentioned here (SEM/EDXA and XRF) and may contribute to the improvement of the equipment used. It is our intent to continue this work and report our findings in a later communication.

ACKNOWLEDGMENTS

The authors gratefully acknowledge the excellent technical assistance of Mr. Carmen DeBlass and Ms. Maureen Daniel for their help with ICP analysis, Mr. Eric Lavine for his XRF work, Mr. Rick Jenkins for computer graphics, Mr. Jonathan Gregory for sample preparation, and Mrs. Nola Murphy for her skillful typing assistance.

REFERENCES

1. Lewis, S. and Matlaga, T., Lead: an age-old problem, new concerns: health and regulatory, *Econ. Mag.*, 1, 1989.
2. Toxicological Profile for Lead, U.S. Public Health Service, Agency for Toxic Substances and Disease Registry, 1988.
3. Lead-Based Paint: Interim Guidelines for Hazard Identification and Abatement in Public and Indian Housing, Office of Public and Indian Housing, 1990.
4. Standard Operating Procedures for Measurement of Lead in Paint Using the Scitec Map-3 X-Ray Fluorescence Spectrometer, U.S. Environmental Protection Agency, Washington, D.C., 1991.
5. Principles and Techniques of Scanning Electron Microscopy, Vol. 5, 1976, 146–149.
6. Kevex Instruments, Inc., Energy-dispersive X-ray microanalysis: an introduction, 1989.
7. Scitec Corporation, Spectrum Analyzer Operations Manual, 1989.

Chapter 19

Characterization and Identification of Lead-Rich Particles: A First Step in Source Apportionment

G. S. Casuccio, M. L. Demyanek, G. R. Dunmyre, B. C. Henderson and I. M. Stewart

CONTENTS

I. INTRODUCTION

The characterization and source apportionment of particulate matter has provided a wealth of information on TSP (total suspended particulates) and PM-10 (particulate matter ≤ 10 μm) over the past 20 years.[1-10] More recently, interest has focused on the use of apportionment techniques to identify sources of lead-rich particles in house dust.[11-13] This is due, in part, to reports that, even at low levels, exposure to lead may seriously affect growth, damage the nervous system, and cause learning disabilities in young children.[14] Furthermore, it has been reported that exposures over extended periods of time can have a cumulative effect.

The magnitude of the problem was emphasized in a 1988 report to Congress from the Agency for Toxic Substances and Disease Registry, which estimated that approximately 2.5 million young children in the U.S. still experience lead poisoning at blood levels above 15 μg/dl.[15] In 1991 the Surgeon General declared lead poisoning the principal environmental health hazard afflicting American children. It is now estimated that one child in six is lead poisoned.

Because of these concerns, numerous studies have been performed to assess the concentration of lead in the indoor environment and to further evaluate the health effects of lead on young children.[16-18] While efforts have progressed in these areas, a widespread assumption had been that plumbing fixtures and deteriorated paint were the primary lead sources in the home. More recently[19] it has been suggested that external sources may have a more significant impact than these internal sources; thus studies are now beginning to focus on determining more precisely the spectrum of sources contributing to particulate lead in house dust. This is of increasing importance because if the lead concentration in the indoor environment is at an elevated level, then it would be of great value to identify the source(s) of the lead, before any remedial action is performed.

Particulate lead can originate from both natural and anthropogenic sources. Naturally occurring lead would typically be present only at low concentrations (i.e., <10 to 100 ppm) in soil throughout most areas of the U.S. However, in certain geologic areas, lead may be present at higher levels.[20] Anthropogenic sources are numerous and varied. Examples include run off from lead-mining activities, emissions from waste incinerators, stack and fugitive emissions from lead smelter and foundry operations, and automobiles that use lead-based fuel additives. It is interesting to note that although the use of leaded gasoline has decreased dramatically over the past decade in the U.S., a recent study has reported strong correlations between elevated lead concentrations and proximity to highways in metropolitan areas.[21] This suggests that historical lead fuel exhaust may still have an impact on the environment. In addition to these sources, a potentially significant source of lead in the indoor environment is lead-based paint.

While typical sources of lead in house dust may include leaded paint, soil, lead-based fuel exhaust, road dust, industrial operations, and incineration, it is difficult, if not impossible, to determine the contribution from various sources, using conventional analytical procedures. For example, although AA (atomic absorption) is extremely sensitive and capable of measuring lead in the ppb (part per billion) range, it gives little indication of the phase of the lead present or of the mineral species with which it may be associated.

SEM (scanning electron microscopy) combined with EDS (energy-dispersive spectroscopy) analysis is able to provide morphological and elemental composition data on individual particles. SEM/EDS techniques are often capable of resolving source attributes on materials that are compositionally similar at the bulk level, but resolvable at the individual particle level. A limiting factor associated with SEM/EDS analysis is that it can be a time-consuming process to obtain data on a sufficient number of particles to assist in the apportionment process.

An analytical technique that overcomes this limitation is CCSEM (computer-controlled scanning electron microscopy). Through computer control, CCSEM is capable of analyzing hundreds of particles per hour while providing information on size, morphology, and composition, in a cost-effective manner. As such, CCSEM has the potential to become an invaluable tool in the characterization, identification, and source apportionment of lead-bearing particles.

The strength of CCSEM comes in identifying particles on the basis of their elemental composition (i.e., intra-elemental relationships) and/or morphology. In situations where a more definitive analysis is required, such as discriminating between lead-oxide and lead-carbonate particles, complementary analytical techniques may be required. The electron microprobe, with its WDS (wavelength-dispersive spectroscopy) capabilities, has the ability to provide assistance in this area. Also, the TEM (transmission electron microscope) is capable of positively identifying lead species through the use of SAED (selective area electron diffraction) analysis in conjunction with EDS.

This paper discusses a multifaceted analytical approach to detecting lead-rich particles in dust and soil samples, and the identification of lead sources based on a combination of chemistry and morphology. Also discussed are the use of novel sample collection methods that not only permit particles to be obtained in their original condition, but are amenable to examination by multiple analytical techniques, to assist in the source apportionment process.

II. CHARACTERIZATION AND IDENTIFICATION METHODS

The first step in a source apportionment study is the collection of samples, and their subsequent analysis. If a source of interest has some unique morphological or compositional characteristics, then it is possible to determine its contribution in a sample, based on the use of source tracers. The apportionment of a source contribution to a sample, through the use of tracers, is often referred to as receptor modeling.[22] In source apportionment or receptor modeling studies, the analytical method that provides the greatest specificity with respect to identification and quantification of source tracers will have the greatest ability to discern the contribution of a source in a sample of interest.[4,8]

Receptor models based on bulk analytical techniques such as AA or XRF (X-ray fluorescence) have been used, with varying degrees of success, to source apportion ambient particulate matter. However, while bulk analytical techniques provide accurate data on elemental concentrations, they give no information on relationships or associations of elements in discrete particles. Furthermore, these methods provide no information on size or morphology associated with a particle. As a result, receptor models based on bulk techniques are often limited in their ability to resolve specific source impact. This is especially true for sources that are compositionally similar.

In order to overcome this limitation, SEM/EDS has been explored as a complementary tool, because of its ability to provide data on size, morphology, and elemental chemistry, on individual particles. Examples of the type of information obtained using SEM/EDS are provided in the figures shown. Figure 1 shows a secondary electron image and elemental spectrum of a fly ash particle. The round (spherical) morphology is indicative of emissions from a high temperature source. In this case the particle is a byproduct from the combustion of coal. It is interesting to note that the elemental chemistry is similar to that of a mixed clay (i.e., silicon/aluminum-rich) and that it is the morphology combined with the chemistry that provides unique information as to its source. A secondary electron image and elemental spectrum of a mixed clay particle is provided in Figure 2. Note the similarities in elemental composition between the fly ash particle in Figure 1 and this particle.

Figure 3 shows a secondary electron and backscattered electron image of a complex agglomerate particle. As can be seen, the secondary electron image provides a topographic perspective of the feature, with a high depth of field. In the backscattered electron image, the brightness of the image is dependent on the

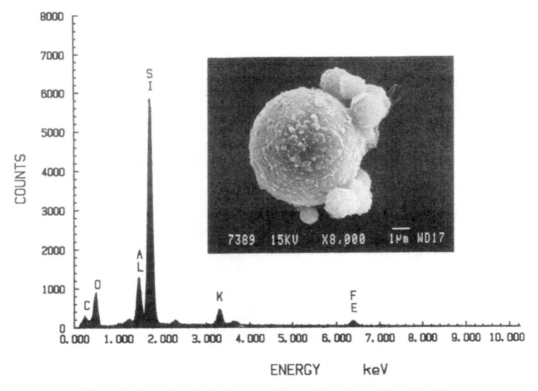

Figure 1   Secondary electron image and elemental spectrum of fly ash particle.

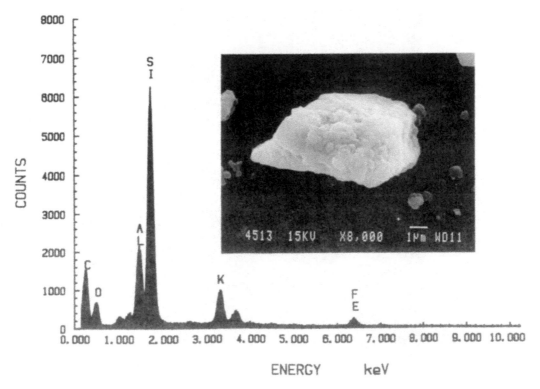

Figure 2   Secondary electron image and elemental spectrum of mixed clay particle.

Secondary Electron Image   **Backscattered Electron Image**

Figure 3   Secondary and backscattered electron image of complex agglomerate particle.

147

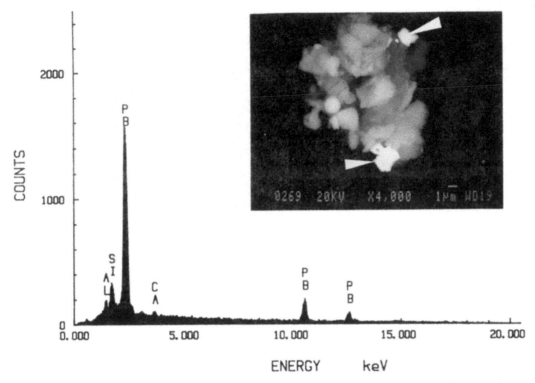

Figure 4 Backscattered electron image and elemental spectrum from a lead-rich area (see arrows).

number of backscattered electrons generated. Higher atomic number elements generate more backscattered electrons than do low atomic number elements, resulting in a brighter image. Thus, the backscattered electron image contains compositional information and permits discrimination between phases containing elements with different atomic numbers. The backscattered electron image clearly shows two bright areas in the particle. The EDS analysis indicates that the bright areas of this particle comprised lead-rich compounds, as shown in Figure 4. Also obvious in the image is a round (spherical) feature, indicative of a high temperature source. The elemental composition of this feature, as provided in Figure 5, is characteristic of fly ash. Figure 6 illustrates that the elemental chemistry associated with the central area of the particle is calcium-rich.

The ability to analyze individual particles in this manner makes SEM/EDS analysis an invaluable tool in source apportionment of particulate matter. A limiting factor of this technology, however, is the amount of time required to provide information of this nature. In order to overcome this limitation, efforts over the past 15 years have focused on the use of the computer to permit an SEM/EDS analysis in an automated and rapid fashion.[23-26] As shown in the figures, the use of secondary and backscattered electron images, along with X-ray data, permits a combined morphological and compositional analysis of a particle. These features are also amenable to computer control, since the electron beam can be monitored and the resultant signals collected and processed simultaneously. For example, the complex feature illustrated in Figures 3 through 6 can be effectively analyzed with CCSEM by rastering the electron beam across the surface of the particle. The resulting X-ray spectrum would, in essence, be a composite of the spectra presented in Figures 4, 5, and 6. Since a digital image of the particle, along with its coordinates, are stored during the analysis, the image of the feature can be reviewed "off line," to provide insight as to whether the lead is present as a discrete particle or part of a larger agglomerate. In addition, the particle can be examined in more detail using SEM/EDS techniques via the stored coordinates, should more information be required. Thus, with appropriate software and hardware, information of this nature can be collected and reviewed in a rapid manner.[8, 27-29]

In summary, it is possible to acquire size, composition, and image data in a totally automatic manner and in sufficient numbers to be representative of the entire sample, within a cost-effective time frame. In this manner, the microscopic data can be directly and reliably related to the bulk properties of the sample.

148

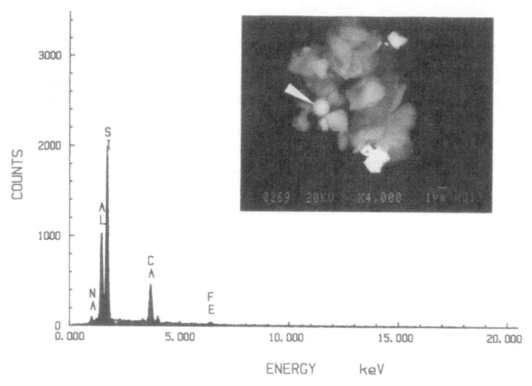

Figure 5 Backscattered electron image and elemental spectrum from a round (spherical) feature (see arrow).

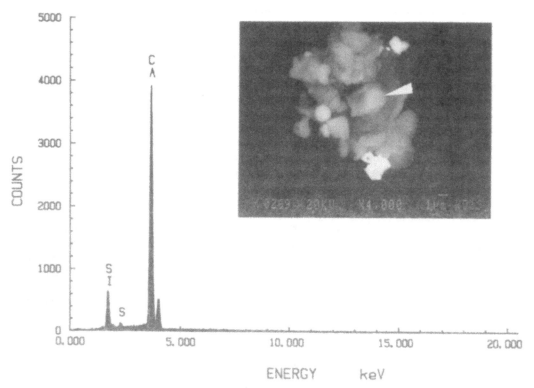

Figure 6 Backscattered electron image and elemental spectrum from central area of the particle (see arrow).

Table 1 Summary of CCSEM Mass Distribution Results (Wt. %): House Dust Sample

		Particle Size Range μm			
<2.5	2.5–5	5–10	10–20	20–40	>40
4	6	8	8	13	61

Table 2 Summary of CCSEM Particle Type Results: House Dust Sample

Particle type	Wt. %	No. of particles
Mixed clay	57	156
Carbon-rich	27	92
Fe-rich	6	5
Pb-bearing	6	27
Ca-rich	3	8
Miscellaneous	1	12

Furthermore, the data can be stored in a manner that makes it available for review and perusal off line on PC systems.[30–32] Thus, CCSEM permits particulate matter to be analyzed in a manner that maintains the strengths of a SEM/EDS analysis while minimizing the amount of time required to acquire the data.

An example of the type of numerical information that can be obtained from a CCSEM analysis is provided in Tables 1 and 2. These tables provide mass distribution and particle type data obtained from the analysis of a house dust sample. The house dust sample was collected by sweeping dust from several areas within the house, into a ziplock plastic bag. Preparation for CCSEM consisted of placing a portion of the dust into acetone and filtering it onto a polycarbonate filter. In Table 1 the mass distribution associated with the total sample is provided and indicates that the majority of the mass of the total sample was associated with particles greater than 40 μm in diameter.

The CCSEM particle type results listed in Table 2 indicate that mixed-clay and carbon-rich particles account for the majority of the sample. Lead-bearing particle types contributed 6% by weight to the overall sample. Because only 300 particles were analyzed on this sample, the number of lead-bearing features was limited and provided minimal source identification information. In an effort to provide additional data on the lead-bearing particles, the sample was reanalyzed with emphasis on particles having a high atomic number. This was accomplished by setting a particle detection threshold, based on the backscattered electron signal, to "ignore" all particles other than those that have higher atomic number constituents, thus generating a more intense signal. Using this parameter to "select" particles, it is possible to scan thousands of particles per hour and thus establish a relatively low detection limit of lead-rich particles. For example, in a sample containing lead at a level of 1000 ppm, only one particle in 1000 may be lead bearing. Analysis for lead using normal CCSEM parameters would be impractical because of the amount of time necessary to analyze sufficient numbers of lead particles. By adjusting the threshold to seek particles containing lead (i.e., based on high atomic number contrast), the 999 particles that do not contain lead would be passed over very quickly, resulting in a great reduction in analysis time.

The use of the backscattered electron signal to assist in the identification and analysis of lead-bearing particles is illustrated in Figure 7. This figure provides secondary and backscattered electron images of a field of view of the house dust sample. The secondary electron image shows numerous particles ranging in size from less than 1 μm to approximately 10 μm. The backscattered electron image, adjusted to enhance the high atomic number particles shows few particles in the image, which indicates that lead-bearing particles, if present, would be few in number. By setting the particle detection threshold to ignore all but the brightest features, only one or perhaps two particles would be analyzed. Analysis of the house dust sample, using these parameters, permitted over 140 particles to be characterized in approximately one half hour. The data indicates that the majority (72%) of the mass associated with the lead-bearing particles occurred in the size ranges less than 5 μm. Review of the lead-bearing features on the off-line PC workstation indicated that most were present as discrete particles. However, many of the lead particles were imbedded within larger particles that resembled paint chips.

By acquiring data in this manner, it is possible not only to provide size distribution information on lead-bearing particles, but also to evaluate the sources of the lead. For example, particles that are rich in lead

Secondary Electron Image   **Backscattered Electron Image**

Figure 7   Secondary and backscattered electron images of house dust sample taken at 2000×, illustrating particle distribution and high atomic number contrast.

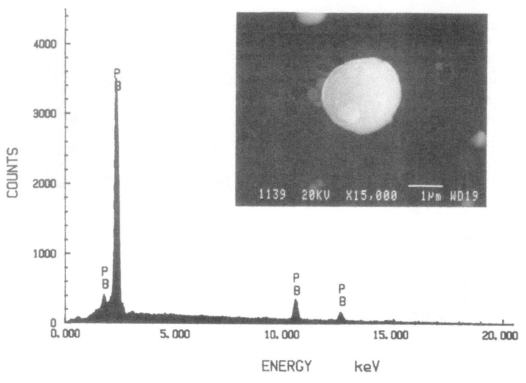

Figure 9 Secondary electron image and elemental spectrum of spherical lead-rich particle from a high temperature source.

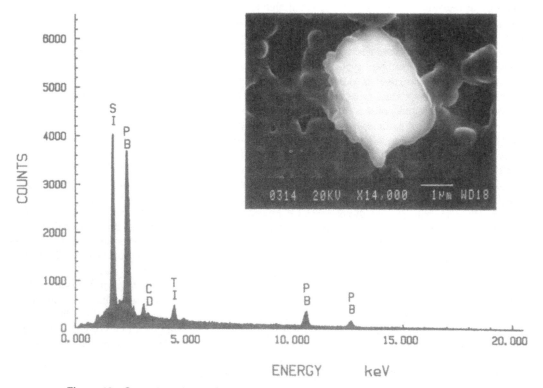

Figure 10 Secondary electron image and elemental spectrum of a lead/silicon-rich particle.

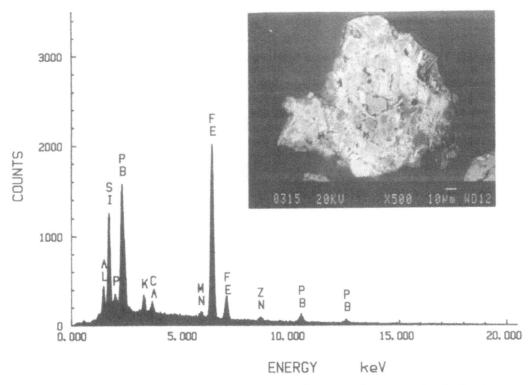

Figure 11   Backscattered electron image and elemental spectrum of a cross-sectioned soil particle.

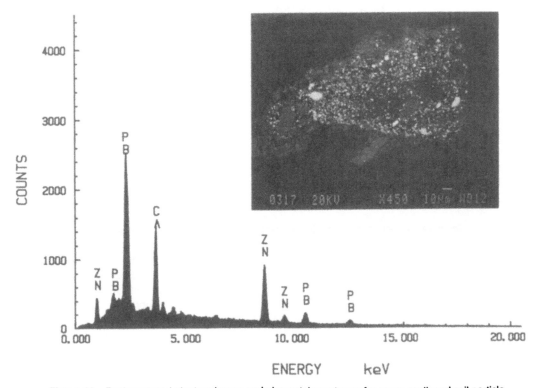

Figure 12   Backscattered electron image and elemental spectrum of a cross-sectioned soil particle.

154

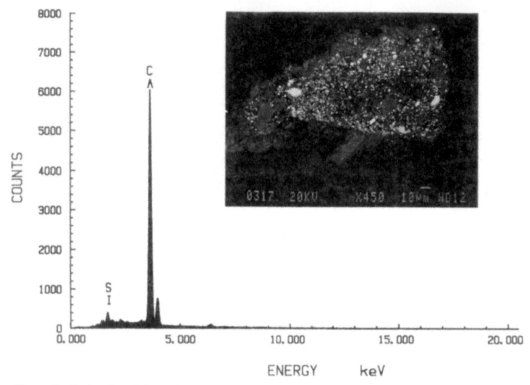

Figure 13 Backscattered electron image and elemental spectrum of "dark" area in cross-sectioned soil particle.

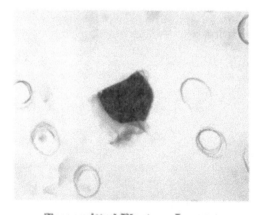

Transmitted Electron Image

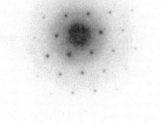

SAED Pattern

Figure 14 Transmitted electron image and SAED pattern of a lead-sulfate particle.

Another powerful tool for the characterization of lead particles is the TEM. The TEM, combined with EDS and SAED capabilities, is very useful in the examination of fine lead-bearing particles (i.e., <1 μm). EDS permits collection of elemental composition of the fine particles. While both SEM and the electron microprobe have this capability, the TEM is unique because of its ability to obtain SAED patterns from crystalline particles. Through measurement of the SAED pattern, it is possible to positively identify the lead compound(s) present. For example, Figure 14 shows a transmitted electron image and an SAED pattern for a lead-rich particle approximately 0.4 μm in size. Measurement and interpretation of the SAED pattern indicates that this is a lead-sulfate particle. A critical study of the EDS pattern indicated that the lead M_α peak at 2.34 keV was indeed stronger, relative to the lead L_α and L_β peaks at 10.6 keV and 12.6 keV, than would have been expected. This increase is attributable to the overlap with the sulfur K_α peak at 2.31 keV.

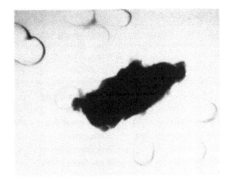
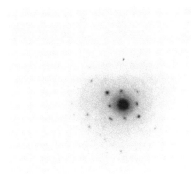

Transmitted Electron Image   **SAED Pattern**

Figure 15 Transmitted electron image and SAED pattern of a lead-oxide particle.

Another example of TEM data is provided in Figure 15. In this figure the particle seen in the transmitted electron image is shown by SAED to be a lead-oxide particle. In addition to having the ability to positively identify lead-rich particles, the TEM has better resolution than the SEM and electron microprobe. Thus, the TEM can be an extremely powerful characterization and source apportionment tool for fine particles.

III. LEAD SAMPLE COLLECTION PRACTICES AND PROCEDURES

Although much attention has been devoted to the development and refinement of analytical protocols to more accurately identify and source apportion lead in house dust samples, the importance of sample collection practices and procedures should not be overlooked. Regardless of the particular method used, it is essential to collect samples that are fully representative of the area of interest. It is equally important to ensure that these samples are suitable for examination by a variety of analytical techniques, if one is interested in gaining the most comprehensive information possible on sources, relative concentrations, and potential bioavailability.

A very effective medium for collecting and subsequently observing lead particles is surface-dust accumulation. Historically, the standard technique used by industrial hygienists to evaluate such surface contamination has been "swipe" or "wipe" sampling. Wipe sampling, as it is most commonly referred to, generally involves the use of a prefabricated filter paper (sometimes referred to as a "Whatman's Filter") or other gauze/absorbant material that may or may not be premoistened with an appropriate solution. The sample medium of choice is then used to wipe a predetermined surface area, with the intent of extracting as much of the surface particulate as possible. Wipe sampling procedures have been used extensively to evaluate a very broad range of indoor and outdoor surfaces, including floors, table/desk tops, ledges, walls, windows (including wells and sills), machinery, and appliances for contaminants such as polychlorinated biphenyls (PCBs), radionuclides, and lead.

Generally, wipe sample data is most useful for comparing relative concentrations between surfaces or for determining whether a contaminant is present or not (qualitative assessment). Major benefits of wipe sampling for lead are the relative ease and expediency with which samples may be collected. For these reasons the September 1990 U.S. Department of Housing and Urban Development (HUD) Interim Guidelines for Lead-Based Paint Abatement in Public and Indian Housing recognize wipe sampling for dust as meeting their criteria for "clearance" sampling after abatement work.[33] However, wipe sampling practices typically used for the evaluation of surface dust, under HUD guidelines, involve using "wipes" (premoistened), rather than filter paper or gauze, to wipe up the dust from a measured surface area. After wiping dust from an approximate 1-ft² area, the wipes are typically stored in centrifuge tubes for transport to the analytical laboratory. The primary disadvantages of wipe sampling for the purpose of characterizing and identifying lead particles are that alteration of the dust distribution takes place such that it cannot be examined microscopically as it was present on the surface, particles tend to be collected in a nonuniform fashion, the wipes themselves are manufactured/packaged in highly variable processes that could significantly alter results, and it is simply not possible to effectively wipe certain surfaces that may be of significant interest (e.g., carpeting, brick). Additionally, loose or damaged surfaces may result in a sample that is not totally representative of the area.

Table 3 A Comparison of Lead Dust Levels Using Wipe and Microvacuum Sampling Techniques[a]

Unit no	Location[b]	Wipe (μgPb/ft²)	Microvacuum (μgPb/ft²)	Substrate surface/condition
1	BR2WW	440	80	Wood/poor
2	LRF	<25	<1	Carpet/fair
3	KF	71	7	Tile/fair
4	BR2F	<25	27	Rubber mat/good
5	BR2F	48	28	Carpet/poor
	LRF	<25	12	Carpet/good
6	BR1WW	1740	3084	Wood/poor
	LRF	<25	3	Wood/good
	KWW	1658	479	Wood/poor
	KF	<25	1	Tile/good
	LRWW	2572	460	Wood/poor
7	KF	<25	<1	Tile/good
	KWW	5928	3677	Wood/poor
8	FPL	41	11	Concrete/good

[a] Analysis by AA flame.

[b] Key to sample location abbreviations: KWS - kitchen window sill, KWW - kitchen window well, KF - kitchen floor, LRF - living room floor, LRWW - living room window well, BR1F - bedroom 1 floor, BR1WW - bedroom 1 window well, BR2F - bedroom 2 floor, BR2WS - bedroom 2 window sill, BR2WW - bedroom 2 window well, FPL - front porch landing.

More recently, "microvacuum" sampling techniques have been developed and used to collect surface dust for assessment purposes.[34] At this point the primary use of microvacuum sampling for lead particles in dust is to provide a semiquantitative index of concentration or to confirm the presence of lead particulate. The technique uses a personal sampling pump calibrated at 2 to 2.5 l/min, a 25- or 37-mm filter cassette with a 0.8 μm pore-size mixed-cellulose ester filter, a slotted nozzle attached to the cassette inlet (to increase capture velocity), and a plastic template (generally 25 cm × 25 cm) to provide an accurate boundary for the surface to be sampled. The filter cassette is attached to the personal sampling pump, with a piece of flexible (Tygon®) tubing. The entire apparatus is then used as a vacuuming device, to extract surface dust from inside the template (25 cm × 25 cm) area. One reported major benefit of using the microvacuum technique for lead assessments is that it has been found to produce results that are believed to show a relationship between lead in dust and blood lead levels in children under 6 years of age.[35]

Still, the microvacuum method alters the dust distribution on the surface of interest, it may alter the particles matrices themselves, and more importantly, many questions remain regarding the true collection efficiency of the method. No findings indicating any strong correlation in "side-by-side" sampling (primarily with wipe samples) have yet been published.

A study was recently conducted by the RJ Lee Group in conjunction with the Georgia Tech Research Institute (GTRI)*, in two large public housing developments. A portion of the study involved side-by-side sampling that was conducted to determine if any correlation existed between the wipe and microvacuum methods. This involved collecting samples in eight different housing units and from a variety of surfaces. Although carpeted surfaces are not typically amenable to wipe sampling, it was attempted for this study. During this side-by-side comparison study, wipe and microvacuum sampling of various surfaces was conducted. The wipe samples were collected following HUD-recommended guidelines, and the microvacuum samples were collected following the procedures outlined in the U.S. EPA Model Curriculum for Lead Abatement Supevisors.[36] Both the wipe and microvacuum samples were analyzed by AA flame spectrometry. A summary of the findings are provided in Table 3.

Due to the limited amount of data presented here, a solid correlation can not be established between microvacuum- and wipe-sample results; however, several interesting points are noted. Where the results are very low (below or close to the detection limits), both the wipe and microvacuum results tracked well. In addition, a review of the data tends to indicate that the microvacuum technique generally produced lower concentrations of lead from the same surfaces, when compared with wipe samples, although one sample (Unit 6, BR1WW) produced a higher result using the microvacuum procedure. It appears from this

* The authors gratefully acknowledge Dave Jacobs, CIH, previously of GTRI, now with the National Center for Lead Safe Housing in Columbia, MD, for his valuable partnership on this project.

Table 4 A Comparison of Lead Dust Levels in Two Large Public Housing Developments Using Side-by-Side Wipe and Adhesive Lift Sampling[a]

Unit no.	Location[b]	Wipe (µgPb/ft²)	Adhesive lift (µgPb/ft²)	Substrate surface/condition
1	LRWW	2572	494	Metal/poor
	FPL	472	1160	Concrete/poor
	LRF	<25	2	Tile/good
	BR1F	<25	<1	Tile/good
2	KWS	<25	12	Wood/fair
	LRF	<25	25	Wood/good
	LRWW	1841	9560	Metal/poor
	BR1F	<25	260	Wood/good
	FPL	<25	25	Concrete/good
3	KWS	<25	<1	Wood/fair
	LRF	<25	19	Wood/poor
	LRWW	874	58	Metal/poor
	BR2WS	356	77	Wood/poor
	FPL	374	446	Concrete/poor
4	KWS	<25	26	Wood/fair
	LRWW	1884	795	Metal/poor
	BR2WS	242	541	Wood/poor
	FPL	49	20	Concrete/fair
5	KWS	<25	7	Wood/poor
	LRF	<25	3	Wood/fair
	LRWW	269	175	Metal/poor
	BR2WW	440	276	Metal/poor
6	KWS	74	23	Wood/good
	LRWW	325	115	Metal/poor
	BR1F	42	2	Wood/fair
	BR2F	<25	23	Wood/good

[a] Analysis by AA Flame.
[b] Key to sample location abbreviations: KWS - kitchen window sill, KWW - kitchen window well, KF - kitchen floor, LRF - living room floor, LRWW - living room window well, BR1F - bedroom 1 floor, BR1WW - bedroom 1 window well, BR2F - bedroom 2 floor, BR2WS - bedroom 2 window sill, BR2WW - bedroom 2 window well, FPL - front porch landing.

data that the relative condition of the substrate may play an important role in determining the potential for elevated lead levels.

As previously mentioned, major disadvantages of both wipe and microvacuum sampling methods are that they produce samples that are not ideal for analysis using multiple analytical techniques, and they involve physical alteration (or at least disturbance) of the dust and particle matrices, as they were originally distributed on the surface of interest. In an effort to overcome these limitations, alternative surface sampling methods have been developed. One useful industrial hygiene tool developed for this purpose is a surface particle adhesive lift sampler that allows for the collection of particulate matter in a very simple manner and that permits the collection of particles *in situ* from a broad range of surfaces (including wood, brick, concrete, etc.). These samplers are also amenable to examination by multiple analytical procedures, including AA, SEM, electron microprobe, and TEM as well as other techniques such as ICP (inductively coupled plasma) and light microscopy (PCM, PLM). If the sample is not too heavily loaded, even CCSEM may be performed using the samplers. It is important to note that any of the surface sampling methods mentioned may alter the original dust accumulation to some extent; however, since the adhesive lift samplers do not involve "swabbing" or wiping the surface, the level of surface disturbance is minimized.

In the same RJ Lee Group/GTRI public housing study referenced previously, the largest portion of the project involved side-by-side sampling to compare lead concentrations, using the traditional wipe technique with adhesive lift sampling technology. Some preliminary results of the comparison study are summarized in Table 4.

For the purpose of illustration, 26 individual sample results, which are representative of the entire study, are summarized in Table 4. This table provides information relative to substrate surface conditions and sample locations. In general, the data show that in approximately 16% of the samples, the adhesive lifts yielded higher lead concentrations from wood, metal (aluminum tracking in windows), and concrete surfaces. In 55% of the samples, both the adhesive lifts and the wipes produced results indicating very low lead concentrations, which would generally not be considered significant from an industrial hygiene standpoint. The wipe results in 20% of the samples indicated higher lead concentrations, as compared to the adhesive lift technology; and in 9% of the side-by-side samples, both the wipes and the adhesive lifts indicated somewhat comparable elevated results.

The variability of the results appears to be dependent on several factors such as the substrate surface condition, the relative homogeneity of the dust distribution, and the actual collection efficiencies of the various techniques. Some of the extreme differences in several of the side-by-side samples can be attributed to substrate surfaces that were nonuniform and in very poor condition (e.g., flaking paint chips) and thus significantly biased certain samples, producing much higher results.

Although work is continuing to further define true collection efficiencies and to perfect adhesive lift sampling technology and its applicability to the full range of analytical techniques, the primary advantages of the technique continue to be the broad applications for industrial hygiene assessments, and the "extra" information that can be gained from individual particles.

IV. SUMMARY

Identifying, characterizing, and apportioning sources of particles in dust accumulations, using integrated analytical techniques, holds a great deal of promise for the fields of materials science, industrial hygiene, and environmental assessment. In dealing with current public health issues such as exposure to lead in the environment (specifically the susceptibility of young children to adverse health effects), gaining a more comprehensive understanding of lead-rich particle chemistry, bioavailability, and sources has become increasingly important.

This paper discussed the importance of source apportionment as it relates to the study of environmental lead. Typical approaches for conducting source apportionment investigations of particulate matter have included dispersion- and receptor-based modeling; however, advancements in sampling technologies and the use of electron microscopy and other analytical approaches in the examination of surface dust accumulations have lent a great deal to this area.

The collection of surface dust samples was discussed due to its importance in obtaining particles that are amenable to analysis by various laboratory procedures. Surface samplers that permit direct preparation for microscopic examination are often preferrable over other surface sampling methods such as wipe and microvacuum sampling, since such methods have the potential to alter the original particle distribution and generally require the dust sample to be redeposited for analysis. Even though the adhesive lift samplers may also alter the particle distribution, the extent of alteration is expected to be less than the other methods, since the surface is not swabbed or vacuumed. In summary, while each sampling technique has its associated advantages and disadvantages, it appears that they may be used in lead studies as complementary tools. Selection of the most appropriate sampler will ultimately depend on the surface to be sampled and the information required.

To explore the use of these samplers as complementary tools, several limited field studies comparing different surface sampling methods (primarily for total lead concentrations) were summarized. These studies were not conclusive enough to generate broad conclusions; however, they did indicate that consideration should be given to the condition of the surface and the type of sampler employed in the investigation. Hopefully, through more detailed side-by-side sampling studies, additional information can be made available on the appropriateness of the various sampling tools discussed in this paper. By examining lead-rich particles in such samples, it may be possible to gain more information on potential sources and relative bioavailability.

REFERENCES

1. Friedlander, S. K., Chemical element balances and identification of air pollution sources, *Environ. Sci. Technol.*, 7, 235, 1973.
2. Gartrell, G. and Friedlander, S. K., Relating particulate pollution to sources: the 1972 aerosol characterization study, *Atm. Environ.* 9, 279, 1975.

3. Lee, R. J., Fasiska, E. J., Janocko, P., McFarland, D., and Penkala, S., Electron beam particle analysis, *Ind. Res. Dev.*, 105, June, 1979.

4. Cooper, J. A. and Watson, J. G., Receptor oriented methods of air particulate source apportionment, *JAPCA*, 30, 1116, 1980.

5. Casuccio, G. S. and Janocko, P. B., Quanitfying blast furnace cast house contribution to TSP using a unique receptor model, presented to the 74th Annu. Meet. of the Air Pollution Control Association, Philadelphia, June 1981.

6. Janocko, P. B., Casuccio, G. S., Dattner, S. L., Johnson, J. L., and Crutcher, E. R., The El Paso airshed; source apportionment using complementary analyses and receptor models, Proc. of the APCA Specialty Conf. on Receptor Models Applied to Contemporary Pollution Problems, Danvers, MA, October 1982.

7. McIntyre, B. L. and Johnson, D. L., A particle class balance receptor model for aerosol apportionment in Syracuse, N.Y., in *Receptor Models Applied to Contemporary Pollution Problems*, SP-48, Air Pollution Control Association, Pittsburgh, 1982.

8. Casuccio, G. S., Janocko, P. B., Lee, R. J., Kelly, J. R., Dattner, S. L., and Mgebroff, J. S., The use of computer controlled scanning electron microscopy in environmental studies, *APCA*. 33, (10), October, 1983.

9. Lucas, J. H. and Casuccio, G. S., The identification of sources of total suspended particulate matter by computer controlled scanning electron microscopy and receptor modeling techniques, presented to the 80th Annu. Meet. of APCA, New York, NY, June 1987.

10. Lucas, J. H., Casuccio, G. S., and Miller, D., Evaluation of PM10 samplers utilizing computer controlled scanning electron microscopy, presented to the APCA/EPA Int. Specialty Conf., San Francisco, CA, February 1988, published in PM-10: Implementation of Standards, Air Pollution Control Association, Philadelphia, 1988, 120–137.

11. Dattner, S. L., Mgebroff, J. S., Casuccio, G. S., and Janocko, P. B., Identifying the sources of TSP and lead in El Paso using microscopy and receptor models, presented to the 76th Annu. Meet. of the Air Pollution Control Association, Atlanta, GA, June 1983.

12. Henderson, B. C., Mershon, W. J., Leger, M., and Cooke, G. A., Comparison of bulk analytical and automated microscopy techniques for the characterization of lead-bearing particulate samples, presented to the American Assoc. for Aerosol Research, Philadelphia, June 1990.

13. Hunt, A., Johnson, D. L., Watt, J. M., and Thornton, I., Characterizing the sources of particulate lead in house dust by automated scanning electron microscopy, *Environ. Sci. Technol.*, 26, (8), 1992.

14. American Academy of Pediatrics. Statement on Childhood Lead Poisoning. Committee on Environmental Hazards, *Pediatrics*, 79, (3), March, 1987.

15. Mushak, P. and Crochetti, A., The nature and extent of lead poisoning in children in the United States. A report to Congress, U.S. Department of Health and Human Services, Agency for Toxic Substances and Disease Registry, Washington, D.C., 1987.

16. Needleman, H., Gannoe, C., Leviton, A., et al., Deficits in psychologic and classroom performance of children with elevated dentin lead levels. *N. Engl. J. of Med.*, 300, 689–695, 1979.

17. Winnike, G., Kramer, G., Brockhaus, A., et al., Neuropsychological studies in children with elevated tooth lead concentration, *Int. Arch. Occupational Environ. Health*, 51, 169–183, 1982.

18. Yule, W., Lansdown, R., Millar, I. B., et al., The relationship between blood lead concentrations, intelligence and attainment in a school population. A Pilot Study, *Dev. Med. Child. Neurol.*, 23, 567–576, 1981.

19. Roberts, J., Budd, W., Camann, D., and Spittler, T., Potential risks from lead in soil and house dust from older Seattle homes, adapted for publication in *The Journal of Air and Waste Management Association*, 1992.

20. Schacklette, H. T. and Boerngen, J. G., Element concentrations in soils and other surficial materials of the conterminous United States. U.S. Geological Survey Professional Paper 1270, Washington, D.C., 1984.

21. Mielke, H., Anderson, J., Berry, K., et al., Lead concentrations in inner city soils as a factor in the child lead problem, *Am. J. Public Health*, 73, (12), 1366–1369, 1983.

22. Hopke, P. K., Ed., *Data Handling in Science and Technology: Receptor Modeling for Air Quality Management*, Vol. 7, Elsevier, New York, 1991.

23. Lee, R. J. and Kelly, J. F., Back-scattered electron imaging for automated particle analysis, in *Microbeam Analysis*, D. E. Newbury, Ed., San Francisco Press, San Francisco, 1979.

24. Scott, W. R. and Chatfield, E. J., A preceision SEM image analysis system with full feature EDXA characterization, *Scanning Electron Microsc.*, 2, 53, 1979.

25. Lee, R. J. and Kelly, J. F., Overview of SEM-based automated image analysis, *Scanning Electron Microsc.*, 1, 303, 1980.

26. Lee, R. J. and Fisher, R. M., Quantitative characterization of particulates by scanning and high voltage electron microscopy, Special Publication 533, National Bureau of Standards, Washington, D.C., 1980.

27. Huffman, G. P., Shah, N., Huggins, F. E., Casuccio, G. S., and Mershon, W. J., Development of computer-controlled scanning electron microscopy (CCSEM) techniques for determining mineral-maceral association, presented at the American Chemical Society, Division of Fuel Meeting, New York, August, 1991.

28. Schwoeble, A. J., Dalley, A. M., Henderson, B. C., and Casuccio, G. S., Computer-controlled SEM and microimaging of fine particles, *J. Metals*, 40 (8), 11–14, 1988.

29. Henderson, B. C., Stewart, I. M., and Casuccio, G. S., Rapid acquisition/storage of electron microscopy images, *Am. Laboratory*, November, 49–52, 1989.

30. Schwoeble, A. J., Lentz, H. P., Mershon, W. J., and Casuccio, G. S., Microimaging and offline microscopy of fine particles and inclusions, *J. Materials Sci. Eng.*, A124, 49–54. 1990.
31. Casuccio, G. S., Schwoeble, A. J., Henderson, B. C., Lee, R. J., Hopke, P. K., and Sverdrup, G. M., The use of computer-controlled SEM and microimaging to assist in airborne particulate characterization, presented to the Fine Particle Society, Santa Clara, CA, July 1988.
32. Schwoeble, A. J., Lentz, H. P., Mershon, W. J., and Casuccio, G. S., Microimaging and offline microscopy of fine particles and inclusions, *J. Materials Sci. Eng.*, A124, 49–54, 1990.
33. U.S. Department of Housing and Urban Development, et al., Lead Based Paint: Interim Guidelines Hazard Identification and Abatement in Public and Indian Housing, Washington, D.C., September 1990.
34. U.S. Environmental Protection Agency, Model Curriculum: Lead Abatement Training for Supervisors (Chapter L), University of Cincinnati, Cincinnati, June 1992.
35. Bornschein, R., Clark, S., Pan, W., and Sucop, P., Midvale Community Lead Study: Chemical Speciation and Bio-Availability and Dietary Exposure of Lead. Society of Environmental Geochemistry and Health, 1991, pp. 149–162.
36. U.S. Environmental Protection Agency, Model Curriculum: Lead Abatement Training for Supervisors (Chapter L), University of Cincinnati, Cincinnati, June 1992.

Chapter 20

Development of a Field-Test Method for the Determination of Lead in Paint and Paint-Contaminated Dust and Soil

P. M. Grohse, K. K. Luk, L. L. Hodson, B. M. Wilson, W. F. Gutknecht, S. L. Harper, M. E. Beard, B. S. Lim, and J. J. Breen

CONTENTS

ABSTRACT

A rapid, simple, inexpensive, and relatively accurate field-test method for the determination of lead (Pb) in paints, dusts, and soils has been developed. The method involves the ultrasonic leaching of 0.1 to 0.5 g of the sample in 5 ml of 25% (v/v) nitric acid for 30 min, followed by colorimetric measurement with a commercially available field-test kit. A variety of actual field samples and several National Institute of Standards and Technology (NIST) standard reference materials (SRMs) were tested using the proposed method. Results were compared with those obtained with a microwave, total-digestion method followed by inductively coupled plasma emission measurement. Lead recovery and method precision were better than 84 and 11%, respectively, for a variety of SRMs and field samples. No apparent interferences were encountered in the ultrasonic/nitric acid sample extracts.

This paper has been reviewed in accordance with the U.S. Environmental Protection Agency peer and administrative review policies and approved for presentation and publication. Mention of trade names or commercial products does not constitute endorsement or recommendation for use.

I. BACKGROUND

The adverse health effects resulting from exposure of young children to environmental lead have received increasing attention in recent years. The major sources of exposure to lead in housing units are thought to be paint, dust, and soil. Lead-based paint is currently a material of concern and a principal medium for lead contamination and exposure. Although less consideration has been given to soil and dust, they are also significant routes of exposure.[1]

Public housing authorities are required, by 1994, to randomly inspect all their housing projects for lead-based paint.[2] The most common test for lead in housing employs the portable X-ray fluorescence (XRF) detector, which gives rapid results and is nondestructive. Inconclusive XRF measurements must be confirmed with field-collected samples in the laboratory, using a more accurate analytical method such as atomic absorption spectrometry (AAS) or inductively coupled plasma emission spectrometry (ICP).[3] Dust and soil samples collected either for risk assessment or postabatement clearance testing must also be returned to the laboratory for analysis. Waiting for laboratory results, however, could result in continued exposure to hazardous levels of lead and delay in reoccupation of a dwelling following abatement. On-site quantitative analysis would eliminate this problem, but this would involve bringing in a mobile laboratory, which would increase costs and require the use of highly trained personnel. A quantitative, chemical test kit that would provide results of acceptable accuracy and precision and could be readily used by typical field testing personnel, would be preferable.

In a previous study five commercially available lead test kits were evaluated.[4] These kits, which are intended principally as qualitative tools, showed a great deal of sample dependency. The kits showed negative responses with available laboratory and real-world samples, at levels of concern (i.e., paint at more than 1 mg Pb/cm^2, dust at more than 200 ppm, and soil at 1000 ppm). The capability of the kits to measure approximately 1 μg of lead in an aliquot of standard solution was demonstrated; however, because of insufficient solubilization and/or negative interferences, positive responses were not obtained with solid, real-world samples containing much more than 1 μg of lead. Several kits for quantitative analysis of lead were also commercially available at the time, but these were not evaluated because they were designed for water analysis and not for solid materials.

In response to this apparent lack of a quantitative field-test kit for analysis of lead in paint, soil, and dust, a study was conducted to identify or develop, if necessary, a quantitative lead extraction procedure and a compatible measurement method.

II. EXPERIMENTAL

A. EXTRACTION STUDIES

In the first phase of the study, a series of experiments were performed to identify suitable lead extraction conditions for paints, dusts, and soils. Parameters studied included extraction-solution composition, sample size, agitation method, and extraction time. Initial work was carried out on paint chips and paint dusts, on the assumption that these would pose the greatest extraction difficulty.

Leachates resulting from the extraction studies were initially subjected to ICP measurement, using a Leeman Labs Plasma Spec I sequential ICP. To correct for any unanticipated differences between standards and samples (and between samples), the use of internal standards and the method of additions were employed. For a comparison check on the ICP measurements, leachates were also measured by direct aspiration AAS.

When the lead concentrations of the test samples were not known, the extraction efficiencies of the candidate methods were determined by analyzing the residue remaining after the extractions. This analysis was performed using a microwave "total digestion" method incorporating nitric (HNO$_3$), hydrofluoric (HF), and hydrochloric (HCl) acids.[5]

Methods for agitating the leaching solution include manual and automatic shaking, static leaching, and ultrasonication. Automated shaking was ruled out because the equipment is too cumbersome for field applications. Manual shaking is labor intensive and was rejected because of low throughput. Therefore, static leaching and two Fisher Scientific ultrasonic baths (the 2.8-l, 100-W and the 1.06-L, 53-W models) were used. The majority of the extractions were performed with the larger unit.

In the initial tests HNO$_3$ solutions ranging from 1 to 50% (v/v) were tested with National Institute of Standards and Technology (NIST) Standard Reference Material (SRM) 1579 (lead in paint). Extraction times ranged between 1 and 24 h for static leaching, and between 15 and 60 min for ultrasonication. Alternative or supplementary extraction media (addition of an HCl extraction step and pretreatment of the samples with alkaline solution) also were evaluated with the goal of improving either (1) the extractability of the sample prior to leaching or (2) the solubilization of the lead compound itself.

The effect of the sample preparation method was tested using oil-based paint films spiked with known levels of lead nitrate or white lead. Paint film samples were leached in one of three configurations: (1) 1-cm^2 intact samples, (2) 1-cm^2 samples cut into 1/16-in. squares, and (3) ground samples crushed to pass 60 to 80 mesh. These tests were repeated on actual real-world paint chips. Initially, between 0.05 and 0.10 g of paint samples were used for the extraction studies. The maximum allowable volume of required extraction solution was also determined.

Once leaching solution, time conditions, and sample preparation methods had been established, the maximum amount of sample to be used for extraction was determined. Following the identification of the most-promising-candidate extraction method, tests were carried out on well-ground and homogenized paint and dust samples. NIST sediment SRMs 2704 and 1645 were used as test soils. Finally, real-world paint-chip samples were subjected to the candidate method.

B. MEASUREMENT STUDIES

In the second part of this study, lead measurement methods that were potentially compatible with the proposed extraction procedure were identified and tested. Because we had determined that ultrasonic

- Dilute sample extract to 100 ml with deionized H_2O
- Add 1 ml preservative (Hach pPb-1 Acid Preservative Solution) and swirl to mix
- Add 1 ml fixer (Hach pPb-2 Fixer Solution) and swirl to mix
- Pass sample through Hach Fast Column Extractor column to retain Pb
- Elute Pb from column using Hach pPb-3 Eluant Solution
- Adjust pH of eluent using Hach pPb-4 Neutralizer Solution
- Add color agent (Hach pPb-5 Indicator Powder Pillow)
- Decolorize 1/2 of above solution as blank using Hach pPb-6 Decolorizer Solution
- Measure sample and blank by colorimetry at 477 nm

Figure 1 Measurement method (highlights): Hach DR 100/700.

leaching produced a usable solution of lead extracted from paint, dust, and soil samples, we were now able to consider measurement methods designed for lead in water. A measurement method was considered an acceptable candidate if it was

1. Portable
2. Selective for lead (i.e., interferences were not significant)
3. Sensitive (should detect solution concentrations of less than 10 µg of Pb per liter)
4. Rapid
5. Inexpensive
6. Simple (not requiring a high level of education/technical background)
7. Compatible with the leachate from the extraction study candidate method
8. Chemically safe to use (i.e., minimum use of hazardous reagents)

Several different measurement methods were considered, including the lead-ion-selective electrode, field portable anodic stripping voltammetry, and colorimetric methods. Preliminary tests with a lead-ion-selective electrode indicated good response for aqueous solutions, but little or no response to lead extracted from paint and dust samples. Anodic stripping voltammetry was rejected because of its complexity.

Although many colorimetric methods for lead are described in the literature, we decided that a commercially available colorimetric lead test kit would be most desirable, since it would be immediately available to testers. A commercial colorimetric analysis system would eliminate the need to purchase and prepare reagents and to purchase a colorimeter. Three quantitative-test kits designed to measure lead in water were purchased for testing: LEADTRAK from Hach,[6] Octet Comparator from LaMotte Chemical,[7] and Lead Test from Chemetrics.[8] The latter two utilized a visual color comparison for quantification; but when they were tested with aqueous standards and paint and dust sample extracts, the color of the sample extracts did not match the color (chroma) of the standards. Therefore, an intensity comparison could not be made. These two kits also used potentially hazardous reagents such as cyanide and carbon tetrachloride.

The LEADTRAK system uses a proprietary colorimetric reaction with a lead ion that reaches maximum color intensity at 477 nm. This kit was judged the most suitable and was used exactly as the manufacturer recommended. Several colorimeters are available for the LEADTRAK system. Tested here were the DR/100 with analog meter readout and the DR/700 with digital readout. The quoted concentration range for both colorimeter systems is 0 to 150 µg/l in drinking water. The method is summarized in Figure 1.

To ensure that the chemistry of the Hach LEADTRACK kit was compatible with the extraction leachates, and specifically to determine the lowest test-sample pH tolerated by the test kit, the same standard (50 µg of Pb per liter) was analyzed in dilute HNO_3 solutions ranging from 0 to 0.3%. A linearity check was performed by analyzing a series of standards throughout the stated analytical range of the colorimeters.

Using the ultrasonic/25% nitric acid leaching method developed in the first part of the study, we used both Hach colorimeters to analyze the paint- and dust-method test samples and SRMs previously measured using ICP and AAS. Three different analysts performed the determinations. Two analysts used both colorimeters to analyze the leachates. Five replicate extractions were performed for each sample. Overall method (extraction/measurement) precision and accuracy were calculated, in addition to operator variability.

Table 1 Effect of Variation in Sample Extraction Weights

Sample	Sample aliquot weight (g)	Ultrasonic leach/ICP measurement (% Pb) N = 2
DUST-1	0.25	0.414
	0.50	0.429
PAINT-1	0.25	0.148
	0.50	0.147
PAINT-2	0.10	0.635
	0.25	0.634
	0.50	0.431
PAINT-3	0.10	3.51
	0.25	3.41
	0.50	1.15
NIST 2704	0.10	0.0157
Buffalo River Sediment	0.25	0.0162
(0.0161)[a]	0.50	0.0143
NIST 1645	0.25	0.0664
River Sediment	0.50	0.0694
(0.0714)[a]		

[a] Certified values in percent.

III. RESULTS AND DISCUSSION

A. EXTRACTION STUDY

Both static and ultrasonic leaching were tested over a variety of extraction times and HNO_3 concentrations. Approximately 0.1-g paint samples were used for these determinations. For the static extraction method, acceptable extraction efficiency (>90%) was achieved after approximately 3 h with two out of five samples of NIST SRM 1579, using 25% (v/v) HNO_3. Using the 100-W ultrasonic bath, greater than 85% efficiency was achieved with only 10% (v/v) HNO_3 after 30 min, for all samples, and greater than 95% efficiency was achieved with 25% (v/v) HNO_3. Assuming that most paint samples would be in the chip form, it appeared unlikely that the static extraction technique would be adequately efficient. Therefore, the remaining studies focused on the ultrasonic extraction method. At this time in the study, 10% HNO_3 appeared to be the most suitable leaching media for subsequent work, because (1) it provided close to 90% lead recovery for samples tested, (2) it is highly desirable to minimize strength of acid solutions used in the field, and (3) extracts having minimum acidity would have the greatest chance of being compatible with the method(s) of measurement ultimately selected.

Using the 10% (v/v) HNO_3 extraction method, the need for a grinding procedure prior to the extraction was evaluated. Latex paint film samples were prepared in-house and extracted in the three configurations described in Section II.A. Results using the 10% (v/v) HNO_3 extraction (30 min) and ICP measurement showed no difference in recovery between intact and ground paint chips. All sample aliquots were in the 0.1-g range. There also appeared to be no difference in efficiency when using 2.5 or 15 ml of extraction solution (10% HNO_3). However, when the grinding procedure was evaluated using real-world, old paint chips, there was an average increase of 13% in the measured lead levels for ground chips, over chips that had been extracted intact.

In an effort to improve the extraction procedure, 1.0 N sodium hydroxide was evaluated as a paint chip "softening" agent (prior to extraction); in a separate experiment, concentrated HCl was added after the 10% HNO_3 extraction step. Results in both cases indicated no improvement in the extraction efficiency for the real-world old paint chip samples. Ultimately, effective recovery was deemed to be more important than avoiding the use of relatively strong acid in the field, or compatibility with potential measurement methods. Therefore, we decided to use 25% (v/v) HNO_3 because this reagent gave better than 95% recovery with the samples tested.

The maximum sample weight for extraction for each matrix that resulted in greater than 90% lead recovery was then determined. Although 0.5-g aliquots appeared to be quantitatively extracted for most samples of each matrix type, some paint samples showed a decline in recovery even above 0.1 g (Table 1). Consequently, a maximum weight of 0.1 g was chosen for paint samples, while 0.25 to 0.5 g was found to be acceptable for dusts and sediments.

Table 2 Determination of Extraction Efficiency Through Extract Residue Analysis[a]

Sample	Extractable lead (µg/g)	Residual lead (µg/g)[b]	Extraction efficiency (%)
Paint			
P711	40,400	330	99.2
P254	2160	59	97.3
P273	7889	25	96.9
P268	52,400	1400	97.4
P355	71,800	600	99.2
P719	288,000	5900	98.0
Dust			
D116	548	14	97.5
D120	1240	34	97.3
D121	1710	8	99.5
D122	850	52	94.2
Sediment			
NIST 2704	160	1	99.4
NIST 1645	653	61	91.4

[a] n = 1.
[b] Calculated relative to original sample weight.

Table 3 Comparison of Total Digestion and Ultrasonication Analysis Results[a]

Sample	Total digestion/ICP	US/ICP[b]	US/AAS[c]
DUST-1	0.496 ± 0.041	0.414 ± 0.017	0.418 ± 0.009
PAINT-1	0.162 ± 0.004	0.160 ± 0.008	0.178 ± 0.012
PAINT-2	0.645 ± 0.023	0.635 ± 0.013	0.642 ± 0.005
PAINT-3	3.60 ± 0.12	3.51 ± 0.12	3.36 ± 0.30
NIST 1579	11.87 ± 0.15	11.40 ± 0.20	12.10 ± 0.30
(11.87%)			

[a] n = 5; % Pb ± standard deviation.
[b] Ultrasonic extraction/ICP measurement.
[c] Ultrasonic extraction/flame AAS measurement.

Extraction efficiency was evaluated by sonicating crushed real-world paint chip samples in 25% (v/v) HNO_3 for 30 min in the 100-W ultrasonic bath, followed by digestion of the undissolved residue, using a more complete microwave method employing HNO_3, HCl, and HF acids. Both the leachate and residue digestate were analyzed by ICP. For every sample, extraction efficiency was greater than 95% (Table 2). Using the same conditions, the smaller, 53-W ultrasonic bath was used to extract samples of the same real-world paint. Extractions with this less powerful unit also yielded recoveries in excess of 95%. Finally, using the larger bath and 25% HNO_3, extraction efficiencies for four "real-world" dust samples and two sediment/soil SRMs were determined, using the same "residue analysis" approach. As shown in Table 2, extraction efficiencies exceeded 94% for dust samples and 91% for sediments.

The extraction method was then tested on three real-world method evaluation paint samples and two dust samples. The preparation and round-robin analyses of these samples were presented in another paper.[9] Five replicate extractions were performed for each sample. Extracts were measured for lead by ICP and AAS. Extraction recoveries for all paints were 93.3% or better. Recovery for one dust sample was 84.3%. The lead level in the other dust sample was too low for an accurate recovery determination. Extraction results were compared to data from the total digestion (HNO_3, HCl, HF) of these samples (Table 3). A statistical pairwise comparison was made of each of the three methods, for each sample.[10] Despite the significant differences among samples (Table 4), the averages of the ultrasonic/ICP and ultrasonic/AAS values differed by less than 5% from total digestion values for the three paint samples and the paint SRM. The difference was 16% for the dust sample. After review of these and earlier test results, a final extraction procedure (Figure 2) was formulated.

B. MEASUREMENT STUDY

Initial tests of the Hach lead test kit (LEADTRACK), a system designed for analysis of lead in drinking water, involved measurement of standard lead solutions and comparison of the results obtained by using

166

Table 4 Pairwise Comparisons of Methods. Test Samples for Which Results Differed at the 95% Confidence Level

	RTI total digestion	US/ICP[a]	US/AAS[b]
RTI total digestion	—	—	—
US/ICP	DUST-1		
	NIST 1579	—	—
US/AA	DUST-1	PAINT-1	
	PAINT-1	NIST 1579	—

[a] Ultrasonic extraction/ICP measurement.
[b] Ultrasonic extraction/flame AAS measurement.

- Weigh 0.1 g paint (0.25 g sediment or dust) in a 50 ml graduated centrifuge tube
- Pipet 5 ml of 25% (v/v) HNO_3 into centrifuge tube and cap
- Sonicate for 30 min (solution will heat to approximately 45°C)
- Remove from bath, allow to cool, and dilute to 50 ml with DI H_2O and mix
- Centrifuge for 5 min at 2,000 rpm or allow to settle for approximately 30 min

Figure 2 Extraction method.

Table 5 Comparison of Atomic Spectroscopic and Hach Measurements of Paint and Dust Leachates[a]

Sample	Atomic spectroscopy (n = 10)	Hach with DR/100 Colorimeter (n =15)	Hach with DR/700 Colorimeter (n = 10)
DUST-1	0.416 ± 0.013	0.409 ± 0.024	0.439 ± 0.027
PAINT-1	0.169 ± 0.010	0.182 ± 0.014	0.179 ± 0.014
PAINT-2	0.639 ± 0.009	0.624 ± 0.023	0.613 ± 0.030
PAINT-3	3.44 ± 0.21	3.24 ± 0.20	3.24 ± 0.19
NIST 1579 (11.87 ± 0.04%)	11.8 ± 0.3	10.7 ± 0.5	10.9 ± 0.6

[a] % Pb ± standard deviation.

both the kit colorimeter and a laboratory-scale UV/Visible spectrophotometer. Results agreed within 10%, demonstrating the potential accuracy of the test-kit measurements. Tests were then performed using NIST SRM 1579 paint extracts, to determine the maximum allowable volume of test solution that could be used with the Hach kit. With a final (diluted) nitric acid concentration of 2.5%, a maximum of 3 ml of the final extract could be used before the buffering capacity of the Hach system was exceeded.

Using a series of aqueous standards, the linearity of the colorimeters was tested. These tests showed the Hach system to yield a linear response over the range of 0 to 150 µg/l. Using a 1-µg/l standard, a series of 11 replicate determinations were made to obtain an estimate of the detection limit. A value of 2.5 µg/l was obtained using three times the standard deviation of these measurements.

Following the preliminary measurements, two different models of Hach colorimeters were used to measure lead concentrations in extracts acquired using the proposed method. Dust and paint samples that had been sieved and homogenized in-house were used. Three operators performed the measurements. Five separate aliquots of each sample were subjected to the ultrasonic extraction method and measurement, using the Hach kit. Data from all operators and replicates for each sample and each colorimeter were pooled and the mean calculated. These data are compared to the pooled ICP and flame AAS measurements of the same leachates (Table 5). All Hach values were within 10% of the atomic spectroscopy values.

A statistical, pairwise comparison was made of each of the three measurement methods, for each sample. The Hach results for PAINT-2, PAINT-3, and the SRM are biased low relative to the atomic spectroscopic results (Table 6), whereas the results are variable for the dust and the other paint. Table 7 presents a breakdown of data from each Hach operator. These data are compared to "total digestion"/ICP results

Table 6 Pairwise Comparison of Methods. Test Samples for Which Results Differed at the 95% Confidence Level

	Atomic spectroscopy	Hach with DR/100 Colorimeter	Hach with DR/700 Colorimeter
Atomic spectroscopy	—	—	—
Hach with DR/100 Colorimeter	PAINT-1		
	PAINT-2		
	PAINT-3	—	—
	NIST 1579		
Hach with DR/700 Colorimeter	DUST-1		
	PAINT-2		
	PAINT-3	DUST-1	—
	NIST 1579		

Table 7 Comparison of Field Method Results with Reference Values[a]

		Field method[c]		
Sample	Ref. value[b]	Operator 1	Operator 2	Operator 3
DUST-1	0.496 ± 0.041	0.419 ± 0.030	0.389 ± 0.017	0.420 ± 0.025
PAINT-1	0.162 ± 0.004	0.182 ± 0.014	0.184 ± 0.012	0.179 ± 0.017
PAINT-2	0.646 ± 0.023	0.604 ± 0.016	0.646 ± 0.021	0.622 ± 0.032
PAINT-3	3.60 ± 0.03	3.27 ± 0.14	3.26 ± 0.33[d]	3.21 ± 0.13
NIST 1579	11.9 ± 0.2	10.5 ± 0.8	10.7 ± 0.4	10.7 ± 0.3

[a] % Pb ± standard deviation.
[b] n = 3.
[c] n = 5.
[d] % RSD = 10.2.

(n = 3 for each sample). The total digestion used microwave heating in the presence of HNO_3, HF, and HCl. Therefore, Table 7 also represents a preliminary measure of the overall field-method accuracy and precision. Average recoveries for the Hach kit field method, calculated relative to the "reference values," exceeded 90% for paints and 82% for the dust sample. In only one instance did the precision over the five replicates of a given sample exceed 10% RSD (10.2%), and generally it was considerably better. A statistical, pairwise comparison of the operator results showed only one difference at the 95% confidence level (the difference between operators 1 and 2 for PAINT-2).

IV. SUMMARY AND CONCLUSION

Due to the current lack of a comprehensive and quantitative lead test kit for paints, dusts, and soils, a field method has been developed that combines a simple extraction technique with a commercially available colorimetric test kit for lead in water. The method is relatively rapid. Weighing and grinding of paint requires about 5 min per sample. Setting up for ultrasonication requires 1 to 2 min per sample. The ultrasonicated samples can be allowed to settle for 30 min, which does not require any labor, or they can be centrifuged, which requires about 5 min per sample. Finally, measurement with the Hach kit requires about 15 min per sample. A number of samples can be processed simultaneously, depending on the capacity of the ultrasonic bath.

The method is capable of detecting 50 μg/g in paint and 20 μg/g in sediments and dusts. Extraction recovery is greater than 85%, and measurement values are within 10% of atomic spectroscopic values. Overall method precision is generally better than 10% RSD. The estimated cost for materials is approximately $5 per analysis (1992); initial outlay for the Hach kit varies, depending on the colorimeter purchased, from approximately $250 for a kit with an analog device to $800 for a kit with a digital device.

Some pretreatment of paint chips appears necessary; a simple 30-s crushing operation with a glass or plastic rod appears to be adequate. Quality-control operations are critical. Check samples should be analyzed periodically to verify calibration curve accuracy and compensate for instrument drift and variations in operator performance. Digital colorimeter units such as the Hach DR/700 are preferable to the analog units, due to the potential variations in "style" in reading the meter. Based on the existing data, it appears that 0.1-g aliquots are most suitable for paint chips and powdered paint, while as much as 0.25 g may be used for house bulk dusts and sediments. The measurement system has a range limited to extracts with between 10 and 150 μg of Pb per liter. For some samples this may be beneficial due to the "diluting out" of any potential matrix or interference effects.

Although the method appears suitable for paints, bulk dusts, and soils, work needs to be performed for dust wipes. In addition, the method should be evaluated for actual field soil samples. The sediments tested in this study were SRMs that were well homogenized and of very small (and easily extractable) particle size. In order to further define the precision and accuracy of the method, it must be tested through a round-robin study, preferably performed in a field environment.

REFERENCES

1. Elwood, P. C., The sources of lead in blood: a critical review, *Sci. Total Environ.*, 52, 1–23, 1986.
2. Lead-Based Paint Poisoning Prevention Act, 42, U.S.C. 4:22 (d)(2)(A), 1971.
3. Lead-Based Paint: Interim Guidelines for Hazard Identification and Abatement in Public and Indian Housing, Department of Housing and Urban Development, Washington, D.C., 1990.
4. Luk, K. K., Hodson, L. L., Smith, D. S., O'Rourke, J. A., and Gutknecht, W. F., Evaluation of lead test kits for analysis of paint, soil and dust, presented at AWMA/EPA Int. Symp. on Measurement of Toxic and Related Air Pollutants, Durham, NC, May 1992.
5. Bao-hou, L, Zhong-quan, Y., and Kai, H., Determination of Si, Al, Mg, Fe, Ti, Mn, Cu, Co and Ni in vanadium-titanium ore by microwave oven digestion, ICP, AA and Chemical analysis methods, Institute of Chemical Industry and Metallurgy, Academy of Sciences of China, Beijing, June 1988.
6. Hach Company, 100 Dayton Ave., P.O. Box 907, Ames, IA 50010.
7. LaMotte Chemical Products Co., P.O. Box 329, Chestertown, MD 21620.
8. Chemetrics, Rt. 28, Calverton, VA 22016–0214.
9. Williams, E. E., Binstock, D. A., Estes. E. D., Neefus, J. D., Meyers, L. E., Gutknecht, W. F., Lim, B. S., Breen, J. J., Harper, S. L., and Beard, M. E., Preparation and evaluation of lead-in-paint and lead-in-dust reference materials, in Chapter 21, this volume.
10. Natrella, M. G., Experimental Statistics, Handbook 91, National Bureau of Standards, Washington, D.C., 1966, pp. 3-23 – 3-30.

Preparation and Evaluation of Lead-Containing Paint and Dust Method Evaluation Materials

E. E. Williams, D. A. Binstock, E. D. Estes, J. D. Neefus, L. E. Myers,
W. F. Gutknecht

CONTENTS

I. INTRODUCTION

A growing concern about the adverse health effects associated with low-level exposure to environmental lead has resulted in increased emphasis on risk assessment and abatement programs at federal, state, and local levels. With increasing demands for sampling and analysis of lead-contaminated materials and the emergence of a laboratory accreditation program for the analysis of environmental lead samples, the need for reference materials for lead in paint, dust, and soil is critical.

Two types of reference materials are important in analytical chemistry quality assurance:

- Standard reference materials (SRMs)
- Performance evaluation materials (PEMs)

Of the two types of reference materials, SRMs are more homogenous and more stringently characterized. The analytical uncertainty for SRMs is usually ≤10% as compared to 10 to 25% for PEMs. SRMs are more costly and, thus, less available for routine quality assurance/quality control (QA/QC) activities. Therefore, the preparation of PEMs in a concentration range appropriate to laboratory analysis is essential.

This paper describes the preparation of lead-based paint and dust performance evaluation materials, called method evaluation materials (MEMs), as prescribed by the U.S. Environmental Protection Agency (EPA)-sponsored Lead Reference Materials Workshop[1] (LRMW) held in May 1991. The design and results of a round-robin study carried out to examine the homogeneity, i.e., the sample-to-sample variation, of these MEMs are also described, along with the bias and precision of five analytical methods used by the laboratories participating in the round-robin study.

II. EXPERIMENTAL

A. PREPARATION OF PAINT METHOD EVALUATION MATERIALS

The samples analyzed for lead in the process of lead-based paint testing are often multiple layers of different kinds of paints that have been embrittled from age and weathering. These materials are expected to respond differently to chemical analysis than new paint materials prepared in the laboratory.[2] Accordingly, it was decided that method evaluation materials should be prepared from real-world paints, because these materials would provide the most meaningful challenge to any given analytical methodology.

Collection of Materials

The Research Triangle Institute (RTI) has participated in a number of projects involving the collection of paint for use as reference materials and, as a result of these projects, has established a large repository of real-world paints. The repository contains paints from interior and exterior walls and trim collected from abatement and demolition projects across the country. The paint chips used for the preparation of the method evaluation materials described in this chapter were collected from a vacant hospital in Athens, Ohio. The hospital was built in the late 19th century and repainted routinely over the duration of its existence (approximately 100 years).

Identification of Appropriate Bulk Materials

Aliquots were removed from bulk quantities of paint chips and subjected to microwave (MW) extraction using a combination of nitric acid (HNO_3) and hydrochloric acid (HCl) and measured using inductively coupled plasma emission spectrometry (ICP),[3] in order to identify materials with concentrations projected as suitable for round-robin analysis. Most of the bulk paint samples collected in Athens were from walls and woodwork painted with multiple layers of pre-1978 paint that contained high levels of lead (5 to 40%). In order to obtain a lower level of lead, layers representing the most recent painting, i.e., the lowest lead concentration, were collected (separated by hand) from one of the bulk samples consisting of chips of multiple-colored layers.

As a result of these initial analyses, two quantities of bulk paint having lead concentrations of 3.8 and 0.36% were selected for use in the preparation of method evaluation materials for the round-robin. The 0.36% material was further refined to approximately 0.15% by manually removing chips determined to contain high levels of lead in order to provide a bulk material having a lower lead concentration. The two bulk samples were designated "High Paint" (3.8%) and "Low Paint" (0.15%).

Grinding

The two bulk samples were ground in a cross-beater mill[4] to prepare materials having particle sizes ≤250 μm. After grinding in the cross-beater mill, the materials were ground in a Retsch[5] grinder to a particle size ≤120 μm.

Blending

The ground paint materials were mixed for 30 min in a Turbula[6] blender. Aliquots were removed from these bulk samples, for initial verification of concentration and homogeneity.

Testing Effects of Sample Size

In an effort to investigate the effects of test sample size on analytical results, aliquots of each bulk sample with weights of 50, 100, and 150 mg were analyzed in triplicate (i.e., three test samples of each of the

Table 1 Dependence of Concentration on Aliquot Size for Paint and Dust Method Evaluation Materials[a]

	Aliquot Size		
Sample	50 mg	100 mg	250 mg
Low Paint	3,600 ± 7.06 (0.196)	3,530 ± 42.4 (1.20)	3,310 ± 28.3 (0.854)
High Paint	36,800 ± 1203 (3.27)	36,200 ± 283 (0.781)	36,000 ± 425 (1.18)
Low Dust	97.4 ± 29.2 (29.9)	79.8 ± 0.42 (0.53)	81.2 ± 0.71 (0.87)
High Dust	4,340 ± 503 (11.6)	4,160 ± 84.9 (2.04)	4,100 ± 6.97 (0.17)

[a] Mean ± standard deviation (% RSD), in micrograms per gram (μg/g); % RSD = percent relative standard deviation.

three aliquot weights for each of the two bulk samples, for a total of 18 tests) using the MW/ICP method noted previously. These weights, though arbitrarily chosen, are within the range commonly used for environmental samples with relatively high analyte concentrations, (i.e., >10 μg/g). The results, shown in Table 1, indicated a decrease in the lead level for the "Low-Paint" sample above 100 mg. To avoid the possibility of problems with the larger aliquot size, an aliquot size of 100 mg was selected and prescribed for further analyses.

B. PREPARATION OF DUST METHOD EVALUATION MATERIALS
Collection
The use of real-world materials for the preparation of dust MEMs was also considered to be appropriate. Dust was collected from households and hotels (vacuum cleaner bags), from street sweepers, and from abatement sites (postabatement, high-efficiency particulate air [HEPA] vacuum bags).

Sterilization
Prior to handling, all dust samples were shipped to Neutron Products, Inc., Dickerson, MD, for sterilization by gamma irradiation. Bags were irradiated for 12 h, receiving a minimum total dose of 2.5 Mrads.

Sieving
The sterilized bags were returned to RTI and sieved individually to remove gross debris and hair, using a Ro-Tap[7] apparatus that consisted of two stacked sieves (2.00 mm and 250 μm).

Identification of Appropriate Bulk Materials
From individual collections of sieved dust, 100-mg aliquots were removed and analyzed using the MW/ICP method.

Blending
Because the yield of sieved dust from one vacuum bag was usually insufficient to provide enough material for thorough testing, batches of sieved dust having concentrations over the range of interest were blended for 30 min in a Turbula[6] blender, to achieve adequate quantities of the desired concentrations. Four aliquots were removed from each batch of blended material and analyzed using the MW/ICP method to determine lead concentration.

As a result of the analyses of both individual and blended bulk dust samples, two samples having lead concentrations of approximately 80 μg/g (Low Dust) and 4100 μg/g (High Dust) were selected for use in the round-robin.

Testing Effects of Sample Size
As in the case of the paint matrix, three different aliquot weights of dust samples (50, 100, and 250 mg) were analyzed in triplicate by the MW/ICP method, to investigate the effects of aliquot size on analytical results. The results, shown in Table 1, indicated consistent concentrations for the three aliquot sizes tested, with an improvement in precision with increasing sample size. In light of these results and because 100 mg of dust approximates a typical field sample collected on a filter or a dust wipe, an aliquot size of 100 mg was selected and prescribed for further analyses.

C. PACKAGING OF SAMPLES
Approximately 150 samples of each of the paint and dust MEMs were prepared by accurately weighing 5 g of the bulk paint and 2 g of the bulk dust, into 20-ml plastic screw-cap bottles. Care was taken to close

the bulk containers and tumble the contents in all directions several times after removing every five to seven samples. The bottles were numbered sequentially to track the loading from the bulk material.

D. FINAL VERIFICATION OF CONCENTRATION

Five bottles were removed at random from each of the four MEM lots (High Paint, Low Paint, High Dust, and Low Dust). Aliquots (100 mg) were removed from each of the five bottles from each lot and analyzed by the MW/ICP method, to determine the lead concentrations of the final bottled materials. The final concentrations of the test samples were as follows:

Low Paint 1,410 ± 44.5 μg/g
High Paint 37,900 ± 500 μg/g
Low Dust   84.2 ± 11.9 μg/g
High Dust 4,670 ± 330 μg/g

III. DESIGN OF THE ROUND-ROBIN STUDY

A. METHODS OF ANALYSIS

Four combinations of extraction and spectrophotometric measurement methods commonly used for lead sample analysis were selected for testing. Either microwave-based (MW) extraction or hotplate-based (HP) extraction, in combination with measurement by atomic absorption spectrometry (AAS) or inductively coupled plasma emission spectrometry (ICP), resulted in four standard methods: MW/AAS, MW/ICP, HP/AAS, and HP/ICP. Laboratory X-ray fluorescence (XRF), although not standardized, was also included in the study, because it was used in the EPA Soil Lead Abatement Demonstration Project Three-City Study.[8]

Laboratories using an extraction method were sent the RTI document, "Standard Operating Procedures for Lead in Paint by Hotplate- or Microwave-Based Acid Digestions and Atomic Absorption or Inductively Coupled Plasma Emission Spectrometry,"[3] and were asked to follow procedures outlined in the document, for microwave (HNO_3/HCl) and hotplate (HNO_3/H_2O_2) methods. Laboratories using laboratory XRF were asked to follow their own procedures for sample preparation and analysis. The EPA Environmental Monitoring Systems Laboratory (EMSL)/Las Vegas document, "Standard Operating Procedures for energy-Dispersive X-ray Fluorescence Analysis of Lead in Urban Soil and Dust Audit Samples,"[9] was sent to the XRF participants, for use as a reference protocol only. Because some of the XRF laboratories did not have dust standards for calibration, all laboratories were provided two audit samples (58 and 2275 μg Pb/gram) of dust from the Three-City Study[8] and were asked to use these audit samples to calibrate for the analysis of the dust test samples.

B. SELECTION OF LABORATORIES

Laboratories were selected both on the basis of their willingness to participate in the round-robin, and the method (or methods) they were willing to perform. The majority of the laboratories invited to participate were involved in an earlier round-robin study of the hotplate extraction method.[10] The remainder were identified through personal inquiries made to individuals active in the area of environmental lead measurement. The goal in selecting laboratories was to have a minimum of eight results for each sample analyzed by each of the five methods, because this would allow acceptable statistical analysis.

C. SUBMISSION OF PAINT AND DUST TEST SAMPLES

The set of paint and dust samples submitted to each laboratory included both method evaluation materials (MEMs) and standard reference materials (SRMs). It was decided to include both low and high levels of both paint and dust materials, to encompass the concentration ranges most relevant to field samples. The National Institute of Standards and Technology (NIST) SRMs were included as blind samples in the set, because the homogeneity and concentrations of these materials are well established. Results of the analyses of the SRMs were expected to provide a means of differentiating problems intrinsic to the laboratories/methods from problems associated with the MEMs.

A test set consisting of 10 samples was sent to each laboratory, for each analytical method performed (i.e., laboratories performing two different methods received two sample sets). The paint samples were labeled P-1 through P-5, and the dust samples, D-1 through D-5. Each test set contained duplicate bottles of both high- and low-lead paint and duplicate bottles of both high- and low-lead dust. Laboratories were unaware of the presence of duplicates; P-1 and P-4, for example, were duplicate low-lead paint samples. The composition of a sample set is shown in Table 2. SRM 1579 was included as the reference material

Table 2 Set of Test Samples Submitted to Each Laboratory

Samples	Source	Concentration (MW/ICP)[a] Mean (µg/g) ± SD (% RSD) n = 5
Low Paint (P-1, P-4)	Athens, OH	1,410 ± 44.5 (3.16)
High Paint (P-3, P-5)	Athens, OH	37,900 ± 500 (1.35)
Paint SRM (P-2)	NIST SRM 1579	118,700 ± 400 (0.34)
Low Dust (D-2, D-4)	Household dust, NC and CA	84 ± 11.9 (14.1)
High Dust (D-1, D-5)	Post-abatement dust, PA	4,670 ± 330 (7.07)
Dust SRM (D-3)	NIST SRM 2711	1,162 ± 31 (2.67)

[a] MW = microwave digestion method; ICP = inductively coupled plasma emission spectrometery.

for paint. The new SRM 2711, although actually a soil, was provided as a surrogate reference material for dust, because of the lack of availability of SRMs for dust and because the particle size of SRM 2711 was considered comparable to dust.

Each laboratory was instructed to tumble every bottle of material prior to aliquoting for analysis. If an extraction technique was used, the laboratory was asked to remove two 100-mg aliquots from each bottle, carry each aliquot through the extraction, and analyze the individual extracts. XRF laboratories were instructed to take two samples from each bottle, that were sufficiently large to prepare "infinitely thick" samples, and analyze each. Thus, each laboratory reported a total of 20 results for each test set. Because of the inclusion of duplicate MEM samples (i.e., two bottles from the same bulk material), each laboratory actually analyzed four samples of the same MEM material. Therefore, the design provided a means of determining repeatability (within-laboratory variation) and reproducibility (between-laboratory variation). The distribution of laboratory participation was as follows:

Procedure	Total number of performances
MW/AAS	7
MW/ICP	9
HP/AAS	9
HP/ICP	10
Lab XRF	7
Total	42

D. SUBMISSION OF RESULTS

Laboratories were asked to report results to RTI on a data reporting form, shown in Figure 1, included with the test samples. A separate form was enclosed for reporting instrumental parameters.

IV. RESULTS AND DISCUSSION

Six combinations of matrix (dust, paint) and level (high, low, SRM) provided data for analysis (see Table 2). A statistical analysis [11-13] of the data was performed to determine the following:

- Mean concentration by method for each of the six test samples
- Consensus concentration for each of the six test samples
- Statistically significant differences between mean concentrations, by method, determined separately for each of the six test samples
- Homogeneity (within-sample variation of the material)
- Repeatability (within-laboratory variance), by method
- Reproducibility (between-laboratory variance), by method

EPA/RTI Round-Robin for Lead in Paint and Dust

Round-Robin No. 002   Lab ID No. _____
_____

Digestion Method   Laboratory
Experience with this Method   years

Analysis Method   Approval Signature:
Experience with this Method   years

Gross Concentration of Lead (ppm)

Sample ID No.	Aliquot 1	Aliquot 2
P-1		
P-2		
P-3		
P-4		
P-5		
D-1		
D-2		
D-3		
D-4		
D-5		
Reagent Blank		

Figure 1   Dust reporting form for EPA/RTI round-robin for lead in paint and dust.

A. CENSORED, MISSING, AND OUTLYING DATA

A total of 33 laboratories reported results for 42 performances, with each performance resulting in duplicate analyses of each of 10 test samples (blind duplicate samples of Low and High Paint and Low and High Dust, plus single blind samples of SRMs 1579 and 2711). One laboratory reported triplicate results, and two results were not reported, for a total of 848 results. Some results, primarily for the low-dust sample, were reported as less than a specific concentration and were removed prior to statistical analysis. A total of 820 results were examined for outliers.

For each of the six combinations of matrix (dust, paint) and level (high, low, SRM), a nominal concentration was obtained as the median of all reported results from the extraction methods. (Laboratory XRF data were excluded because (1) these results were of lower values than the results for the extraction methods, and (2) the extraction methods were performed according to specific standard procedures, whereas the XRF analyses were not.) A recovery was then calculated for each individual extraction method result, as the ratio of the reported concentration divided by the nominal concentration. Using recoveries between 0.35 and 2, the average and standard deviation of recovery was calculated separately for each of the 30 method(5)-by-matrix(2)-by-level(3) combinations. This restriction to recoveries between 0.35 and 2 was a prescreen intended to remove grosser outliers that could cause distortion of the means and standard deviations. A recovery score was then calculated for the recovery of each reported result, by subtracting the average recovery from the individual calculated recovery and dividing by the standard deviation of recovery for the given combination. Any measurement whose absolute score exceeded 2.76 was excluded as an outlier. This corresponded to the upper and lower one half of the 1% of a normal distribution. This two-step procedure resulted in the exclusion of 28 observations as outliers.

B. CONSENSUS VALUES

Consensus values for each of the six samples were calculated as the simple average of the method means for the extraction methods only. (Outliers and censored data were excluded from the determination of the method means.) Tables 3 and 4 show the means (by method), consensus values, and standard deviations. (Standard deviations for the consensus values are pooled standard deviations.) XRF values were excluded from calculation of the consensus values, but method means for the XRF procedure were included in Tables 3 and 4 to allow for a simple comparison of results. Recoveries, expressed as the ratio of means (by method) to consensus values, are given in Table 5.

C. SAMPLE HOMOGENEITY

The MEM samples were supplied to the laboratories as blind duplicates. For these samples, it is possible to test for homogeneity of the parent stocks, using two-way analysis of variance of logs; treating sample-to-sample variation, analysis, and their interaction as random effects. That is, laboratories using the same method, and replicate samples selected from the same parent stock, such as D-2 and D-4, were both viewed as random selections from a normally distributed population of the same. The assumption of random effects is appropriate in order to generalize results to a larger population of laboratories. This model was fit separately to all 20 combinations of method(5)-by-matrix(2)-by-level(2), which involved non-SRM samples.

A preliminary test for the absence of interaction or interdependence between sample and laboratory analysis, indicated that this assumption was reasonable. Only one of 20 interaction tests was significant at the 5% level, with this data set (Low Dust, MW/AAS, $0.025 < p < 0.5$). This is the expected number of rejections, by chance alone, under the null hypothesis of no interaction. Accepting the hypothesis of no interaction means that the contributions of sampling and analysis, to the total variance, can be thought of as additive.

Using a two-way analysis of variance, the relative standard deviations (RSDs) for the samples, equivalent to the percentage difference between samples, were calculated. These values are presented in Table 6. Only one case (Low Dust, HP/ICP) was determined to have a significant difference between samples (8.9%). In all other cases the sample-to-sample differences were zero (16 out of 20 cases) or not significant. On the average, over the 20 cases the sampling component of variance accounted for 1.37% of the total variance. A 95% upper confidence limit for the sampling component of variance was below 2.5%. The data indicate that 95% of all 0.1-g samples selected from the bulk materials are within 5% (between 95 and 105%) of the consensus values.

The conclusion is that bulk sample materials prepared by RTI are homogeneous. Furthermore, sample-to-sample variation is not significant relative to the variation of the methods.

Table 3 Method Mean and Consensus Values for Round-Robin Paint Samples

Matrix/ sample no.	Consensus value[a] (µg/g) ± SD[b] (% RSD)	Method[c]	Method mean (µg/g) ± SD (% RSD)
High Paint	37,632 ± 861	MW/AAS	41,281 ± 1,274 (3.1)
(P-3, P-5)	(2.3)	HP/AAS	36,921 ± 713 (1.9)
		MW/ICP	36,654 ± 672 (1.8)
		HP/ICP	35,670 ± 796 (2.2)
		Lab XRF	27,404 ± 1,567 (5.7)
Low Paint	1,690 ± 63	MW/AAS	1,896 ± 63 (3.3)
(P-1, P-4)	(3.8)	HP/AAS	1,661 ± 74 (4.5)
		MW/ICP	1,603 ± 45 (2.8)
		HP/ICP	1,600 ± 66 (4.1)
		Lab XRF	1,034 ± 76 (7.4)
Paint SRM	109,859 ± 6,521	MW/AAS	122,432 ± 6,507 (5.3)
(P-2)	(6.0)	HP/AAS	104,340 ± 8,681 (8.3)
NIST 1579		MW/ICP	118,281 ± 2,476 (2.1)
Certified value:		HP/ICP	94,382 ± 7,021 (7.4)
118,700 ± 400		Lab XRF	112,721 ± 13,259 (11.8)

[a] Lab XRF excluded from consensus value determination.
[b] Pooled standard deviations.
[c] Legend: MW = microwave method (EPA/AREAL), HP = hotplate method (NIOSH 7082), ICP = inductively coupled plasma emission spectrometry, AAS = atomic absorption spectrometry, XRF = X-ray fluorescence, SRM = standard reference material.

Table 4 Method Mean and Consensus Values for Round-Robin Dust Samples

Matrix/ sample no.	Consensus value[a] (µg/g) ± SD[b] (% RSD)	Method[c]	Method mean (µg/g) ± SD (% RSD)
High Dust	4550 ± 120	MW/AAS	4847 ± 127 (2.6)
(D-1, D-5)	(2.7)	HP/AAS	4677 ± 103 (2.2)
		MW/ICP	4281 ± 113 (2.6)
		HP/ICP	4397 ± 133 (3.0)
		Lab XRF	2485 ± 117 (4.7)
Low Dust	104 ± 6	MW/AAS	114 ± 6 (5.3)
(D-2, D-4)	(5.8)	HP/AAS	108 ± 7 (5.3)
		MW/ICP	98 ± 3 (3.1)
		HP/ICP	98 ± 9 (9.2)
		Lab XRF	93 ± 8 (8.6)
Dust SRM	1186 ± 44	MW/AAS	1327 ± 72 (5.4)
(D-2)	(3.8)	HP/AAS	1173 ± 32 (2.7)
NIST 2711		MW/ICP	1133 ± 24 (2.1)
Certified value:		HP/ICP	1112 ± 42 (3.8)
1162 ± 31		Lab XRF	1029 ± 33 (3.2)

[a] Lab XRF excluded from consensus value determination.
[b] Pooled standard deviations.
[c] Legend: MW = microwave method (EPA/AREAL), HP = hotplate method (NIOSH 7082), ICP = inductively coupled plasma emission spectrometry, AAS = atomic absorption spectrometry, XRF = X-ray fluorescence, SRM = standard reference material.

D. REPEATABILITY AND REPRODUCIBILITY

The repeatability (within-laboratory) and reproducibility (between-laboratory) standard deviations are based on a one-way analysis of variance of log recoveries, ignoring sample-to-sample differences that were shown to be nonsignificant (i.e., these differences are absorbed into the estimates of repeatability and reproducibility). Values for repeatability and reproducibility (%) are presented in Table 6. Laboratory XRF has the best repeatability (i.e., lowest percentage variation) for all six samples. This result is highly significant and may be due to the possibility that the log transformation did not sufficiently stabilize the variances and that Lab XRF is actually operating at a different apparent level than the other methods on some of the samples.

Table 5 Analytical Recovery by Method[a] Relative to Round-Robin Consensus Values (%)

Method	Paint			Dust		
	High	Low	SRM	High	Low	SRM
MW/AAS	110	112	111	107	110	112
MW/ICP	97.4	94.9	108	94.1	94.2	95.5
HP/AAS	98.1	98.3	95.0	103	104	98.9
HP/ICP	94.8	94.7	85.9	96.6	94.2	93.8

[a] Lab XRF recoveries were not determined because these results were excluded from the determination of consensus values.

Table 6 Estimates of Sample-to-Sample Variation, Within-Lab Variation (Repeatability), and Between-Lab Variation (Reproducibility)

Matrix	Statistics	Methods				
		MW/AAS	HP/AAS	MW/ICP	HP/ICP	Lab XRF
Low Paint	Mean (µg/g)	1896	1661	1603	1600	1034
	Sample RSD (%)	4.2	0	0	2.2	0
	Repeatability (%)	11.5	12.4	11.9	9.7	3.4
	Reproducibility (%)	13.3	17.7	13.3	16.2	18.3
High Paint	Mean (µg/g)	41281	36921	36654	35670	27404
	Sample RSD (%)	0	0	0	0	0
	Repeatability (%)	5.6	4.9	3.8	4.5	3.3
	Reproducibility (%)	9.5	7.1	6.5	8.2	15.7
Low Dust	Mean (µg/g)	114	108	98	98	93
	Sample RSD (%)	0	0	0	8.9	0
	Repeatability (%)	18.3	12.2	16.0	24.5	8.6
	Reproducibility (%)	20.2	20.6	16.5	35.3	22.2
High Dust	Mean (µg/g)	4847	4677	4281	4397	2485
	Sample RSD (%)	0	3.5	0	0	0
	Repeatability (%)	6.2	6.2	9.6	11.5	3.7
	Reproducibility (%)	8.9	8.9	10.6	13.7	13.2
Paint SRM	Mean (µg/g)	122432	104340	118281	94382	112721
	Repeatability (%)	7.2	6.2	4.4	12.5	1.3
	Reproducibility (%)	14.8	30.2	7.1	29.0	32.4
Dust SRM	Mean (µg/g)	1327	1173	1133	1112	1029
	Repeatability (%)	3.2	3.7	5.1	3.2	2.5
	Reproducibility (%)	14.2	8.9	7.5	12.7	8.7

At the same time, Lab XRF was fairly consistent in repeatability across all levels. No other consistencies could be recognized.

The most important single measure of method performance is reproducibility, because it reflects inter-laboratory, as well all within-laboratory, variability. Laboratory XRF has the worst reproducibility (highest variation) for all three paint samples. MW/ICP has the best reproducibility (lowest variation) on five of the six samples.

It is desirable to have constant repeatability and reproducibility across all concentration levels for a given method. However, as Table 6 shows, repeatability and reproducibility vary with concentration. Nevertheless, regressions of repeatability and reproducibility (expressed as µg/g) vs. sample concentration (µg/g) for each method (shown in Figures 2 and 3) provide a useful estimation of method variability at a specific concentration level. Data from paint and dust matrices were pooled to provide estimates of the performance of methods over the entire linear range of measurement. This range was considered to be appropriate to the concentrations of real-world paint and dust samples submitted to laboratories for analysis. If either the paint or dust matrix had been examined separately, the concentration range would have been limited. The plots are forced through zero so that slopes (representing the percentage change in repeatability or reproducibility for each unit change in concentration) may provide a visual comparison of methods in a selected concentration range. The plots do not attempt to model performance at minimum detection. Another approach is to pool variance data by methods over all matrices and concentrations, into an analysis of variance (one-way disregarding concentration level). Table 7 represents values of repeatability and

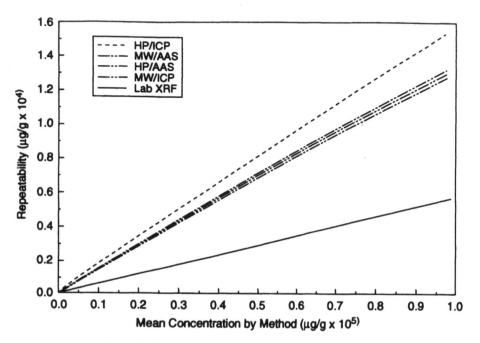

Figure 2 Repeatability vs. lead concentration, by method.

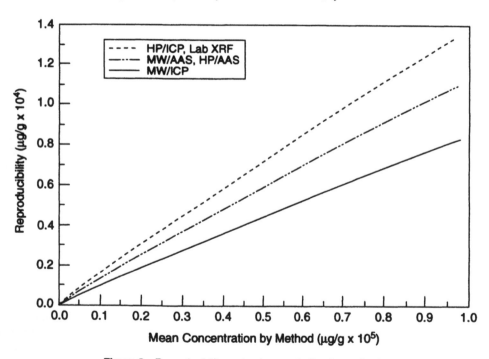

Figure 3 Reproducibility vs. lead concentration, by method.

reproducibility, calculated using this approach. Again, it should be noted that these are approximate estimates because repeatability and reproducibility do vary with concentration.

The differences in reproducibility for analysis by laboratory XRF and the extraction methods are attributed to the lack of standard procedures for Laboratory XRF analysis, as opposed to use of standard procedures for the extraction procedures. A quadratic tendency is evident in the laboratory-specific recovery plots for Lab XRF analyses (excluding SRM samples). This suggests that instrument calibrations may have been carried out with an inadequate number of standards.

Table 7 Estimates of Repeatability and Reproducibility by Method (% Variation)

Method	Repeatability	Reproducibility
MW/AAS	10.7	13.7
HP/AAS	9.7	17.2
MW/ICP	10.5	11.7
HP/ICP	12.9	21.0
Lab XRF	4.8	19.4

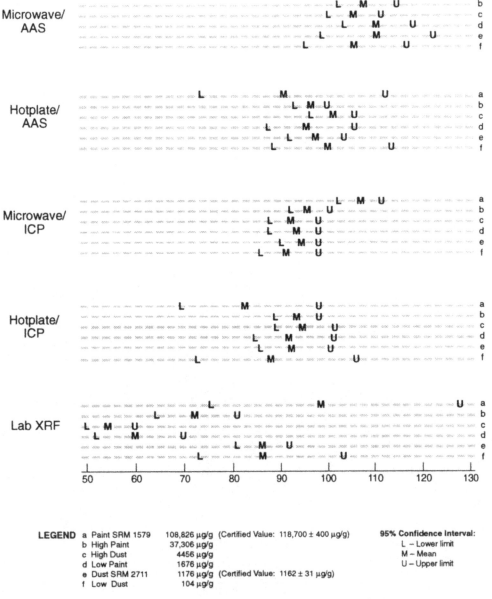

Figure 4 95% confidence interval for the geometric mean recovery (%), by method.

Figure 4 shows the 95% confidence intervals of the geometric mean recoveries for the five methods used in this study. The six horizontal lines associated with each method correspond (from top to bottom) with concentrations—SRM 1579, High Paint, High Dust, Low Paint, SRM 2711, and Low Dust. L, M, and U correspond to the lower end of the 95% confidence level, the mean value, and the upper end of the 95% confidence level, respectively.

E. PAIRWISE COMPARISON OF METHOD MEANS

Pairwise comparisons of method means within each of the six samples were performed using ordinary nonsimultaneous t tests at the 5% significance level. There are 10 possible paired comparisons of methods within each of the six samples (60 total comparisons), so that three false rejections of the hypothesis of no difference would be expected by chance alone. Table 8, which summarizes the results of the pairwise comparisons, indicates that no differences were declared in connection with the low-dust sample, and only two differences were declared for the paint SRM samples. Of 28 declared differences, 26 involve MW/AAS and Laboratory XRF. These results are consistent with the order (or ranking) of the methods, with respect to mean values (see Tables 3 and 4). MW/AAS has the highest mean for five or six test materials. The chance of this happening was 0.000064 if all the methods were equivalent. It is significant that Lab XRF has the lowest mean for five of the six samples.

F. COMPARISONS OF AAS AND ICP RESULTS

Previous work had shown low recoveries for NIST SRM 1579 with ICP analysis, while no such effects were observed for AAS measurements.[10] Because of this bias, attributed to matrix-caused signal suppression, laboratories had been instructed to dilute solutions for ICP analysis into the range of 1 to 10 µg Pb/ml. Despite this procedure, results determined by the MW/AAS method were higher than results from the MW/ICP method, by 3 to 18%, as noted in Tables 3, 4, and 5.

The difference in MW/AAS and MW/ICP results was examined in a preliminary study that showed that the MW/ICP results obtained using a total-digestion method (HNO_3, HCl, HF) agreed closely with the round-robin MW/ICP results, but were consistently lower than the MW/AAS results. Therefore, an explanation for the positive bias of the AAS results, relative to the ICP results observed in the round-robin, was proposed as the lack of background correction by a number of AAS laboratories. The existence of matrix effects not eliminated by the required dilution into the 1 to 10 µg Pb/ml range was also proposed as an explanation for ICP signal suppression. The potential contribution of matrix effects to ICP signal suppression was not investigated further in this study.

V. SUMMARY AND CONCLUSION

The round-robin study showed that the methods of preparation yielded homogeneous materials with lead concentrations in the targeted ranges. The hypothesis of homogeneity was accepted in 19 of 20 cases. One rejection in 20 is expected by chance alone when testing at the 95% confidence level. In 16 of 20 cases the sampling component of variance estimate was zero. In the other four cases the sampling component was 10% or less of the total. On the average, the sampling component accounted for 1.37% of the total variance.

The five analytical methods produced somewhat different results, on the average. AAS results were consistently higher than ICP results. Explanations include the lack of background correction for AAS measurements and/or ICP signal suppression attributed to matrix effects, even though sample dilution was required. Laboratory XRF results were decidedly lower than those of the extraction methods, while MW/AAS tended to produce the highest results. An examination of significant differences in analytical results showed, in fact, that MW/AAS and Laboratory XRF produced the most significant differences in results. Laboratory XRF results showed an apparent negative bias relative to the AAS and ICP methods, attributable to the lack of standardized procedures for sample preparation and analysis. Furthermore, a quadratic tendency was evident in the laboratory-specific recovery plots for Laboratory XRF (excluding SRM samples). This suggests that the calibrations varied from laboratory to laboratory and that an inadequate number of standards was used for calibration.

Laboratory XRF was significantly more repeatable than were the extraction/atomic spectroscopic methods. MW/AAS, HP/AAS, and MW/ICP had very similar repeatability characteristics, while HP/ICP appeared less repeatable. The performance of the HP/ICP procedure may be attributed to variation in hotplate temperature, possible loss due to spattering or "bumping," and variation in the technique by which volumes are reduced during the extraction process. MW/ICP had the best reproducibility for five of the six samples.

181

Table 8 MEMs and SRMs Identified to Differ Significantly by Sample-Specific, Pairwise Method Comparison

| | Method | | | |
	MW/AAS	HP/AAS	MW/ICP	HP/ICP
HP/AAS	Low Paint High Paint Dust SRM	—	—	—
MW/ICP	Low Paint High Paint High Dust Dust SRM	High Dust		—
HP/ICP	Low Paint High Paint Paint SRM High Dust Dust SRM	None	Paint SRM	—
Lab XRF	Low Paint High Paint High Dust Dust SRM	Low Paint High Paint High Dust Dust SRM	Low Paint High Paint High Dust	Low Paint High Paint High Dust

Based on the results of this study, MW/ICP was considered to be a method of choice because it showed good reproducibility (total system coefficient of variation less than 12%) and gave the least-variable recovery across concentrations (Figure 4).

REFERENCES

1. Williams, E. E., Grohse, P. M., Neefus, J. D., and Gutknecht, W. F., A Report on the Lead Reference Materials Workshop, EPA Contract No. 68D10009, U.S. Environmental Protection Agency, Washington, D.C., 1991.
2. Welcher, F. J., Ed., *Scott's Standard Methods of Chemical Analysis,* 6th ed. IIB, D. Van Nostrand, New York, 1963, 1680–1719.
3. Binstock, D. A., Hardison, D. L., Grohse, P. M., and Gutknecht, W. F., Standard Operating Procedures for Lead in Paint by Hotplate- or microwave-based Acid Digestions and Atomic Absorption or Inductively Coupled Plasma Emission Spectrometry, NTIS Publication No. PB 92–114172, EPA Contract No. 68–02–4550, U.S. Environmental Protection Agency, Washington, D.C., 1991.
4. Cross Beater Mill, Model SK1 (Dietz), Serial No. 71475, Glen Mills, Inc., Maywood, NJ.
5. Retsch Grinder, Model ZM1, Serial No. 33060, Oriden, Brinkman Instruments Co., Westbury, NY.
6. Turbula Blender, Model T2C, Serial No. 910880, Glen Mills, Inc., Maywood, NJ.
7. Ro-Tap Generator, Model No. 5KH35JN3132T, Rotary Model No. L143, Lid Model No. L45, Tapper Model No. L42, W.S. Tyler Ro-Tap, Fisher Scientific, Pittsburgh, PA.
8. U.S. Environmental Protection Agency, Office of Health and Environmental Assessment, Environmental Criteria and Assessment Office, Urban soil lead abatement demonstration project. EPA/600/AS-93-001, Volumes 1–4. U.S. EPA, Research Triangle Park, NC, 1993.
9. Boyer, D. M. and Hillman, D. C., Draft report, Standard Operating Procedures for Energy Dispersive X-ray Fluorescence Analysis of Lead in Urban Soil and Dust Audit Samples, Contract No. 68–CO–0049, U.S. Environmental Protection Agency, Washington, D.C., 1992.
10. Binstock, D. A., Hardison, D. L., White, J., and Grohse, P. M., Evaluation of atomic spectroscopic methods for determination of lead in paint, dust and soil, in Proc. of the 1991 U.S. EPA/AWMA Int. Symp., Measurement of Toxic and Related Air Pollutants, Air and Waste Management Association, Pittsburgh, 1991, 1058–1070.
11. Kleinbaum and Kuppa, *Applied Regression Analysis and Other Multivariable Methods,* Duxbury Press, North Scituate, MA, 1978.
12. Steiner, E. H., Planning and analysis of results of collaborative tests, in *Statistical Manual of the Association of Official Analytical Chemists,* Youden, N.J. and Steiner, E. H., Eds., Association of Official Analytical Chemistry, Arlington, VA, 1975.
13. Miller, R., *Simultaneous Statistical Interference,* Springer Verlag, New York, 1981; also SAS User's Guide: Statistics, Version 6, Cary, NC 1987.

Chapter 22

NIST-SRM 2579 Lead Paint Films for Portable X-Ray Fluorescence Analyzers

P. A. Pella, M. McKnight, K. E. Murphy, R. D. Vocke, E. Byrd, J. R. DeVoe, J. S. Kane, E. S. Lagergren, S. B. Schiller, and A. F. Marlow

CONTENTS

I. INTRODUCTION

The adverse health effects resulting from exposure of young children and adults to sources of lead in the environment is receiving increased attention. Studies have suggested that even exposure to low levels of lead can result in impairment of the central nervous system, mental retardation, and behavioral disorders. The principal sources of bioavailable lead are paint, dust, and soil, with lead-based paint being the major source of high-dose lead poisoning in the U.S.[1]

The U.S. Department of Housing and Urban Development (HUD) estimates that currently 57 million homes in the U.S. contain lead-based paint,[2] which poses a serious health threat, especially to young children. In an attempt to address this health issue, the Lead-Based Paint Poisoning Prevention Act, as amended by Section 566 of the Housing and Community Development Act of 1987,[3] requires testing of all painted surfaces in public and Indian housing, by 1994.

The need for lead paint reference materials for calibration of analytical methods for quantitative lead analysis was recognized by the National Institute of Standards and Technology (NIST) as early as 1970. In 1973 NIST issued Standard Reference Material (SRM) 1579, a powdered, lead-based paint collected from old housing undergoing renovation.[4] This material was recertified as SRM 1579a in 1991. In 1977 a set of lead paint reference materials was prepared for HUD by NIST, for assessing the operational performance of portable X-ray fluorescence (XRF) analyzers. The use of portable XRF analyzers to test painted surfaces is recommended as part of HUD's interim guidelines for lead-based paint hazard identification and abatement;[5] however, the reference materials prepared in 1977 are no longer available. Therefore, a new lead paint film set, SRM 2579, was produced, in cooperation with HUD, to fulfill this continuing need and is described in this chapter. SRM 2579 consists of a set of five 7.6 × 10.2-cm (3 × 4-in) mylar sheets, four of which are coated with a single paint layer of different lead content. The paint layers and the mylar sheets are 0.04 mm and 0.2 mm thick, respectively. The fifth sheet is coated with a lead-free lacquer layer of the same thickness as the lead paint layers and is included as a blank. Each level was also colored with a different lead-free organic tint, to make them easily distinguishable when used in the field. All sheets are covered with a clear, thin, plastic laminate to protect the paint layers from abrasion. An automated process was used to prepare these paint films where a flowing paint stripe of fixed width was allowed to spread continuously at a controlled thickness on a mylar substrate fed from a large roll. After the paint dried, the mylar substrate was cut into sheets from which the SRM was produced. Known amounts of a lead chromate concentrate were diluted with a commercial paint vehicle to obtain the desired lead concentrations. SRM 2579 was fabricated by Munsell Color, Newburgh, NY, in accordance with NIST specifications. The lead concentration levels, in units of mass per unit area, relate to regulatory levels for lead abatement purposes.

II. CERTIFICATION STRATEGY

An important requirement of SRM 2579 is that the uniformity of the lead-paint layer on the substrate must be acceptable for its intended use. Prototype samples were prepared by Munsell so that we could characterize the uniformity of the paint layer as typically produced in the fabrication process. Samples from initial runs showed that there was a systematic, nonuniformity in the paint layer across the width of the paint stripe, as measured by wavelength-dispersive X-ray spectrometry (WDXRF). In some of the prototype samples, for example, the paint layer was thicker in the center of the paint stripe than at the edges. By reducing the width of the paint stripe in the fabrication process, the magnitude of radial nonuniformity was sufficiently small enough to provide material suitable for preparing this SRM.

The original intent for certification of the lead content of this SRM was to use isotope dilution mass spectrometry (IDMS) only. However, since it was necessary to characterize the nonuniformity of the paint layer, a large number of samples (e.g., 120) were measured by WDXRF for each level. IDMS measurements were made on six of these, representing the two lowest, two in the middle range, and two highest, X-ray counts for each level. A calibration line (WDXRF counts vs. IDMS concentrations) was then fit for each level. From each calibration line, a median value of X-ray counts was used to derive a robust central lead value together with conservative uncertainties, for each level.

III. XRF CHARACTERIZATION OF PAINT UNIFORMITY

A production run for each level consisted of 300 separate painted sheets where each sheet was sequentially labeled to identify its place in the production run. Each sheet was painted with a stripe (15 cm wide by 30 cm long) that was centered on the sheet. Of the 300 sheets, 30 sheets that were representative of the beginning, middle, and end of each run were selected for measurement by WDXRF. Each of the painted stripes on the 30 sheets was subdivided into a grid pattern of 2.5 cm × 2.5 cm squares along 12 rows (length) and 6 columns (width). The squares were individually labeled so that any WDXRF measurement could be referenced to its parent sheet and its location on that sheet. A 2.5 cm × 2.5 cm square was chosen for measurement because it represents the smallest area normally measured by portable XRF units. From one row on each sheet, selected at random, four squares were selected from the six columns, for a total of 120 squares for each level. At least one of the four squares taken from each sheet was from columns nearer the edges of the paint stripe, to better characterize the uniformity of the paint layer.

Figure 1 shows box plots of XRF counts plotted vs. the column (width) of the painted stripe for each level. In a box plot the 75th and 25th percentiles of the data are portrayed by the top and bottom of a rectangle. Vertical lines extend from the ends of the box to the most extreme data point, or 1.5 times the difference between the 75th and 25th percentiles, whichever is smaller. Any extreme value falling beyond these lines is plotted as a point and may indicate an outlier. Each box plot in Figure 1 summarizes measurements on approximately 20 samples for that column at the given level. The horizontal line connects the median value obtained for each column. As shown in the figure, variations in X-ray counts across the width of the stripe occur in addition to a number of outliers at each level (see, for example, one measurement of 13.75 kCps [counts per second × 10^3] on a sample from column 6 of level I). These outliers were included in the statistical treatment of the data, to obtain the certified values and corresponding uncertainties, and described in Section V.

IV. QUANTITATION OF THE LEAD CONTENT

IDMS measures a concentration that has units of weight per unit weight. The certified values for this SRM, however, must be expressed in units of mass per unit area, for use with X-ray analyzers. Therefore, it was necessary to accurately measure the area of each square sample measured by IDMS, for conversion to the proper units. First, eight points, two along each edge of the sample square, were digitized using an optical microscope at 30× magnification. The area of each of the square samples was then computed using CAD graphics software. When the lead concentrations of these squares were measured by IDMS on a weight-per-unit-weight basis and then converted to a weight per unit area, using the measured areas, the relative standard deviation of the averages increased by approximately 0.5% in all levels. This is most likely a reflection of the uncertainty of measurement of the actual paint surface area within each square sample.

An IDMS determination of lead concentration requires two separate analyses to fully characterize a sample, because lead is one of several elements that can show considerable variation in the natural

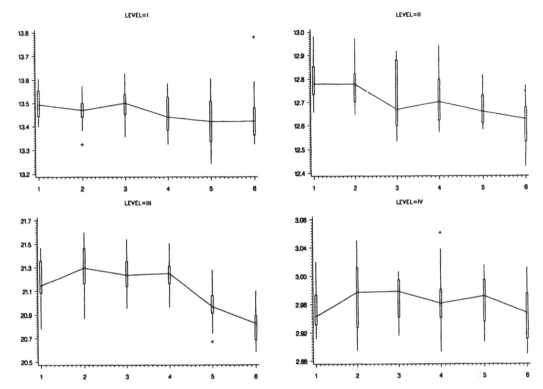

Figure 1 X-ray counts in kCps (counts per second × 10³) (y-axis) vs. column number (x-axis) for each level, I, II, III, IV, respectively.

abundance of its four isotopes, from one material to another. In order to measure the natural isotopic composition of lead in the paint film, four unspiked sample squares, one from each level, were wet ashed and electrochemically purified. The chemical purification step was necessary to obtain a stable, reproducible Pb^+ ion beam in the mass spectrometer. The average $^{208}Pb/^{206}Pb$ ratio of the four levels was measured with a relative standard deviation of 0.02% and indicated that the lead in the different levels were isotopically indistinguishable within measurement precision.

Lead concentrations were determined on six (2.5 cm × 2.5 cm) square samples selected from each level previously measured by WDXRF. Each square represented a unique point for developing a calibration curve between WDXRF counts vs. IDMS concentration, for that level. Each film was weighed by difference into clean teflon beakers, and an appropriate amount of isotopically enriched ^{206}Pb spike solution was added gravimetrically. The sample and spike were isotopically equilibrated by wet ashing with HNO_3, HCl, and $HClO_4$. The lead from these solutions was separated electrochemically and then redissolved in dilute acid, for introduction into the mass spectrometer.

The lead isotope ratios were measured on a thermal ionization solid source isotope ratio mass spectrometer. The samples were loaded with silica gel and phosphoric acid onto a single, outgassed Re filament. Typical sample loadings were approximately 100 ng of lead. SRM 981 was run as a control standard, to allow the isotopic ratios to be corrected for isotopic fractionation in the thermal source. Duplicate mass-spectrometric analyses of the same sample solution typically showed less than a 0.05% relative difference.

The data obtained from the perturbed lead ratios of the spiked samples, together with the isotopic composition data from the unspiked sample, were used to calculate a concentration value. The overall uncertainty in the measurement for each sample consisted of uncertainties in the measurement of the ratios, the calibration of the spike concentration, and uncertainties in the fractionation and blank corrections. The combined uncertainty estimate of the lead concentration for each level was no greater than 0.2% on a per weight basis.

Table 1 Certified Values for Lead in NIST SRM 2579

Level	Code	Certified value (mg/cm²)	Estimated uncertainty (mg/cm²)
I	Yellow	3.53	0.24
II	Orange	1.63	0.08
III	Red	1.02	0.04
IV	Green	0.29	0.01
Blank	Clear	<0.0001	

V. CERTIFIED VALUES AND UNCERTAINTIES

The certified lead concentrations and their corresponding uncertainties are given in Table 1. As mentioned previously in Section III, outliers were observed in the data shown in Figure 1 for levels I, III, and IV. The presence of outliers indicates that the data are not normally distributed. Therefore, a distribution-free tolerance interval was used here to generate conservative estimates of uncertainty. Typically, the lower and upper limits of this interval would be the lowest and highest measurements, respectively, but an additional component, due to the nonuniformity of the paint layers, was required. The tolerance interval for each level was obtained as follows. First, the data in Figure 1 for levels I and II were fitted as a linear function of column number, and the data for levels III and IV were fitted as a quadratic function of column number. The residuals from the respective fits reflected the random component of variation and are the data from which the tolerance intervals were derived. Then, in addition to the tolerance interval, a column bias component was estimated as one half the difference between the maximum and minimum least-squares-fitted values over the columns. The final lower tolerance limit was calculated as the median value for counts minus the absolute value of the minimum residual minus the column bias, while the final upper tolerance limit was calculated as the median value for counts plus the maximum residual plus the column bias.

The interval in counts is then converted to one in concentration units, using a calibration line obtained by least-squares fitting of XRF counts to IDMS concentrations. Errors in calibration were accounted for by an appropriate intersection of the count limits to 95% confidence bands placed on the calibration line (see Figure 2). The certified lead concentration was obtained by calibrating the median value of counts to concentration. Note that this interval is not symmetric about the certified value. Therefore, a conservative estimate of uncertainty was obtained as the larger of the difference of the upper limit and the certified value, or the difference of the certified value and the lower limit. In the absence of systematic error, the certified value plus or minus its uncertainty will contain 95% of the true lead concentrations, with 95% confidence.

VI. QUALITY ASSURANCE

To ensure that the previously tested production-run samples were representative of the final SRM samples that were delivered to NIST, a quality assurance test was performed. A random sample of 20 films from each of the four levels was selected to provide a check on the tolerance limits for each level. For levels I, II, and IV, all 20 measurements fell within the limits at the respective level. For level III, however, 3 of the 20 measurements fell outside the upper limit. With a 95% confidence, 95% coverage tolerance interval one would expect only one of twenty measurements to fall outside the limits. As a result, it was decided to measure each individual level III sample in order to remove any sample that exceeded the preselected tolerance limits. Because the size of the SRM samples did not permit direct measurements to be taken in the X-ray spectrometer used previously, measurements were made with a commercial energy-dispersive X-ray spectrometer (EDXRF) equipped with a sample changer appropriately modified for this purpose. As a result, the tolerance interval computed from the original XRF counts needed to be converted to one in EDXRF counts. This was done by recalibrating samples measured previously (20 of the original 120 samples were selected, spanning the range of lead content for level III films). From a least-squares fit of the EDXRF counts vs. counts obtained previously, the tolerance limits were converted to EDXRF counts and also adjusted for error. An adjustment was also made based on the measured X-ray attenuation (about 1.5% relative) of the lead Lα line by the laminate. The result of the EDXRF measurements was that about 20% of level III samples were found to exceed the preselected tolerance limits and were rejected.

In conclusion, SRM 2579 paint films on mylar should prove to be more durable for use in the field, than previous films prepared in 1977, and is an important factor for checking the operational performance of lead-paint portable XRF analyzers over a long period of time. The magnitude of the uncertainty for each

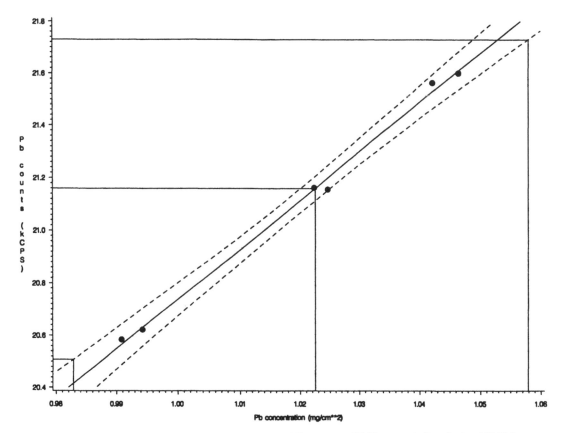

Figure 2 Calibration line with 95% confidence bands for XRF counts vs. IDMS concentrations for level III. Tolerance limits and median value for counts intersect the calibration bands and line to give corresponding tolerance limits and certified lead concentration.

certified level in this SRM is considered small enough for its intended use and reflects the current state-of-the-art in fabrication.

ACKNOWLEDGMENT

The authors wish to thank Ron Morony of HUD for the financial support for the development of this SRM under IAG DU-100I91–0000026.

REFERENCES

1. Comprehensive and Workable Plan for the Abatement of Lead-Based Paint in Privately Owned Housing, report to Congress, U.S. Department of Housing and Urban Development, Washington, D.C., 1990.
2. Lead-Based Paint Poisoning and Prevention Act, as amended by section 566 of the Housing and Community Development Act of 1987, United States Code, Title 42, the Public Health and Welfare.
3. Lead-Based Paint: Interim Guidelines for Hazard Identification and Abatement in Public and Indian Housing, Office of Public and Indian Housing, Department of Housing and Urban Development, Washington, D.C., 1990.
4. Greifer, B., Maienthal, E.J., Rains, T.C., and Rasberry, S.D., Development of NBS Standard Reference Material: No. 1579, Powdered Lead-Based Paint, NBS Special Publication 260–45, U.S. Department of Commerce, National Bureau of Standards, Washington, D.C., 1973.
5. Lead-Based Paint: Interim Guidelines for Hazard Identification and Abatement in Public and Indian Housing, Office of Public and Indian Housing, Department of Housing and Urban Development, Washington, D.C., 1990.

Chapter 23

Vacuum Sampling of Settled Dust for Lead Analysis

B. S. Lim, J. G. Schwemberger, P. Constant, and K. Bauer

CONTENTS

I. INTRODUCTION

In Section 566 of the Housing and Development Act of 1987, Congress required that U.S. Department of Housing and Urban Development (HUD) report to Congress their findings and recommendations from a demonstration of different strategies for abating lead-based paint (LBP) in HUD-owned, single-family housing.

The Environmental Protection Agency (EPA) Office of Pollution Prevention and Toxics (OPPT) is currently conducting a study, the Comprehensive Abatement Performance Study (CAPS), at some HUD demonstration houses, to assess the performance of the abatements after the houses have been reoccupied. The performance of the abatements will be addressed by measuring the levels of lead in dust and soil. To prepare for this study, EPA conducted a pilot study in Denver in 1991.[1] One of the objectives of the pilot study was to test and assess the performance of the sampling and analysis protocols. In the pilot, field personnel observed that the dust collection device used in HUD's National Survey had limitations; namely, inadequate collection efficiency and versatility. The dust of interest in the National Survey was loose surface dust that is transferable from hand to mouth. However, for the pilot study, dust was defined as all loose material on the surface within the designated area from which a sample was to be collected, resulting in dust with much smaller particle sizes.

An exploratory investigation was undertaken to (1) determine the capabilities of the original dust collector with the blue nozzle used on HUD's National Survey,[2] (2) identify the critical parameters needed to be considered in the modification of the blue-nozzle dust collector for the EPA CAPS, and (3) fabricate the modified blue nozzle or design and fabricate a new dust collector.

II. DUST COLLECTOR DESIGN

The newly designed dust collector must meet the following performance and operational requirements, to satisfy the pilot study objectives:

1. Ability to collect a sample of dust particulates $\leq$ 250 to 2000 μm in size, from 1-ft^2 area in 2 min or less
2. Overall average dust collection efficiency of 85% or more for carpets, wood floors, linoleum-covered floors, concrete, window sills, and window channels
3. Low cost, lightweight, portable, easy to use, 110-vac powered, and capable of collecting up to 2 g of dust

The dust collector that uses the blue nozzle is shown in Figure 1. This collector has three major units: the nozzle, the vacuum pump, and the filter cassette, which collects the dust. Tubing interconnects the three

192

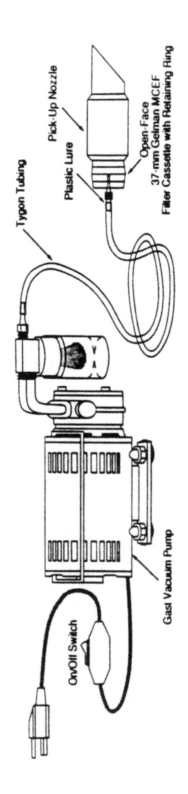

Blue Nozzle Dust Collector

Figure 1 Schematic of dust collector used in pilot study.

major components, and a vacuum gauge is connected to the pump, to provide a means to monitor the vacuum. Two additional dust collectors were designed. These new designs are shown in Figures 2 and 3. Figure 2 is a schematic drawing of a vacuum-driven dust collector that has a 37-mm cassette with a preloaded, 0.8-μm, cellulose-ester membrane as a modification of the blue nozzle dust collector. Figure 3 is a schematic of a vacuum-driven cyclone separator dust collector.[3] The cyclone dust collector is made of PVC pipe and pipe fittings. Figure 4 is an engineering blueprint of the newly designed cyclone dust collector. The cassette is located in the cassette holder plug, which screws into the bottom of the cyclone sampler case. The sampler case is a pipe reducer with an end cap attached to the 4 1/2-in. portion. The nozzle is a short length of 1-in. PCV pipe connected to the inlet of the cyclone sampler case, via a 90-degree elbow. The discharge unit is a 1 1/2-in. coupler inserted into a machined hole at the top of the cyclone sampler case (4 1/2-in. end cap). A 1-in. reducer is placed into the discharge end of the coupler, to accommodate a 1-in. short piece of pipe. A 110-vac, 60-Hz, 2-amp, commercially available vacuum source is connected to the cyclone discharge end, via the appropriate size pipe and fittings. This dust collector utilizes the same-type filter cassette as the in-line dust collector. It is located at the bottom of the cyclone collector. Since the filter actually serves no useful function in this design, the filter cassette can be replaced by some other appropriate collection unit.

III. TEST OF DESIGN PROTOTYPES

The in-line dust collector operates on the principle of impaction. The air enters the nozzle and flows through the filter cassette. The particulate matter in the air stream impacts onto the filter in the cassette, and the air is discharged from the outlet of the cassette. The pore size of the filter is the determining factor for the size of particles discharged from the outlet of the cassette. The particles are submicron in size: the size of particulate matter that most likely remains airborne.

The cyclone collector operates on a different principle from that of the in-line dust collector, which is an impactor. The dust enters the sampler body tangentially at a relatively high velocity and proceeds in a rotary direction. The air within the body forms a vortex, travels up, and is discharged through the top of the sampler case. This air is discharged at a relatively low velocity. The dust moves downward within the collector's body and is discharged at the lower end of the cone, into a collector container.

The forces that separate the dust from the discharging air are centrifugal and gravitational. This separation is possible because the air velocity is reduced considerably after it enters the body of the cyclone, thus allowing the forces to separate the particles from the air stream. Some very fine particulate matter is discharged from the device. The amount of the fine particulate discharged is a function of the amount picked up and the design of the cyclone collector.

IV. EXPERIMENTAL DESIGN

Three factors were considered in the experimental design for the laboratory test to estimate dust collection efficiency collector type, particle size, and surface sampled.

One blue nozzle, two in-line collectors, and two cyclones, for a total of five dust collectors, were tested.

Sand, soil, and paint chips were collected and prepared for the dust-collector test. The sand and soil were separated into particulate ranges by using a series of sieves. Paint chips were collected from residences in the Kansas City area. Most of them were 1/4 in. in size. Less than 10% were from 1 to 2 in.^{2} in area. The chips were quite brittle and broke easily. They were placed into a clear plastic bag and hand crushed. Each portion by weight of each of the different size ranges of sand and soil were taken to prepare three composites. Paint chips (~0.9 g) were added to each of the following composites for a particular run.

Three types of composite materials covering the range of less than 250 to 2000 μm were sampled: Composite C_1 (particles of size <250 μm plus paint chips), Composite C_2 (particles of size >250, but <2000 μm plus paint chips), and Composite C_3 (<2000 μm plus paint chips).

Five types of surfaces were used: wood floor, linoleum, concrete, carpet, and window sill.

A full factorial design requiring $5 \times 3 \times 5 = 75$ unique runs was selected. Two replicates of each combination were performed so that some measure of the reproducibility of the dust collection procedure could be estimated. Thus, a total of 150 runs followed, with two exceptions: one additional run was made on the in-line dust collector, and one was made on the cyclone dust collector. The former was needed because of incorrect sampling, and the latter was to verify performance of cyclone on a concrete surface that provides a collection efficiency of over 100%. The runs were scrambled so that the two prototypes of

194

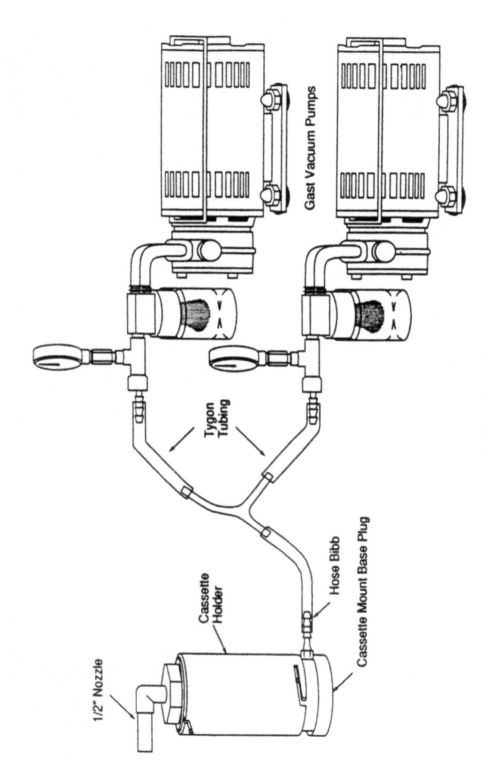

1/2" Nozzle

Cassette
Holder

Tygon
Tubing

Gast Vacuum Pumps

Hose Bibb

Cassette Mount Base Plug

In-Line Dust Collector

Figure 2 Schematic of vacuum dust collector no. 1.

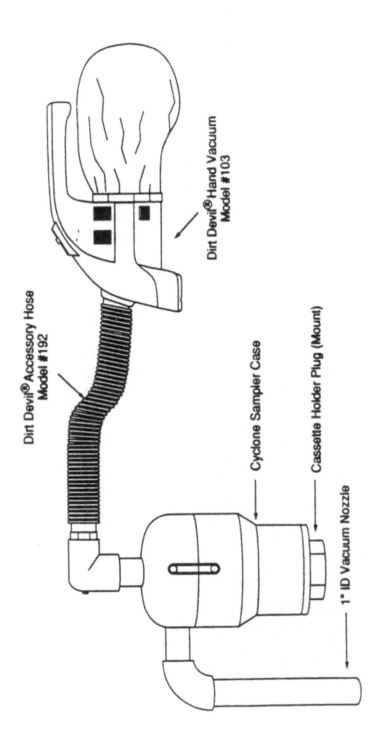

Figure 3 Schematic of vacuum dust collector no. 2.

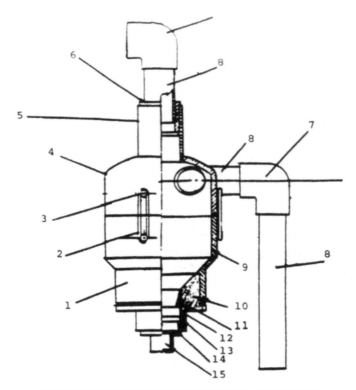

CEMENT NOVA WELD "P" BY GENOVA	
1	4" x 3" PVC PIPE RED. COUP. MOD. COUP.
2	1/8" x 1 3/8" "O" RING
3	FASTENING PLUG
4	4" PVC PIPE CAP. MOD.
5	1 1/4" PVC PIPE COUP.
6	1 1/4" x 1" PVC BUSHING
7	1" PVC 30o ELBOW
8	1" PVC PIPE 160 PSI 73°F 73°F
9	
10	3" x 1 1/2" PVC BUSH. MOD.
11	SPACER RING
12	SPACER RING
13	1 1/2" FIP x 1 1/2" PVC ADAP. MOD.
14	1/8" x 1 5/8" "O" RING
15	1 1/2" PVC PIPE PLUG MOD.

Figure 4 Engineering blueprint of cyclone dust collector.

each design were not run consecutively throughout the test. This was done to eliminate any operator bias in trying to improve the vacuuming technique with repeated runs.

V. EXECUTION OF TEST

Five principle steps were taken in performing the test: (1) weighing the cassettes before they were used to collect dust, (2) applying dust to a surface, (3) collecting the dust from a surface, (4) weighing the dust sample, and (5) recording data.

The dust was prepared by measuring out a 0.9-g aliquot of the dust composite to be used for the run and added to these paint chips to bring the aliquot to approximately 1 g. The filter cassette was placed into the dust collector, and the aliquot of dust was applied to the 1-ft^2 inscribed area of the surface (wood floor, linoleum, concrete, and carpet). Dust samples were taken from a defined 1-ft^2 area of the surface in overlapping passes, starting from top left, moving to the right, returning to the left less than one nozzle diameter below. This was continued until the right bottom corner was reached. The process was repeated starting at the lower left-hand corner going upward.

After the dust was collected with the cyclone dust collector and the a.c. power was turned off, the operator tapped the outside surface of the cyclone sampler case several times with a metal rod ($\sim$1/4-in. diameter). This operation caused any fine particulate that had collected onto the interior surface of the case to fall into the filter cassette.

At the completion of the run, the cassette was removed from the dust collector and given to the analyst, who weighed the cassette and recorded its weight on the appropriate form.

VI. TEST RESULTS

A total of 152 runs were performed. These included two replicate runs for each of the 75 unique combinations, plus two additional runs. For each run the collection efficiency (%) was calculated as

$$\text{Efficiency (\%)} = \frac{(\text{Cassette Wt. After Use} - \text{Cassette Wt. Before Use})}{\text{Wt. of Composite Applied to Surface}}$$

Carpet fibers were vacuumed into the cassettes during some carpet runs. In these cases, after a cassette was weighed with dust and the accompanying carpet fibers, the fibers were removed by hand, and the cassette was reweighed. An adjusted efficiency was calculated based on that reduced weight. Both unadjusted and adjusted efficiency results are shown in Figures 5, 6, and 7 for these runs.

The absolute and relative variations due to replication, for the three types of dust collectors, are shown in Table 1. These replication errors are for wood floor, linoleum, concrete, and carpet samples.

VII. INTERPRETATION OF STATISTICAL RESULTS

The results from the dust collection tests were analyzed by analysis of variance. A 5% significance level was assumed for all significance tests and comparisons. The analysis of variances was performed to estimate the effects of collection type, surface type, dust composite type, and their interactions, on the collection efficiency. In addition, the two prototype cyclone dust collectors were analyzed for differences in their dust collection efficiency, and it was found that they were not different. The results of the statistical analysis are summarized in Table 2.

A. BLUE-NOZZLE DUST COLLECTOR

Among the three dust collectors tested, the blue nozzle shows the greatest variability in collection efficiency, when applied to a variety of surfaces and dust composite types. It does not achieve the minimum required 85% collection efficiency, regardless of the surface and size of dust composite particle sizes. This is shown by the fact that the lower 95% confidence limit to the mean efficiency is below 85% in all test cases.

B. IN-LINE DUST COLLECTOR

The in-line dust collector's efficiency varies significantly between carpet and all other surfaces, including the window sill. This holds true for all dust particle sizes. It does not achieve the minimum required 85%

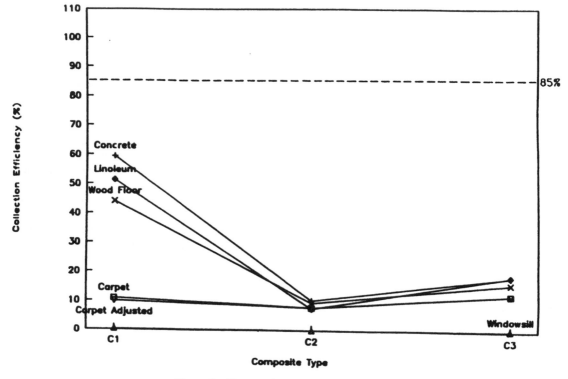

Figure 5 Blue-nozzle dust collector efficiency.

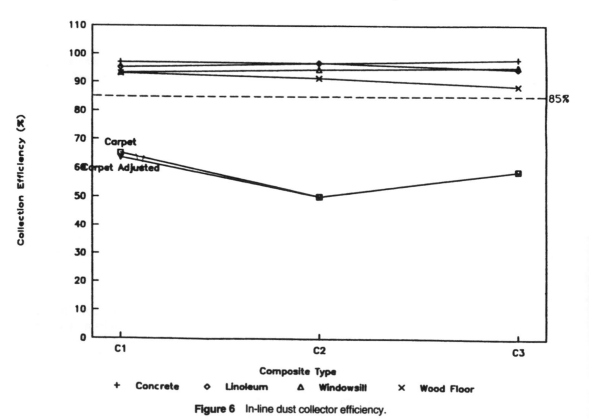

Figure 6 In-line dust collector efficiency.

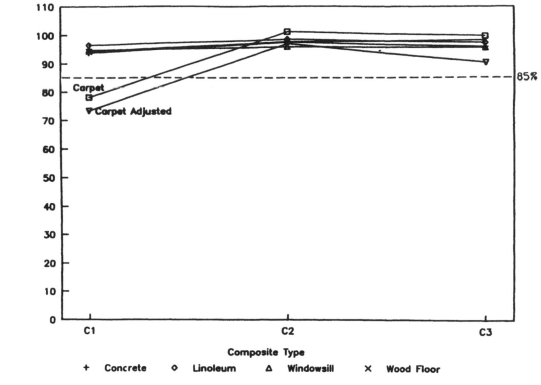

Figure 7 Cyclone dust collector efficiency.

Table 1 Absolute and Relative (%) Replication Errors

Collector type	Without adjustment		With adjustment	
Blue-nozzle	2.75	15.5%	2.61	14.8%
In-line	3.93	4.5%	3.94	4.5%
Cyclone	2.93	2.3%	2.25	2.4%

collection efficiency on carpet, regardless of dust composite particle sizes. However, for all smooth surfaces and the window sill, the in-line dust collector achieves a high average efficiency of 94.5%, regardless of dust particle sizes, thus significantly exceeding the required minimum of 85%.

C. CYCLONE DUST COLLECTOR

Overall, the cyclone dust collector's performance slightly exceeds that of the in-line dust collector. Except for small-sized particles on carpet, the efficiency of the cyclone dust collector significantly exceeds the required minimum of 85%. Excluding carpet, the smallest lower 95% confidence limit to the mean is 94.0%. For small particles, i.e., composite C_1-particles of size of less than 250 μm plus paint chips, the cyclone dust collector's mean efficiency of 78.2% does not meet the required minimum of 85% on carpet.

VIII. CONCLUSIONS

The principal conclusions are (1) the blue-nozzle dust collector is not suitable for the CAPS because of its low dust-collection efficiency; (2) the in-line dust collector is not adequate for the CAPS because the time required to collect a sample of dust from 1 ft^2 exceeds 2 min (low-rate surface coverage) due to the small nozzle inlet opening; and (3) the newly designed cyclone dust collector is suitable for the study.

Further studies are needed to characterize its performance as well as the performance of other vacuum samplers.

Table 2 Collection Efficiency (%) by Dust Collector and Surface Type

| Collector type | Surface type | Collector meets 85% requirement? | Number of runs | Mean | Standard error of mean | Collection Efficiency (%)[a] | | | | |
| | | | | | | 95% Confidence limits to the mean | | Minimum | Maximum | Relative standard deviation (%) |
						Lower	Upper			
Blue nozzle	Concrete	No	6	29.3	9.70	4.40	54.3	10.2	60.8	81.0
	Carpet	No	6	10.3	1.26	7.06	13.6	6.84	14.4	30.0
	Carpet (adjusted)[b]	No	6	10.0	1.10	7.19	12.8	6.84	12.6	26.9
	Linoleum	No	6	25.7	8.41	4.12	47.4	4.29	53.4	80.0
	Wood floor	No	6	23.1	6.75	5.79	40.5	8.81	44.8	71.4
	Window sill	No	6	0.21	0.06	0.04	0.37	0.02	0.43	74.9
In-line	Concrete	Yes	12	97.3	0.54	96.1	98.5	92.4	99.5	1.9
	Carpet	No	12	58.0	3.20	51.0	65.1	37.7	73.2	19.1
	Carpet (adjusted)	No	12	57.5	3.11	50.7	64.4	37.7	72.1	18.8
	Linoleum	Yes	12	95.5	0.93	93.5	97.6	86.9	99.4	3.4
	Wood floor	Yes	13	91.0	1.05	88.8	93.3	84.0	96.6	4.1
	Window sill	Yes	12	94.4	0.50	93.3	95.5	92.4	98.2	1.8
Cyclone	Concrete	Yes	13	96.0	0.77	94.3	97.6	92.5	104.1	2.9
	Carpet	Yes	12	93.1	3.41	85.6	100.6	70.3	105.8	12.7
	Carpet (adjusted)	No	12	87.0	3.21	79.9	94.1	65.4	97.9	12.8
	Linoleum	Yes	12	97.5	0.28	96.9	98.1	96.1	99.4	1.0
	Wood floor	Yes	12	96.8	0.68	95.3	98.3	92.4	99.2	2.4
	Window sill	Yes	12	95.5	0.42	94.6	96.4	93.0	98.0	1.5

[a] Statistics are calculated across prototypes, replicates, and dust composite types.
[b] Adjustment is made for carpet fiber weight collected on cassettes.

REFERENCES

1. Comprehensive Abatement Performance Pilot Study, Vol. 1, Results of Lead Data Analyses: Draft Final Report, U.S. Environmental Protection Agency, Washington, D.C., 1992.
2. Analysis of Soil and Dust Samples for Lead, U.S. Environmental Protection Agency, Washington, D.C., 1991.
3. Constant, P. and Bauer, K., Engineering Study to Explore Improvements in Vacuum Dust Collection, Contract No. 68-DO-0137, U.S. Environmental Protection Agency, Washington, D.C., 1992.

Chapter 24

Analysis of Factors Contributing to Lead in Household Dust: Accounting for Measurement Error

B. Price and E. C. Baird, III

CONTENTS

I. INTRODUCTION

Lead in household dust (PbD), lead in soil (PbS), and lead-based paint (LBP) have been noted as the most direct sources of Pb causing elevated blood levels (PbB) in children.[1-2] PbS has been associated with industrial emissions, leaded gasoline emissions, and deteriorated exterior LBP. PbD has been associated with PbS tracked or blown inside and LBP on window frames or walls that have deteriorated or have been disturbed during renovation or repair. Research efforts to identify sources of Pb that are the most active contributors to human exposure rely on statistical analysis of Pb concentration data collected for various media. Measurement error is an inherent characteristic of Pb concentration data, which if ignored in statistical analysis may lead to incorrect inferences.

This paper is a review of previous research conducted by Price Associates, Inc. (PAI)[3] that addressed the effects of measurement error in statistical analysis of relationships between PbD, LBP, and other variables that are potential sources of Pb exposure. Its purpose is to alert researchers conducting Pb exposure studies, to the following: (1) the magnitude of measurement error in Pb concentration measurements can be substantial, and (2) ignoring measurement error in statistical analysis may lead to incorrect conclusions.

The remainder of the paper is divided into four parts. First, results obtained in the previously conducted review addressing relationships among PbD, LBP, and other Pb exposure variables are summarized. The results described are the types of results that typically would be affected by measurement error. Second, characteristics of measurement error and its effects are presented. Next, a model for Pb concentration measurements, that incorporates measurement error, is outlined. The concluding section contains a discussion of general recommendations for addressing measurement error in statistical analysis of Pb variables.

II. OVERVIEW OF RESULTS FROM PRIOR STUDIES: RELATIONSHIPS BETWEEN PbD AND SOURCES OF Pb

A review of prior studies of Pb exposure was conducted by PAI,[3] with two objectives in mind: (1) to identify data that could be used to analyze the relationships between PbD and sources of Pb, and (2) to identify quality control data that would provide a basis for assessing measurement error in statistical analyses of PbD vs. sources of Pb. Concerning the first objective, only one study[4] was identified that directly analyzed the effects of Pb sources on PbD. A few studies[5-8] were identified that included frequency or correlation tables of PbD measurements and variables representing environmental sources of Pb. The raw measurement data from these studies were not readily available; therefore, a complete reanalysis of

204

Table 1 Classification of U.S. Privately Owned Dwelling Units, by Lead Content on Painted Surfaces and Lead in Household Dust

Category	PbD⁺	PbD⁻
	# DUs in (1000s)	
Any LBP⁺	9,950	47,420
Interior LBP⁺ only	671	10,013
Exterior LBP⁺ only	2,546	15,423
LBP⁻	723	19,084
Intact	6,724	51,787
LBP⁺	6,214	37,336
LBP⁻	510	14,449
Not intact	3,949	14,719
LBP⁺	3,736	10,084
LBP⁻	213	4,635

Notes: LBP⁺ = painted surface with Pb content greater than 1 mg/cm², by XRF.
LBP⁻ = painted surface with Pb content less than 1 mg/cm², by XRF.
PbD⁺ = household dust with Pb content above HUD guideline limits (see Reference 7, pp. 3–12).
PbD⁻ = household dust with Pb content below HUD guideline limits.

Source: Comprehensive and Workable Plan for the Abatement of Lead-Based Paint in Privately Owned Housing: Report to Congress, Office of Policy Development and Research, U.S. Department of Housing and Urban Development, Washington, D.C., 1990, 3–13 and 3–16.

Table 2 Odds Ratios Relating Lead in Household Dust to Lead Content on Painted Surfaces, in U.S. Privately Owned Dwelling Units

Presence of LBP with Pb above 1mg/cm² (LBP⁺)	Odds ratio for PbD above HUD limits (PbD⁺)
Any LBP⁺	5.54
Interior only	1.77
Exterior only	4.36
Intact	4.71
Not intact	8.06

Notes: Odds ratios calculated from data in Table 1.
LBP⁺ = painted surface with Pb content greater than 1 mg/cm², by XRF.
PbD⁺ = household dust with Pb content above HUD guideline limits (see Reference 7, pp. 3–12).

statistical relationships based on the underlying data was outside the scope of the review. Also, little, if any, information was provided in the reports concerning measurement error. The frequencies and correlations, however, allowed a limited investigation of relationships between PbD and Pb source variables. Concerning the second objective, quality control (QC) data useful for assessing measurement error were reported by Westat,[8] McKnight et al.,[9] and MRI.[10]

A. ANALYSIS OF FREQUENCY DATA

Among the studies reviewed, the most completely documented frequency data were collected in a national survey of privately owned housing built prior to 1980, conducted for HUD.[7] Referred to subsequently as the "National Survey," data were collected using a probability sample of 284 dwelling units (DUs) from a universe of approximately 77 million DUs. Measurements were obtained on the Pb content of interior and exterior paint, condition of painted surfaces, PbS, and PbD. Details of the sampling plan and measurement methods are described in HUD,[7] Westat,[8] and MRI.[10]

HUD[7] provides tables showing national estimates of DUs with LBP above the HUD limit of 1 mg/cm² (subsequently denoted as LBP⁺) and PbD above the HUD Pb concentration limits for household dust (subsequently referred to as PbD⁺). The LBP data also are classified as "interior LBP⁺ only" vs. "exterior LBP⁺ only," and "intact" surfaces vs. "not intact" or damaged surfaces. National totals projected from the survey data are shown in Table 1.

To investigate relationships between PbD levels and suggested predictors of PbD, the totals in Table 1 were translated into odds ratios (Table 2). An odds ratio is an index for summarizing count data to show the strength of a relationship between two variables. The "odds" of finding a DU with PbD⁺ is the ratio of the proportion of DUs with PbD⁺ to the proportion of DUs with PbD⁻. To investigate the relationship between PbD and LBP, odds would be estimated separately for DUs with LBP⁺ and for DUs with LBP⁻. The ratio of these odds provides an indication of the likelihood of DUs with PbD⁺ where the Pb concentration in paint is above the HUD limit (LBP⁺), relative to the likelihood of DUs with PbB⁺ and LBP⁻. An odds ratio approximately equal to one indicates that the presence of LPB⁺ in a DU does not increase the likelihood that the DU also will have PbB⁺. Stated another way, an odds ratio approximately equal to one would be interpreted as evidence of no relationship between the Pb content of painted surfaces and PbD. An odds ratio greater than one indicates a positive relationship between the variables (i.e., PbD⁺ is more likely in DUs with LBP⁺ than in DUs with LBP⁻).

The odds ratios in Table 2 show a positive relationship, in general, between LBP and PbD. The relationship is more pronounced where LBP⁺ is on exterior surfaces only, a ratio of 4.36, vs. 1.77 for LBP⁺ on interior surfaces only. The relationship between LBP and PbD also is more pronounced where surfaces with LBP⁺ are damaged ("not intact"), a ratio of 8.06, vs. 4.71 where painted surfaces are not damaged ("intact"). Statistical tests to compare the odds ratios to one and to each other were not conducted, because these tests are based on the observed counts, which were not reported in HUD.[7]

B. REGRESSION ANALYSIS

A number of the studies[4-8] reviewed used regression analysis to investigate relationships among Pb exposure variables. Only one of the studies, Bornschein,[4] directly analyzed the relationship of PbD vs. environmental sources of Pb. Reports describing the other studies, although not providing raw measurement data, recorded correlations among Pb variables. As described by PAI,[3] the correlations were used to estimate regression relationships with PbD as the response variable.

In summary, the regression results pointed to the following variables as predictors of PbD: condition of painted surfaces; LBP hazard index (combines Pb content of a painted surface with an index of surface condition); exterior LBP; window-sill LBP; PbS near the DU and PbS in general; and building age. These findings are intended to be treated as preliminary until a statistical analysis, using the raw data and accounting for measurement error, is completed. Although information on measurement error was not provided in the study reports, the potential impacts of including measurement error in the analysis are suggested in the example presented in the next section.

III. MEASUREMENT ERROR AND ITS EFFECTS ON STATISTICAL INFERENCE

All measurements of Pb exposure variables potentially are subject to measurement error. Measurement methods generally have some degree of inherent bias and exhibit variability in replicate measurements of a single sample. Even where measurement bias is negligible, measurement variability alone can have a significant impact on statistical analysis results. Measurement error may lead to misclassification of DUs: for example, with respect the Pb content of paint. Since Pb content measurements vary around the true Pb content, a measurement may indicate that a DU is LBP⁺ when in fact the DU is LBP⁻, or vice versa. Datasets that contain misclassified DUs are likely to lead to a biased estimate of the proportion of DUs with LBP⁺ and, subsequently, to biased estimates of odds ratios and other statistical summaries of count data. In regression analysis, which is one of the primary statistical methods used for assessing sources of Pb, ignoring measurement error may result in biased estimates of regression coefficients. More importantly, standard errors for the estimated regression coefficients may be understated, and as a consequence, the regression results may incorrectly identify significant predictors of Pb exposure.

A. COUNT DATA

Measurement error affecting LBP and PbD measurements may lead to the misclassification of DUs with respect to LBP⁺/LBP⁻ and PbD⁺/PbD⁻ categories. Estimates of proportions and, in turn, odds ratios are likely to be biased if the observed data, unadjusted for misclassification error, were used.

The following example demonstrates how measurement error leads to misclassification. To simplify the example, consider a study intending to estimate the proportion of DUs with LBP⁺ (i.e., the proportion of DUs with painted surfaces having Pb concentration in excess of 1 mg/cm²). Surfaces in a sample of n randomly selected DUs would be classified as LBP⁺ or LBP⁻ based on spectrum X-ray fluorescence (XRF) analyzer measurements. A DU would be classified as LBP⁺ if any surface in the DU had an XRF-measured Pb concentration of at least 1 mg/cm²: the HUD LBP limit. XRF measurements typically exhibit some

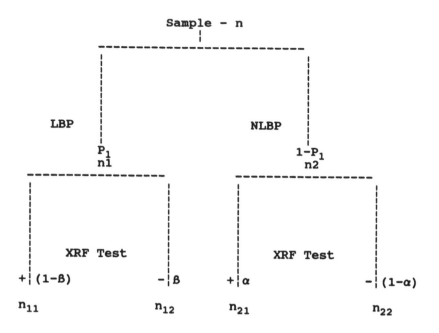

Figure 1 Distribution of survey responses resulting from misclassification due to XRF measurement error.

degree of bias and some degree of inherent variability, in repeated measurements. As a consequence, XRF measurements may result in both false-positive and false-negative classifications. A false positive occurs if a DU with Pb less than 1 mg/cm² on a painted surface is classified, on the basis of XRF measurements of that surface, as LBP⁺. A false negative occurs if a DU with Pb more than 1 mg/cm² is classified, by XRF measurements, as LBP⁻. The probability of a false-positive classification, subsequently denoted by α, and the probability of a false-negative classification, subsequently denoted by β, are both determined by the degree of measurement error in the XRF measurement method. The misclassification probabilities and resulting count data contaminated by misclassified DUs are the causes of biased estimates of proportions and subsequent biased statistical summaries such as odds ratios.

The misclassification scheme described above is depicted in Figure 1. The figure indicates that the true proportion of LBP⁺ DUs is P_1. The figure also shows n_1 DUs in the sample of n, that are truly LBP⁺. XRF testing, however, identifies only n_{11} of these DUs as LBP⁺, misclassifying the remainder as LBP⁻ (false-negative classifications). XRF testing also identifies n_{21} of the n_2 DUs that are truly LBP⁻ as LBP⁺ (false-positive classifications). Ignoring misclassification, the estimate of P_1 would be the ratio of the number of DUs that tested as LBP⁺ divided by the total number of DUs tested, or $(n_{11} + n_{21})/n$. The correct estimate would be n_1/n; however, the value of n_1 cannot be known, because of misclassification.

A scheme for producing estimates of proportions that are approximately unbiased where data are subject to misclassification due to measurement error is described in the Appendix to this report. The scheme is designed to address data that would be classified according to two variables, LBP and PbD, in order to

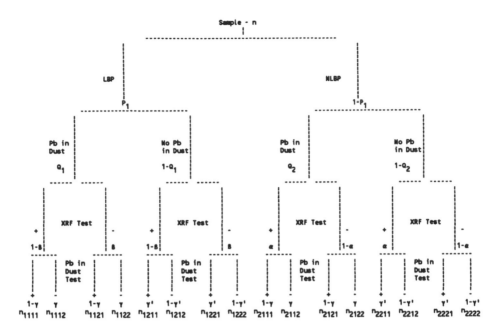

Figure 2 Distribution of survey responses resulting from misclassification due to XRF and Pb in dust measurement error.

analyze the relationship between LBP and PbD. The potential for misclassification in this case is greater than in the situation described above, due to the utilization of two measurement methods in the analysis: one for LBP, and one for PbD. The misclassification possibilities for this case are shown in Figure 2, where misclassification rates are denoted by α, β, γ, and γ', for LBP and PbD, respectively.

Applying this scheme to the estimation of P_1 alone, the bias in estimates of P_1 would be $\alpha - (\alpha - \beta) \cdot P_1$, absent any adjustment for measurement error. The National Survey report (HUD[7]) estimated P_1 to be 0.74 (74%). HUD[7] did not utilize information on measurement error, to assess the magnitude of the misclassification probabilities, α and β. If, for example, both probabilities were equal to 0.05, the estimate of P_1, adjusted for measurement error bias, would be would be 0.77 (77%), not 74%. If both misclassification probabilities were 0.10, the adjusted estimate would be 0.80 (80%). It is more likely, however, that the HUD estimate of P_1 is larger than the true value for the following reasons. Each DU in the study was classified as LBP+ if at least one of approximately 10 surfaces tested showed an XRF reading greater than 1 mg/cm2. Testing multiple surfaces to classify a DU magnifies the false-positive classification probability and reduces the false-negative classification probability. If α and β were both 0.05 for a single measurement, the corresponding probabilities for misclassifying a DU become 0.40 and 1×10^{-13}, respectively, when 10 surfaces are measured. The corresponding estimate of P_1, adjusted for measurement error, would be 0.56 (56%), not 0.74 (74%), as reported in HUD.[7] If α and β both were 0.10, the estimate of P_1 adjusted for measurement error would be 0.26 (26%).

B. REGRESSION ANALYSIS AND MEASUREMENT ERROR

Regression analysis methods have been widely used to investigate the significance of sources of Pb exposure. The results of a limited investigation using regression analysis for relating PbD to PbS, LBP, and other Pb environmental variables were described earlier. Measurement error was not accounted for in those analyses. Ignoring measurement error in the explanatory variables may result in biased estimates of regression coefficients, and understated estimates of standard errors. The consequences may be misidentification of significant sources of PbD.

Table 3 Regression Results for PbD vs. Pb Exposure Variables Showing the Effects of Measurement Error in Explanatory Variables[a]

| Variable | Zero Meas. Error | | Case 1 | | Case 2 | | Case 3 | | Case 4 | |
	Coef	t	Coef	t	Coef	t	Coef	t	Coef	t
Const	5.78	—	5.25	—	5.24	—	1.19	—	1.19	—
Age	–0.18	–3.30	–0.18	–3.33	–0.18	–3.33	–0.23	–2.28	–0.23	–2.28
NPbS	0.23	2.13	0.24	2.19	0.24	2.19	0.32	1.83	0.32	1.82
FPbS	0.29	1.63	0.28	1.53	0.29	1.52	0.18	0.65	0.18	0.65
PLBP	0.95	1.32	1.09	1.39	1.09	1.39	2.10	1.15	2.10	1.15
ELBP	1.08	1.85	1.19	1.90	1.19	1.90	1.98	1.38	1.98	1.37
KLBP	2.37	1.91	2.83	1.86	2.83	1.86	6.51	1.05	6.52	1.05
BLBP	1.07	1.38	1.15	1.37	1.15	1.36	1.49	1.15	1.49	1.15

[a] Results based on hypothetical dataset. Refer to text for description.

Variables are as follows: PbD — Pb concentration in household dust; Age — last two digits of year DU was built; NPbS — Pb concentration in soil near the DU (μg/g); FPbS — Pb concentration in soil at a distance from the DU (μg/g); PLBP — XRF for Pb in porch paint (mg/cm²); ELBP — XRF for Pb in exterior paint (mg/cm²); KLBP — XRF for Pb in kitchen paint (mg/cm²); BLBP — XRF for Pb in bedroom paint (mg/cm²).

Cases are as follows: Case 1 — $\sigma_{\ln(PbS)} = 0.50$; $\sigma_{\ln(LBP)} = 0.25$; Case 2 — $\sigma_{\ln(PbS)} = 0.60$; $\sigma_{\ln(LBP)} = 0.25$; Case 3 — $\sigma_{\ln(PbS)} = 0.50$; $\sigma_{\ln(LBP)} = 0.50$; Case 4 — $\sigma_{\ln(PbS)} = 0.60$; $\sigma_{\ln(LBP)} = 0.50$.

Estimation methods for regression coefficients when explanatory variables are subject to measurement error are discussed by Fuller.[11] The method described in Section 2.2 of Fuller, which incorporates QC data estimates of measurement error into the estimation of regression coefficients, is utilized below with a hypothetical dataset, to demonstrate the potential impact of ignoring measurement error. Results for a hypothetical dataset are discussed because the studies reviewed did not provide sufficient information to develop estimates of measurement error for the real data. In addition, a complete reanalysis of the studies was outside the scope of the review project.

The hypothetical dataset, consisting of 40 observations, reflects characteristics of the data presented by Stark.[6] The regression model is

$$\ln(PbD) = \beta_0 + \beta_1 \cdot Age + \beta_2 \cdot \ln(NPBS) + \beta_3 \cdot \ln(FPbS)$$
$$+ \beta_4 \cdot \ln(PLBP) + \beta_5 \cdot \ln(ELBP) + \beta_6 \cdot \ln(KLBP)$$
$$+ \beta_7 \cdot \ln(BLBP) + \varepsilon$$

where

PbD = Pb concentration in household dust (μg/g),
Age = last two digits of year DU was built,
NPbS = Pb concentration in soil near the DU (μg/g),
FPbS = Pb concentration in soil at a distance from the DU (μg/g),
PLBP = XRF for Pb in porch paint (mg/cm²),
ELBP = XRF for Pb in exterior paint (mg/cm²),
KLBP = XRF for Pb in kitchen paint (mg/cm²),
BLBP = XRF for Pb in bedroom paint (mg/cm²), and
ε = random error from a normal distribution.

Ignoring measurement error, the estimated model has an R^2 value of 0.384. The statistically significant explanatory variables based on a p value of 0.05 (t = 1.7) are Age, NPbS, ELBP, and KLBP. (The regression results are displayed in Table 3.)

Table 3 also displays regression results that incorporate measurement error. Four cases are considered based on estimates of measurement error standard deviations derived from the National Survey Quality Control (QC) data. For purposes of the example, measurement bias is assumed to be zero. The measurement error standard deviations used to define the four cases are $\sigma_{\ln(PbS)} = 0.50$ or 0.60, and $\sigma_{\ln(LBP)} = 0.25$ or

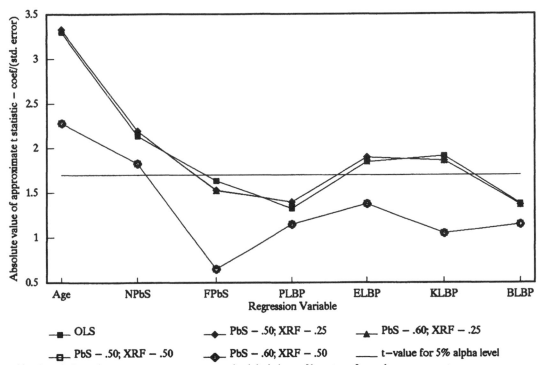

Numbers in legend are measurement error standard deviations of log–transformed measurements.
Results are based on hypothetical data consisting of 40 observations. Refer to text for variable definitions.

Figure 3 Statistical conclusions based on t-tests in regression analysis may be incorrect if measurement error in explanatory variables is not taken into account.

0.50. These values reflect the estimates derived from QC data reported by Westat[8] and MRI,[10] which are developed in Section IV. The measurement error standard deviations are treated as estimates in the regression calculations (see Fuller,[11] Section 2.2), based on 60 QC samples each for PbS and LBP.

As expected, accounting for measurement error leads to decreased statistical significance of the explanatory variables. In the most extreme case considered in Table 3, Case 4, only building age (Age) and soil Pb near the DU (NPbS) show large t values. The t values for each of the four cases also are plotted (Figure 3) showing how changes in measurement error variability affects statistical conclusions for these data.

IV. MEASUREMENT ERROR MODEL

QC data consisting of measurements derived from blanks, standard samples (i.e., samples with known concentrations), and split samples provide information necessary to assess measurement error. Quality control data also may be used to estimate measurement error parameters that subsequently may be employed in the types of statistical analyses described above. Among various studies concerning factors leading to Pb exposure reviewed for this chapter, HUD's National Survey contains the best documented QC data. QC data for LBP measurements collected in the National Survey are summarized by Westat.[8] QC data for PbD and PbS are reported by MRI.[10] Brief descriptions of the use of these data to estimate measurement error parameters follow in the next two sections.

A. SPECTRUM X-RAY FLUORESCENCE ANALYZER (XRF) MEASUREMENT ERROR

McKnight et al.[9] reported estimates of bias and variability for the spectrum XRF measurement method, based on a limited study of prepared surfaces with known Pb concentrations. The estimates are bias = 0.10 mg/cm^2 and standard deviation = 0.30 mg/cm^2. Westat[8] included a larger QC database for the XRF method. Bias was analyzed by developing calibration curves relating XRF measurements to Pb in paint concentrations of prepared samples. These data cover four substrates (wood, steel, drywall, and concrete), each at two concentrations, 0.60 mg/cm^2 and 2.99 mg/cm^2. Westat[8] developed calibration equations for

Table 4 Bias and Variability of Adjusted XRF Measurements

	Substrate									
	Wood		Steel		Drywall		Concrete 1		Concrete 2	
True Pb	Mean	SD	Mean	SD	Mean (mg/cm²)	SD	Mean	SD	Mean	SD
0.00	0.60	0.00	0.12	0.16	0.60	0.00	0.70	0.26	0.60	0.00
0.40	0.61	0.04	0.41	0.27	0.62	0.05	0.81	0.40	0.60	0.00
0.75	0.78	0.16	0.74	0.29	0.77	0.17	1.02	0.54	0.61	0.09
1.00	1.01	0.20	1.02	0.30	1.00	0.20	1.16	0.61	0.64	0.17
1.25	1.28	0.22	1.26	0.31	1.27	0.24	1.33	0.68	0.69	0.26
2.00	2.04	0.27	2.02	0.34	2.02	0.28	2.06	0.82	1.46	0.77
3.00	3.07	0.39	3.01	0.41	3.07	0.39	3.09	0.87	3.13	1.01
4.00	4.11	0.51	4.02	0.53	4.08	0.51	4.16	0.95	4.92	1.32

Note: Bias is estimated as the difference between the mean and the true Pb concentration.

Source: Westat, Inc., Draft Report: National Survey of Lead-Based Paint in Housing, Part II, prepared for Office of Policy Development and Research, U.S. Department of Housing and Urban Development, Washington, D.C., 1991, Table 5–2.

each of eight instruments used in the National Survey and for each substrate. These 32 equations may be used to estimate bias-adjusted values for every XRF measurement obtained in the National Survey. The statistical characteristics of these adjusted XRF measurements, obtained by Westat[8] using simulation, are reproduced in Table 4. The standard deviations in the table provide estimates of measurement error variability for the XRF method. Based on Table 4 the range of XRF measurement variability quantified as the coefficient of variation (i.e., CV equal to the standard deviation divided by the mean), is approximately 10% (0.10) to slightly over 100% (1.00).

B. PbS AND PbD MEASUREMENT ERROR

Soil samples collected in the National Survey were analyzed by inductively coupled plasma atomic emission spectrometry (ICP), and dust samples were analyzed by graphite furnace atomic absorption spectrometry (GFAA). (Refer to MRI[10] for details concerning these analytical methods.) Three laboratories participated in the analytical program: the Core laboratory at Aurora, CO, the Core laboratory at Casper, Wyoming, and the MRI laboratory. The QC program included performance check samples (PCSs) to assess bias (bias and recovery are used interchangeably and have the same meaning throughout this discussion), and split samples to assess measurement variability. As reported in MRI,[10] no dust reference materials were available. Soil samples were used, in place of dust samples, as PCSs when field dust samples were being analyzed. Also, field dust samples could not be split; therefore, no data concerning variability of dust measurements were generated.

A simple measurement model may be used to describe measurement recovery and variability as follows:

$$X = \Theta C + \sigma_s C Z_s + \sigma_\varepsilon \Theta C Z_\varepsilon \qquad (1)$$

where

> X = measured Pb concentration,
> C = true Pb concentration,
> Θ = recovery $(0 < \Theta \leq 1)$,
> σ_s = measure of variability across replicate samples of a given soil,
> Z_s = standard normal variate,
> σ_ε = measure of variability for replicate analyses of a given sample, and
> Z_ε = standard normal variate.

The mean and variance for measurements are

$$E(X) = \Theta C \qquad (2)$$
$$Var(X) = (\sigma_s^2 + \Theta^2 \sigma_\varepsilon^2) C^2 \qquad (3)$$

A transformation to the logarithmic scale is indicated to simplify the model: in particular, to eliminate the dependence of the variance on the true Pb concentration. The transformation is $Y = \ln(X)$. Then

$$E(Y) = \ln(\Theta C) = \ln(\Theta) + \ln(C) \tag{4}$$

and, based on a first order approximation,

$$\text{Var}(Y) = \sigma_s^2/\Theta^2 + \sigma_\varepsilon^2 \tag{5}$$

This model leads to simple estimates of the measurement error parameters, σ_ε and Θ, in terms of the logarithmically transformed data.

An estimate of σ_ε^2, measurement error variance, is obtained as follows. The two parts of a split sample may be represented as

$$Y_1 = \ln(\Theta) + \ln(C) + \sigma_s Z_s/\Theta + \sigma_\varepsilon Z_{\varepsilon 1} \tag{6}$$
$$Y_2 = \ln(\Theta) + \ln(C) + \sigma_s Z_s/\Theta + \sigma_\varepsilon Z_{\varepsilon 2}. \tag{7}$$

The difference is

$$Y_2 - Y_1 = \sigma_\varepsilon(Z_{\varepsilon 1} - Z_{\varepsilon 2}) \tag{8}$$

and the variance of the difference is

$$\text{Var}(Y_2 - Y_1) = 2\sigma_\varepsilon^2 \tag{9}$$

An estimate of σ_ε^2, therefore, is obtained by estimating the variance of $Y_2 - Y_1$ and dividing that estimate by 2:

$$\text{est}(\sigma_\varepsilon^2) = \text{est}(\sigma_{Y2-Y1}^2)/2 \tag{10}$$

From data reported by MRI[10] for the Core laboratory at Casper, $\text{est}(\sigma_\varepsilon^2) = 0.252$, based on 62 split samples. For the MRI laboratory, $\text{est}(\sigma_\varepsilon^2) = 0.356$, based on 43 split samples. The F test for equality of these measurement error variances has a p value of 0.107.

An estimate of recovery, Θ, may be derived from PCSs by averaging the logarithms of the ratios of PCS measurements divided by their known true concentrations. As described by MRI,[10] PCS Pb concentrations were set at three levels: 30, 347, and 2100 μg/g. Recovery results are summarized in Tables 5 and 6. Statistical tests for differences in recoveries across concentrations and laboratories also are recorded in the tables. If recovery is constant across concentrations and laboratories, all raw measurements could be subjected to a simple recovery adjustment prior to statistical analysis. If laboratories have different recoveries, the adjustment would be laboratory specific. If recovery depends on concentration, however, recovery adjustment would be virtually impossible, and the analytical method should be altered to reduce this dependence.

For ICP (Table 5), a difference in recovery between the Core laboratory at Casper and the MRI laboratory, for the low concentration, is indicated (i.e., comparison of categories one and four in the table, $\alpha = 0.01$). Recovery differences (for example, $\alpha = 0.10$), recovery at the low concentration would be interpreted as different than recovery at the high concentration in the MRI laboratory ($\alpha = 0.10$). None of the statistical tests for other recovery comparisons, whether between laboratories or across concentrations, suggest differences.

For GFAA, which was used to analyze dust samples for Pb, PCSs were prepared from soil, not dust. Although these PCSs may be indicative of the general performance of the GFAA analytical process, they provide little direct information on the magnitude of recovery or recovery differences across concentrations and laboratories, for determining PbD concentrations. Dust samples were analyzed in three laboratories: Core, Aurora; Core, Casper; and MRI. Most of the PCS data were from the two Core laboratories. Table 6 shows the recovery data and statistical test results for recovery differences. Differences are indicated between the high and low concentrations at both Core laboratories. Also, differences in recoveries are

Table 5 Soil Quality Control Data: ICP Recovery Analysis

Data category	Lab	Level	N	Mean recovery
1	CC	L	21	89.4%
2	CC	M	18	86.4%
3	CC	H	22	85.6%
4	MRI	L	14	110.9%
5	MRI	M	15	96.9%
6	MRI	H	14	94.6%

Recovery Comparisons Based on Log-Transformed Data	
Comparison categories	S-Test statistic
1,2	−0.236
1,3	−0.519
2,3	−0.259
4,5	−2.798
4,6	−3.295*
5,6	−0.553
1,4	5.385***
2,5	2.557
3,6	2.255

Notes: CC = Core laboratory in Casper, WY.
MRI = Midwest Research Institute.
L = 30 µg/g; M = 347 µg/g; H = 2100 µg/g.
S-Test refers to the S-Method for all contrasts (see Reference 12).
Statistical significance levels: * = 0.10; ** = 0.05; *** = 0.01.
One data point from the CC lab was omitted from this analysis. The true concentration was 2100 µg/g; the measured value was 36 µg/g.

indicated between the two Core laboratories, at the medium and high concentrations. The MRI laboratory analyzed only three PCSs at each concentration, providing little statistical power to assess potential recovery differences. No recovery comparisons involving the MRI laboratory were found to be statistically significant.

Both analytical methods, ICP and GFAA, exhibit statistically significant recovery differences that are concentration dependent. An approach to adjust measurements for recovery differences where recovery depends on concentration is not apparent. Differences that are statistically significant, however, may not be materially significant. The magnitude of recovery differences identified may not affect conclusions based on application of the measurements in question. Determining whether or not recovery differences have a material effect in an application depends on characteristics of the particular application and must be analyzed on a case-by-case basis.

V. DISCUSSION AND RECOMMENDATIONS

Measurement error, consisting of both bias (recovery) and variability, is an inherent component of Pb concentration data. Quality assurance procedures, which rely on QC samples to monitor data quality, are employed to limit, to the extent possible, bias and variability in analytical measurements. QA procedures, however, cannot eliminate measurement error, and depending on its magnitude, measurement error may have a substantial effect on the results of statistical analyses.

Typically measurement error is ignored in statistical analysis. Measurement error was formally addressed in only one of the studies (Westat[8]) reviewed for this paper. The examples that have been considered here indicate how measurement error can affect the results of statistical analyses used to study relationships among Pb exposure variables. Measurement error may be responsible for misclassification of DUs with respect to levels of Pb concentrations, which in turn leads to biased estimates of proportions and other indices used to summarize relationships that rely on estimates of proportions, such as odds ratios. Ignoring measurement error in regression variables can lead to errors in identifying significant predictors from a proposed set of explanatory variables.

Table 6 Dust Quality Control Data: GFAA Recovery Analysis

Data category	Lab	Level	N	Mean recovery
1	CA	L	37	119.9%
2	CA	M	40	115.7%
3	CA	H	33	96.5%
4	CC	L	34	104.0%
5	CC	M	32	90.7%
6	CC	H	33	83.5%
7	MRI	L	3	109.6%
8	MRI	M	3	113.4%
9	MRI	H	3	104.1%

Recovery Comparisons Based on Log-Transformed Data

Comparison categories	S-Test statistic
1,2	–0.580
1,3	–3.996**
2,3	–3.506
4,5	–3.129
4,6	–5.329***
5,6	–2.143
7,8	0.233
7,9	–0.379
8,9	–0.612
1,4	–2.860
2,5	–5.556***
3,6	–4.163**
1,7	–0.607
2,8	–0.069
3,9	1.693
4,7	0.524
5,8	2.114
6,9	2.170

Notes: Soil samples were used as performance check samples.

CA = Core laboratory in Aurora, CO; CC = Core laboratory in Casper, WY; MRI = Midwest Research Institute.

L = 30 µg/g; M = 347 µg/g; H = 2100 µg/g.

S-Test refers to the S-Method for all contrasts (see Reference 12).

Statistical significance levels: * = 0.10; ** = 0.05; *** = 0.01.

Two data points from the CA lab were omitted from this analysis. The true concentrations were 2100 µg/g; the measured values were 164 and 271 µg/g.

Few final substantive conclusions can be provided concerning the study data reviewed for this report. Generally, a thorough reanalysis of the data those studies were based on was beyond the scope of the review. Also, information, such as QC data, that would allow estimation of measurement error parameters typically were not available. It is apparent from the examples, however, that the impact of measurement error, if not accounted for, can be substantial.

Statistical methods that correct for measurement error have been described in this paper. The statistical performance characteristics of these methods, however, have not been studied. The estimation procedure proposed for proportions in the analysis of count data depends on estimates of misclassification probabilities. The regression procedure depends on estimates of measurement error variances. These estimates may be obtained from QC data applied with the measurement model presented as Equation 1. The performance characteristics of the two proposed statistical methods will depend, to a great degree, on the accuracy and precision of the measurement error parameter estimates.

The most direct approach to evaluating the performance of the two statistical methods would be by Monte Carlo simulation. Evaluation efforts have received minimal attention to date. Notwithstanding the absence of performance information, the proposed methods may be used currently as part of a sensitivity analysis to assess the possible effects of measurement error on statistical conclusions.

Appendix

ESTIMATION METHOD FOR PROPORTIONS

Referring to Figure 2, which reflects two tests, there are 16 outcomes with probabilities as follows:

$$P_1 \cdot Q_1 \cdot (1-\beta) \cdot (1-\gamma) \qquad P_1 \cdot Q_1 \cdot (1-\beta) \cdot \gamma$$
$$P_1 \cdot Q_1 \cdot \beta \cdot (1-\gamma) \qquad P_1 \cdot Q_1 \cdot \beta \cdot \gamma$$
$$P_1 \cdot (1-Q_1) \cdot (1-\beta) \cdot \gamma' \qquad P_1 \cdot (1-Q_1) \cdot (1-\beta) \cdot (1-\gamma')$$
$$P_1 \cdot (1-Q_1) \cdot \beta \cdot \gamma' \qquad P_1 \cdot (1-Q_1) \cdot \beta \cdot (1-\gamma')$$
$$(1-P_1) \cdot Q_2 \cdot \alpha \cdot (1-\gamma) \qquad (1-P_1) \cdot Q_2 \cdot \alpha \cdot \gamma$$
$$(1-P_1) \cdot Q_2 \cdot (1-\alpha) \cdot (1-\gamma) \qquad (1-P_1) \cdot Q_2 \cdot (1-\alpha) \cdot \gamma$$
$$(1-P_1) \cdot (1-Q_2) \cdot \alpha \cdot \gamma' \qquad (1-P_1) \cdot (1-Q_2) \cdot \alpha \cdot (1-\gamma')$$
$$(1-P_1) \cdot (1-Q_2) \cdot (1-\alpha) \cdot \gamma' \qquad (1-P_1) \cdot (1-Q_2) \cdot (1-\alpha) \cdot (1-\gamma')$$

There are four observable outcomes based on the tests: LBP and Pb in dust; LBP and no Pb in dust; no LBP and Pb in dust; no LBP and no Pb in dust. The probabilities of these outcomes, denoted by Θ_1, Θ_2, Θ_3, and Θ_4, respectively, are

$$\Theta_1 = P_1Q_1(1-\beta)(1-\gamma) + P_1(1-Q_1)(1-\beta)\gamma' + (1-P_1)Q_2\alpha(1-\gamma) + (1-P_1)(1-Q_2)\alpha\gamma'$$
$$\Theta_2 = P_1Q_1(1-\beta)\gamma + P_1(1-Q_1)(1-\beta)(1-\gamma') + (1-P_1)Q_2\alpha\gamma + (1-P_1)(1-Q_2)\alpha(1-\gamma')$$
$$\Theta_3 = P_1Q_1\beta(1-\gamma) + P_1(1-Q_1)\beta\gamma' + (1-P_1)Q_2(1-\alpha)(1-\gamma) + (1-P_1)(1-Q_2)(1-\alpha)\gamma'$$
$$\Theta_4 = P_1Q_1\beta\gamma + P_1(1-Q_1)\beta(1-\gamma') + (1-P_1)Q_2(1-\alpha)\gamma + (1-P_1)(1-Q_2)(1-\alpha)(1-\gamma')$$

where:

P_1 = true proportion of DUs with LBP,

Q_1 = true proportion of DUs with LBP and Pb in dust,

Q_2 = true proportion of DUs with no LBP and Pb in dust,

α = XRF test false-positive rate,

β = XRF test false-negative rate,

γ = Pb in dust test false-negative rate, and

γ' = Pb in dust test false-positive rate.

Ignoring misclassification due to measurement error, the estimates of P_1, Q_1, and Q_2 are $(n_{++} + n_{+-})/n$, $n_{++}/(n_{++} + n_{+-})$, and $n_{-+}/(n_{-+} + n_{--})$, respectively, where:

n_{++} = observed number of DUs with LBP and Pb in dust,

n_{+-} = observed number of DUs with LBP and no Pb in dust,

n_{-+} = observed number of DUs with no LBP and Pb in dust, and

n_{--} = observed number of DUs with no LBP and no PB in dust.

The $n_{..}$'s are defined in Figure 2. Although each $n_{..}$ is the sum of four components, the components are not observable. Only the sums are observable.

The alternative estimates, which account for misclassification due to measurement error, are obtained by solving the following set of equations for P_1, Q_1, and Q_2:

$$n_{++}/n = \Theta_1$$
$$n_{+-}/n = \Theta_2$$
$$n_{-+}/n = \Theta_3$$
$$n_{--}/n = \Theta_4$$
$$\Theta_1 + \Theta_2 + \Theta_3 + \Theta_4 = 1$$

If the solutions for Q_1 or Q_2 are negative, they are replaced with zero.

REFERENCES

1. U.S. Environmental Protection Agency Strategy for Reducing Lead Exposures, U.S. Environmental Protection Agency, Washington, D.C., 1991.
2. USHHS Strategic Plan for the Elimination of Childhood Lead Poisoning, developed for the Risk Management Subcommittee, Committee to Coordinate Environmental Health and Related Programs, U.S. Department of Health and Human Services, Washington, D.C., 1991.
3. Price Associates, Inc., Review of Studies Addressing Relationships Between Lead in Household Dust, Lead in Soils, and Lead Based Paint, prepared for USEPA Exposure Evaluation Division, Washington, D.C., 1992.
4. Bornschein, R.L., Succoup, P.A., Kraft, K.M., Clark, C.S., Peace, B., and Hammond, P.B., Exterior surface dust lead, interior house dust lead and childhood lead exposure in an urban environment, in Health: Proc. of the University of Missouri's 20th Annu. Conf., Hemphill, D.D., Ed., 1986.
5. Rabinowitz, M., Levinton, A., Needleman, H., Bellinger, D., and Waternaux, C., Environmental correlates of infant blood levels in Boston, *Environ. Res.*, 38, 77, 1985.
6. Stark, A.D., Quah, R.F., Meigs, J.W., and DeLouise, E.R., The relationship of environmental lead to blood-lead levels in children, *Environ. Res.*, 27, 372, 1982.
7. Comprehensive and Workable Plan for the Abatement of Lead-Based Paint in Privately Owned Housing: Report to Congress, Office of Policy Development and Research, U.S. Department of Housing and Urban Development, Washington, D.C., 1990.
8. Westat, Inc., Draft Report: National Survey of Lead-Based Paint in Housing, Parts I and II, prepared for Office of Policy Development and Research, U.S. Department of Housing and Urban Development, Washington, D.C., 1991.
9. McKnight, M.E., Byrd, W.E., and Roberts, W.E., Measuring Lead Concentrations in Paint Using a Portable Spectrum Analyzer X-Ray Fluorescence Device, National Institute of Standards and Technology, U.S. Department of Commerce, Washington, D.C., 1990.
10. MRI, Analysis of Soil and Dust Samples for Lead (Pb), Final Report. Prepared by Midwest Research Institute for the Field Studies Branch, Exposure Evaluation Division, Office of Toxic Substances, U.S. Environmental Protection Agency, and Division of Policy Development Research, Office of Public and Indian Housing, U.S. Department of Housing and Urban Development, Washington, D.C., 1991.
11. Fuller, W.A., *Measurement Error Models*, John Wiley & Sons, New York, 1987.
12. Scheffe, H., *The Analysis of Variance*, John Wiley & Sons, New York, 1959.

Chapter 25

Multielement Analysis of Lead-Based Paint Abatement Data

J. G. Kinateder, S. W. Rust, and J. G. Schwemberger

CONTENTS

I. INTRODUCTION

This chapter presents the results of a multielement analysis of data obtained during the Comprehensive Abatement Performance (CAP) Pilot Study. A separate report deals exclusively with the statistical analysis of observed levels of lead.[1] Six houses with differing abatement history were sampled with the primary intent of determining the relationship of lead in dust and soil, with abatement history. Two houses were abated by enclosure or encapsulation methods; two were abated by removal methods; and two were control houses identified as being free from lead-based paint, by previous XRF testing. Along with the determinations of lead obtained in the study, levels of four other metals were measured: cadmium, chromium, titanium, and zinc. These metals are known to be used in the composition of paint. The purpose of measuring the levels of these other metals in the samples was to cluster samples into groups that appear to have come from similar sources, with the ultimate goal of identifying prominent sources of lead found in household dust.

The major objectives addressed in the analysis of the multielement data from the pilot were to

1. Characterize the levels of lead, cadmium, chromium, titanium, and zinc in household dust and soil
2. Characterize the effects of renovation and abatement on the concentrations of lead, cadmium, chromium, titanium, and zinc in household dust and soil
3. Investigate the relationship between concentrations of lead, cadmium, chromium, titanium, and zinc in the different types of samples collected, and specifically between concentrations of these elements in household dust and exterior soil, air ducts, and bedcover/rug/upholstery

The intention of this initial multielement data examination was to perform a series of analyses, to determine whether a more detailed analysis would be fruitful. With data for only six housing units, most relationships were not strongly detectable. Therefore, it was decided that a further, more detailed analysis of the pilot multielement data would be less productive than a more detailed examination of multielement data that might be obtained during the full study.

Section 2 describes the data. Section 3 describes the analyses performed and their results. Section 4 provides conclusions.

II. DATA

By design, 18 regular vacuum dust samples were to be collected from each house. For each of the six dust-sample types (window channel, window stool, air duct, floor, bedcover/rug/upholstery, and entryway), two different rooms were targeted for sampling. Soil samples were to be collected at six different locations at each house: one just outside the front and back entryways, two at different locations on the foundation, and two at different locations on the property boundary.

1-56670-113-9/95/$0.00+$.50

Table 1 Component Abbreviations used in Tables and Figures

Media	Mnemonic	Component
Vacuum dust samples	ARD	Air ducts
	BRU	Bed/rug/upholstery
	EWY (-I)	Entryway (-inside)
	FLR	Floor
	WCH	Window channel
	WST	Window stool
Soil samples	BDY	Boundary
	EWY (-O)	Entryway (-outside)
	FDN	Foundation

Multielement analysis was obtained for most of the vacuum dust and soil core samples collected during the pilot study. Wipe dust samples were also collected, but multielement data were not measured on these samples. Only element concentrations ($\mu g/g$) were analyzed for this report. Element loadings ($\mu g/ft^2$) are available for analysis at a later date.

Table 1 contains a description of the acronyms used in the remaining tables and figures, to denote various components sampled in the study.

Due to the general lack of room level effects found in the analysis of the CAP pilot lead data, the basic experimental unit considered in the multielement data analysis is the house. House geometric average concentrations of the five elements were the basic quantities used in the statistical analyses. These are presented in Table 2 by sample type for each of the five elements. Also included in the tables is the number of samples for which all five elements had concentration measurements. Table 3 displays the abatement and renovation history of each of the six houses sampled. Renovation is described in Section III.B.

III. ANALYSIS RESULTS

The analysis is divided into three parts corresponding to the three major objectives introduced above. Section A contains a characterization of the concentration levels of the different elements in the various sample types; Section B describes the estimated effects of abatement and renovation; and Section C examines the relationships among the elements and sample types.

A. CHARACTERIZATION OF ELEMENT LEVELS
Levels of each of the five elements observed varied by sample type. Figures 1a through 1e display house geometric mean sample concentrations by sample type for cadmium, chromium, titanium, lead, and zinc, respectively. These figures display all the data used in the analysis. Grand means of each element across all houses, by sample type, are displayed in Table 4. These were obtained by taking the geometric mean of the house mean log concentrations in Table 2, for each sample type and element. Each house where a sample was taken (for a particular sample type) is given equal weight in these averages. In Table 4 each average listed is followed by its log standard deviation. This represents a measure of the between-house variation for that response, without controlling for abatement or renovation effects. These grand means are also displayed in Figures 1a through 1e as circles.

Of the metals considered, the highest geometric mean concentrations were observed for zinc, in the indoor samples. For the outdoor samples, the highest levels were observed for titanium. Of the different components sampled, lead concentrations were observed to be highest in air duct, window stool, and window channel samples. The greatest variation in (log) geometric means across sample types was observed for lead; the least was observed for chromium and titanium. Notice the somewhat parallel nature of these variations across sample types for the different elements, particularly between lead and zinc.

B. ABATEMENT AND RENOVATION EFFECTS
The impact of abatement and renovation on the multielement data was assessed by fitting a statistical model containing both renovation and abatement effects, to the data in Table 2. The model fitted was

$$C_j = m + aI_j + rR_j + E_j \qquad j = 1, \ldots, 6$$

The parameters for this model are defined in Table 5.

Table 2 Unit Geometric Mean Concentrations by Sample Type

Sample type	House	Samples taken in unit	Geometric mean concentrations (µg/g)				
			Pb	Cd	Cr	Ti	Zn
WCH	33	1	7259	29.7	39	659	13767
	43	2	1176	25.3	28	354	2080
	80	3	2922	23.1	77	561	3429
	51	3	829	7.7	31	464	1097
WST	33	4	424	—	100	420	1353
	43	3	523	43.4	47	365	2208
	17	6	369	139.8	38	416	2779
	19	3	140	13.7	49	299	765
	80	5	3828	18.2	106	407	5219
	51	4	1863	7.9	27	365	1588
ARD	33	2	871	35.9	46	189	16482
	43	1	1141	11.0	164	407	7785
	17	2	513	200.3	51	270	4537
	19	1	626	23.8	145	351	1466
	80	3	863	6.6	48	169	2059
FLR	33	7	130	35.9	204	290	645
	43	7	221	—	41	265	1313
	17	7	166	9.1	29	151	567
	19	5	172	7.9	65	145	572
	80	7	305	5.7	40	206	608
	51	6	1224	6.4	21	179	1236
BRU	33	1	117	25.5	69	388	446
	43	2	141	6.1	39	260	1939
	17	1	67	10.0	44	85	572
	19	2	483	10.4	123	166	821
	80	2	151	5.2	35	179	302
EWY-I	33	2	107	19.7	221	567	469
	43	2	395	6.1	30	478	1261
	17	2	270	15.0	35	308	513
	19	2	192	7.9	40	265	567
	80	2	276	5.9	31	340	572
	51	2	1604	9.0	29	250	1437
EWY-O	33	3	79	4.0	91	652	161
	43	3	337	3.6	29	493	344
	17	2	161	25.8	100	534	299
	19	3	73	2.5	24	433	233
	80	3	380	8.6	32	503	428
	51	2	672	4.1	23	334	403
FDN	33	3	147	3.6	32	503	268
	43	3	247	4.4	37	633	608
	17	3	69	2.6	39	399	308
	19	2	108	3.1	23	330	257
	80	3	513	8.6	26	372	503
	51	3	602	3.7	19	354	407
BDY	33	2	86	2.1	23	376	136
	43	2	133	2.1	20	384	140
	17	3	59	2.5	42	584	130
	19	2	44	1.9	16	233	118
	80	2	324	7.6	24	376	395
	51	3	324	2.9	20	293	252

Table 3 Abatement and Renovation History by House

House	Interior abatement history	Exterior abatement history	Renovation
17	Abated: Removal	Abated: E/E	
19	Control	Control	Partial
33	Control	Control	
43	Abated: Removal	Abated: Removal	
51	Abated: E/E	Abated: Removal	Full
80	Abated: E/E	Abated: E/E	

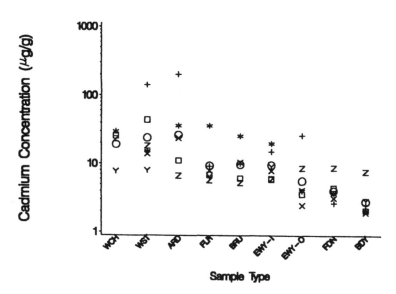

Figure 1a Cadmium concentration vs. sample type (geometric house mean).

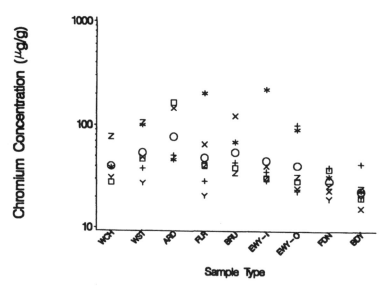

Figure 1b Chromium concentration vs. sample type (geometric house mean).
Legend: * 17, □ 19, + 33, X 43, Z 51, Y 80, 0 Mean.

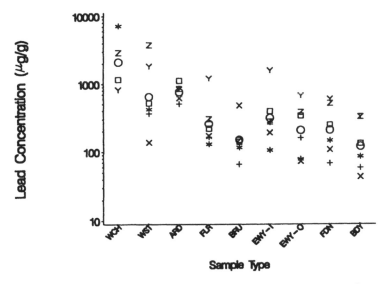

Figure 1c   Lead concentration vs. sample type (geometric house mean).

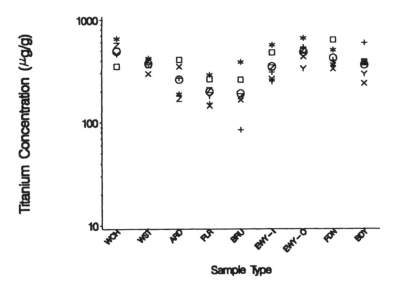

Figure 1d   Titanium concentration vs. sample type (geometric house mean).

House 51 was assigned an R_j value of 1, indicating "full renovation;" and House 19, a value of 0.5, indicating "partial renovation." The other four houses were assigned R_j values of zero, indicating that no renovation was being performed.

In the analysis of the lead data, the method of abatement (E/E or removal) was also considered as a factor in the statistical model. No significant effect was found, and therefore, it was not included in the final lead model. For consistency, no method effect is included in the multielement model either.

Estimates of the model parameters are reported in Tables 6, 7, and 8. Table 6 contains estimates and log standard errors of the geometric mean concentration of each element in unrenovated, control houses, by sample type. Tables 7 and 8 contain estimates and log standard errors of the multiplicative effects of renovation and abatement, respectively, again by sample type. In Tables 7 and 8 a multiplicative effect of 1.0 implies no effect. A multiplicative effect less than 1.0 indicates that lower levels were observed in renovated (abated) houses, while a multiplicative effect greater than 1.0 indicates that higher concentrations were observed in renovated (abated) houses. Those multiplicative effects that are significantly different from 1.0, at the 0.05 significance level, are indicated by a footnote.

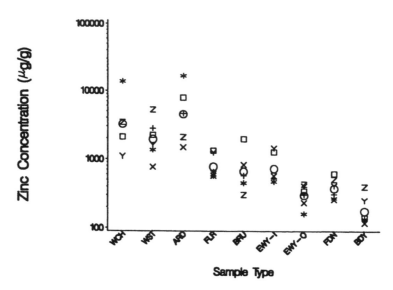

Figure 1e Zinc concentration vs. sample type (geometric house mean).

Table 4 Geometric Mean Concentration and Log Standard Deviation Across Houses by Sample Type

Sample medium	Sample type	No. of units sampled	Lead Geo mean (µg/g)	Log	Cadmium Geo mean (µg/g)	Log	Chromium Geo mean (µg/g)	Log	Titanium Geo mean (µg/g)	Log	Zinc Geo mean (µg/g)	Log
Vacuum	WCH	4	2121	0.97	19.1	0.61	40.0	0.46	497	0.27	3229	1.07
	WST	6	658	1.20	23.8	1.03	54.0	0.54	376	0.13	1939	0.66
	ARD	5	772	0.31	26.3	1.32	77.4	0.64	262	0.38	4447	0.98
	FLR	6	259	0.81	9.29	0.68	48.4	0.80	200	0.29	772	0.39
	BRU	5	151	0.72	9.67	0.62	55.1	0.52	190	0.57	658	0.70
	EWY-I	6	314	0.91	9.48	0.49	45.6	0.79	350	0.33	720	0.49
Soil	EWY-O	6	208	0.90	5.64	0.85	40.8	0.67	482	0.23	295	0.37
	FDN	6	208	0.87	4.01	0.41	28.7	0.28	419	0.24	372	0.35
	BDY	6	120	0.85	2.77	0.51	22.8	0.34	361	0.30	175	0.48

Table 5 Definition of Model Parameters

C_j	represents the observed average log-concentration in house j
m	represents the average log-concentration in unrenovated control houses
a	represents the added effect of abatement
I_j	= 1 if house j was abated; 0 if house j was a control house
r	represents the added effect of a full renovation
R_j	is the degree of renovation house j was undergoing at the time of sampling
E_j	represents the variation from house to house

Figures 2a, 2b, and 2c display companion block charts of the estimates in Table 6 (portrayed on a log scale). Figure 2a displays the estimated unrenovated control house (log) means for air ducts, window stools, and window channels. Figure 2b displays the corresponding estimates for bed/rug/upholstery, entryway, and floor samples. Figure 2c shows the estimates for soil samples (boundary, entryway, and foundation). Air ducts, window stools, and window channels typically had the highest baseline levels of lead and zinc. Soil samples had the lowest concentrations of these elements. Notice the relatively similar behavior of these estimates across the different elements within each of the three sample groups. Figure 2a depicts a lower titanium level, than lead or zinc level, for air ducts and window channels. Levels portray a general rise as one moves from lead to titanium to zinc, for window stools in Figure 2a and for floors, entryways, and bed/rug/upholstery samples in Figure 2b. In contrast, titanium was the element with the highest concentration in each of the soil samples. This is portrayed in Figure 2c. The relationship among levels of the

Table 6 Model Estimates and Log Standard Errors of Geometric Mean
Concentrations in Unrenovated Control Houses

Sample medium	Sample type	n	Lead Geo mean (µg/g)	Log	Cadmium Geo mean (µg/g)	Log	Chromium Geo mean (µg/g)	Log	Titanium Geo mean (µg/g)	Log	Zinc Geo mean (µg/g)	Log
Dust	WCH	4	7259	0.64	29.6	0.07	38.8	0.72	658	0.33[a]	1376	0.35
	WST	6	226	1.17	21.5	0.89	86.4	0.46	368	0.13	1224	0.37
	ARD	5	871	0.41	35.8	1.84	46.5	0.69	188	0.44	1648	0.67
	FLR	6	102	0.33	19.1	0.59	141	0.36	221	0.33	555	0.40
	BRU	5	116	0.45	25.5	0.34	68.7	0.12	387	0.57	445	0.94
	EWY-I	6	96.5	0.19	13.0	0.57	108	0.63	445	0.28	437	0.40
Soil	EWY-O	6	62.8	0.43	3.85	0.81	59.7	0.68	601	0.08	183	0.19
	FDN	6	102	0.89	3.59	0.48	31.5	0.18	441	0.26	270	0.29
	BDY	6	535	0.81	2.09	0.57	20.9	0.33	336	0.22	120	0.52

[a]Indicates effect was significant at $p = .05$ level.

different elements for window stools appears more similar to those for floors, entryways, and bed/rug/upholstery samples than to those for window channels and air ducts. This is likely a reflection of the general composition of these dust samples.

Close attention should be given to the log standard errors of the estimates reported in Tables 6, 7, and 8. Most of these are very large in comparison to the (logarithm of the multiplicative) estimates. Note that a total of 90 statistical tests were performed in the analysis, supporting the results in Tables 7 and 8. Each test was performed at the 5% level. Therefore, even if there were no effects of abatement or renovation on any of these element concentrations, we would still expect four or five sample type/element/factor combinations to be significant. A total of eight combinations were found to be significant. Of these, lead was the element involved in four cases; each of the other four elements was involved in one of the four remaining cases. Entryways were involved in five cases of significance (three soil and two dust). Floors were involved in two cases, and one case of significance was observed for window channels.

Thus, the number of statistically significant results found was small relative to the number of tests performed. This makes it difficult, and perhaps inadvisable, to draw general conclusions from the estimates reported in Tables 7 and 8.

C. RELATIONSHIPS AMONG THE ELEMENTS
Bivariate Relationships (Correlations)
Displays portraying the bivariate relationships among the five elements are provided for interior entryway dust in Figure 3a and for exterior entryway soil samples in Figure 3b. (For a complete set of these graphs, see the EPA report on this study.[2]) For each sample type, house log concentrations for each element are plotted against each of the other elements. Ellipses are drawn on each plot containing 95% of the estimated bivariate distribution. Those plots for which the ellipse is narrow represent pairs of elements for which there was a strong observed correlation. Pairs of elements that are negatively correlated have an ellipse with major axis running from upper left to lower right. The magnitude of the correlation can be inferred from the shape of the ellipse, by comparing it to the key in Figure 3c. For distributions with known correlations of 90, 60, 30, and 0%, comparable ellipses are displayed.

On the plots in Figures 3a and 3b, each house is identified with a different symbol. This allows one to determine whether certain houses have similar characteristics with respect to the various elements and/or sample types.

It can be seen from the figures that the correlations among the elements in the interior entryway dust are similar to those in the exterior entryway soil (except for the lead/cadmium and cadmium/zinc relationships). This could be an indication that much of the interior dust near the entryway is tracked in from outside. Considering the remaining seven sample types (not displayed here), we made the following observations:

- No pair of elements was consistently highly correlated.
- Zinc was the only element consistently positively correlated with lead; this correlation was most pronounced on entryway and window channel vacuum samples, and boundary and entryway soil samples.

Table 7 Estimates and Log Standard Errors of Multiplicative Renovation Effects

Sample medium	Sample type	Lead		Cadmium		Chromium		Titanium		Zinc	
		Multiplicative effect of renovation	Log standard error	Multiplicative effect of renovation	Log standard error	Multiplicative effect of renovation	Log standard error	Multiplicative effect of renovation	Log standard error	Multiplicative effect of renovation	Log standard error
Dust	WCH	0.45	0.79	0.32[a]	0.08	0.67	0.88	1.04	0.41	0.41	0.43
	WST	1.34	1.25	0.21	0.95	0.43	0.49	0.85	0.14	0.47	0.40
	ARD	0.51	1.15	0.43	5.21	9.89	1.96	3.48	1.24	0.01	1.90
	FLR	4.67[a]	0.35	0.60	0.63	0.44	0.39	0.74	0.35	1.45	0.43
	BRU	17.08	1.28	0.17	0.96	3.19	0.33	0.18	1.62	3.35	2.67
	EWY-I	4.87[a]	0.21	0.84	0.61	0.56	0.67	0.58	0.30	1.91	0.43
Soil	EWY-O	2.12	0.46	0.43	0.86	0.38	0.72	0.62[a]	0.09	1.25	0.20
	FDN	2.29	0.95	0.78	0.52	0.57	0.19	0.72	0.27	0.90	0.31
	BDY	1.73	0.87	0.84	0.61	0.68	0.36	0.62	0.24	1.21	0.55

[a] Indicates effect was significant at $p = .05$ level.

Table 8 Estimates and Log Standard Errors of Multiplicative Abatement Effects

Sample medium	Sample type	Lead		Cadmium		Chromium		Titanium		Zinc	
		Multiplicative effect of abatement	Log standard error	Multiplicative effect of abatement	Log standard error	Multiplicative effect of abatement	Log standard error	Multiplicative effect of abatement	Log standard error	Multiplicative effect of abatement	Log standard error
Dust	WCH	0.26	0.79	0.82	0.08	1.19	0.88	0.68	0.41	0.19	0.43
	WST	4.45	1.01	2.11	0.77	0.68	0.40	1.09	0.11	2.63	0.32
	ARD	0.91	0.47	0.68	2.13	1.60	0.80	1.41	0.51	0.25	0.78
	FLR	2.27	0.29	0.41	0.51	0.27[a]	0.31	0.95	0.29	1.42	0.35
	BRU	0.96	0.52	0.27	0.39	0.56	0.14	0.41	0.66	1.55	1.09
	EWY-I	3.25[a]	0.17	0.67	0.49	0.33	0.54	0.86	0.25	1.65	0.35
Soil	EWY-O	4.51[a]	0.37	2.41	0.70	0.81	0.59	0.86	0.07	1.89[a]	0.16
	FDN	2.13	0.77	1.30	0.42	1.07	0.16	1.05	0.22	1.68	0.25
	BDY	2.77	0.70	1.62	0.49	1.33	0.29	1.33	0.19	1.63	0.45

[a] Indicates effect was significant at $p = .05$ level.

- Chromium and titanium were consistently positively correlated.
- Lead and cadmium were negatively correlated for five of the six dust sample types (all but window channel), and positively correlated for all three of the soil sample types.

Multivariate Relationships (Principal Components)

Multivariate relationships among the elements were also investigated. For the estimated model parameters displayed in Tables 6, 7, and 8, a principal components analysis was performed on the variation of the control house averages, abatement effects, and renovation effects, across the nine sample types. The purpose of this analysis was not only to identify consistent patterns in the composition of dust across different sample types (unrenovated control house analysis), but also to determine whether abatement or renovation impacts different components in different ways. The analysis was performed on a log scale.

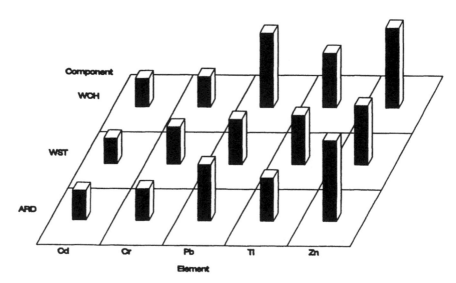

Figure 2a Estimated unrenovated control unit geometric mean concentrations for window and air duct dust samples (log scale).

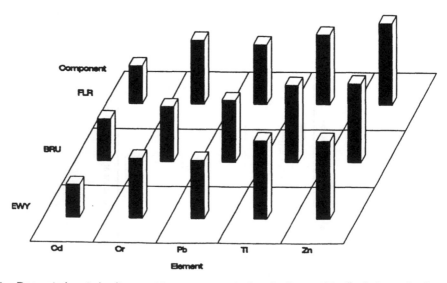

Figure 2b Renovated control unit geometric mean concentrations for floor and textile dust samples (log scale).

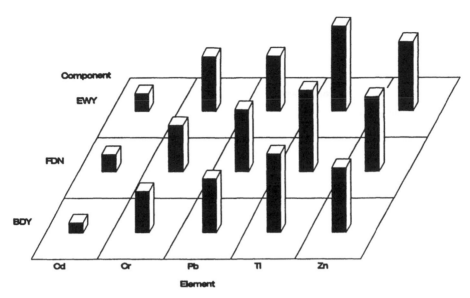

Figure 2c   Estimated unrenovated control unit geometric mean concentrations for soil samples (log scale).

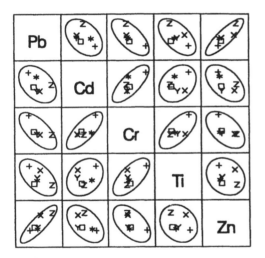

Figure 3a   Entryway dust house mean correlation scatterplot.

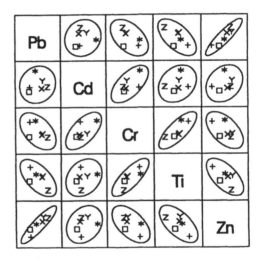

Figure 3b   Entryway soil house mean correlation scatterplot.

	90%	60%	30%	0%
Z1	⬭	⬭	⬭	◯

Figure 3c   Key to correlation scatterplots. *Legend:* * 17, □ 19, + 33, X 43, Z 51, Y 80.

The numerical results of the principal components analyses, and plots of the first two principal components, are displayed in Table 9 and Figure 4. Table 9 displays estimates of the first two principal components, followed by the cumulative proportion of explained variation. Figure 4 displays the relationship of the first two principal components (the directions in which the greatest variability were observed).

Two principal components generally accounted for at least 80% of the variability. This means that although there were five elements measured (lead, cadmium, chromium, titanium, and zinc), most of the

Table 9 Principal Components for Control Home Averages, Abatement Effects, and Renovation Effects

Response	Principal component	Principal component coefficients					Cumulative explained variability
		Pb	Cd	Cr	Ti	Zn	
Unrenovated control unit means	1	0.53	0.57	0.16	–0.15	0.59	0.53
	2	0.41	–0.24	–0.68	0.53	0.18	0.81
Abatement effect	1	0.61	0.48	–0.26	0.08	0.57	0.52
	2	0.05	0.37	0.65	0.65	–0.16	0.87
Renovation effect	1	0.48	–0.02	–0.38	–0.56	0.56	0.52
	2	0.48	–0.55	0.57	0.29	0.25	0.84

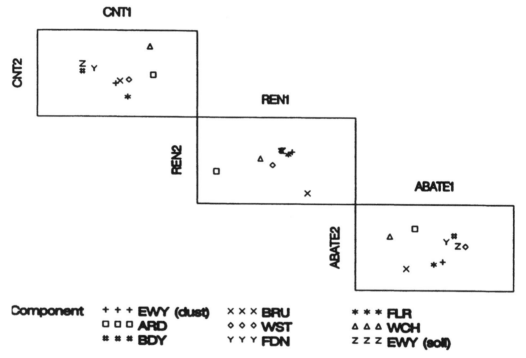

Component   + + + EWY (dust)   × × × BRU   * * * FLR
            □ □ □ ARD   ◇ ◇ ◇ WST   △ △ △ WCH
            # # # BDY   Y Y Y FDN   z z z EWY (soil)

Figure 4 First two principal components plotted vs. each other for unrenovated mean log-concetrations, renovated effects, and abatement effects.

variation among the nine sample types occurs within a two-dimensional space (i.e., two linear combinations of the five element concentrations). This is not very significant in light of the fact that there were between four and six houses sampled for each sample type.

For control house averages, the three soil sample types are grouped into one cluster; floor, entryway, window stool, bed/rug/upholstery dust sample types form another cluster; and window channels and air ducts stand alone. For the renovation effect, all samples are grouped into one cluster, except for air ducts and bed/rug/upholstery, which stand alone. One must recognize that air ducts and bed/rug/upholstery were not sampled in the fully renovated house. Therefore, the estimated impact of renovation on these sample types is less meaningful than on the other sample types, which were sampled in the fully renovated house.

For the abatement effects there were no clear clusters or outlying sample types, but the three soil sample types appear close together.in Figure 4. This may be an indication that even after abatement, the composition of the soil near the foundation is similar to that of the soil at the boundary.

Notice that the clusters apparent in Figure 4 (especially the unrenovated control houses) are consistent with the groupings identified in Figures 2a, 2b, and 2c. The composition of the dust observed in window stools is more similar to that of the dust on floors and textiles than it is to the composition of the dust in window channels and air ducts.

IV. SUMMARY

The most prevalent element in indoor dust (of those considered) was zinc. In soil, titanium had the greatest concentrations. Of the four nonlead elements measured, concentrations of zinc were most positively correlated with lead, both within sample types and across sample types. Cadmium and lead levels were negatively correlated indoors and positively correlated outdoors (except for window stools, for which a positive correlation was observed).

The relative levels of the elements in the following groups of sample types were similar:

- Soil samples
- Floor, entryway, bedcover/rug/upholstery, and window stools
- Air ducts and window channels

Except for lead, there were no strong differences between levels of the different elements in abated, renovated, and control houses. However, five of the eight sample type/element/factor combinations observed to be significant involved entryways (two, dust; three, soil). Lead and zinc levels were higher in abated and renovated homes.

Floor and entryway dust was similar in composition for the five elements except titanium. The concentration of titanium in the dust at entryways was more similar to that observed in soil than to other floor samples, suggesting that the composition of the interior dust at the entryway is influenced by the exterior soil.

REFERENCES

1. Battelle Memorial Institute and Midwest Research Institute, Comprehensive Abatement Performance Pilot Study, Vol. I, Results of Lead Data Analyses, EPA 747-R-93–007, under Contracts No. 68-D2–0139 and 68-D0–0137, U.S. Environmental Protection Agency, Washington, D.C., 1993.
2. Battelle Memorial Institute and Midwest Research Institute, Comprehensive Abatement Performance Pilot Study, Vol. II, Multi-Element Data Analysis, EPA 747-R-93–007, under Contracts No. 68-D2–0139 and 68-D0–0137, U.S. Environmental Protection Agency, Washington, D.C., 1993.

Chapter 26

Determination of Environmental Lead, Using Compositing of House Dust Samples

M. R. Farfel and C. A. Rohde

CONTENTS

I. INTRODUCTION

House dust is a major link in the pathway of lead exposure in children, via the hand-to-mouth route of ingestion.[1,2] Nearly 10 million of the 57 million private dwellings in the U.S estimated to contain lead-based paint are occupied by children under the age of 7 years. Of particular concern are the estimated 3.8 million dwellings, occupied by young children, that have elevated lead levels in settled dust and/or deteriorating lead paint.[3] Assessing exposure to lead-contaminated dust in large numbers of dwellings is a complex and costly process. Compositing of samples may provide a practical and affordable means of determining levels of lead in settled dust.

The EPA-supported "Repair and Maintenance Pilot Study"[4] provided an opportunity to study the properties of compositing. Our objectives in this preliminary work were to investigate (1) the relationship between weighted averages of individual samples and the values obtained from composite samples, and (2) the relationship between composite samples and maximum and minimum values of the individual samples. Sources of variability for individual and composite samples will be explored in a companion paper in which the theory of composite samples[5] will be investigated using these data.

II. METHODOLOGY

A. STUDY DESIGN

The compositing study was designed to produce vacuum dust-lead data from four rooms in each of four houses. Four individual samples and one composite sample of settled dust were collected from each of the rooms. A total of 64 individual vacuum dust readings were planned from 16 rooms, along with 16 vacuum dust composite readings. To allow for a wide range of dust-lead levels and housing conditions, samples were collected from two dwellings previously abated of lead-based paint, and two older unabated dwellings.

Front and back rooms on the first and second floors were selected for study. In each of the four rooms, four individual vacuum dust samples (1 ft² each, equivalent to 929 cm²) were collected. Each 1-ft² sample was collected at a randomly selected site immediately to the left or to the right of each of the four 1-ft² subareas sampled as part of the floor composite sample.

1-56670-113-9/95/$0.00+$.50

B. METHODS

Dust samples were collected with an in-line filter device that consisted of a 37-mm polystyrene cassette (Gelman GN-4) containing a mixed cellulose ester filter (0.8-μm pore size) and support pad attached, by Tygon® tubing, to an air mover (GAST rotary vane vacuum pump, 16 l/min). The filter cassette cover was removed and replaced with a Teflon® nozzle (34-mm ID) cut at a 45 degree angle. Surfaces were vacuumed in an orthogonal raster pattern with overlapping passes at a collection rate of 2 min/ft². A template (1 ft²) was used to define the sampling area. This sampling device was used in a recent national survey of lead-containing paint and dust in privately owned U.S. housing.[3]

In the laboratory the filters were rinsed into prepared 50 ml Pyrex beakers, using deionized water. The dust rinsate was dried (110°C) and cooled. After the beaker was reweighed, the sample was digested on a hotplate, using concentrated nitric acid and hydrogen peroxide according to EPA SW846 Method 3050 (final dilution volume 25 ml). All dust samples were first analyzed by flame atomic absorption spectro-photometry (flame-AAS, Instrument Laboratory, Model 551). Samples with low lead concentrations (<LOQ) were analyzed by graphite furnace-AAS (Perkin Elmer, Model 5100). All analytical work was performed at the Kennedy Krieger Institute Trace Metals Laboratory.

C. DATA ANALYSIS

After eliminating readings due to sample loss and low-weight samples, 58 individual readings of Pb loading and concentrations and 15 corresponding room composite readings were available for data analysis. In order to compare weighted averages and composite samples, on a uniform scale, only those rooms where there were four individual samples and a composite sample were used for data analysis. This resulted in 11 data points for statistical analysis. Weighted averages of dust lead concentrations were calculated using the individual sample weights as the weighting factor. Weighted averages of the dust lead loadings were calculated using surface area sampled as the weighting factor. The statistical package GLIM was used for all statistical analyses.[6]

D. HYPOTHESES

The hypotheses tested in this study are expressed below in null form:

H₁: There is no relationship between the weighted average of lead-dust loading estimates (PbD) from individual samples, and the PbD values obtained from composites.

H₂: There is no relationship between the maximum of lead-dust loading estimates (PbD) from individual samples, and the PbD values obtained from composites.

H₃: There is no relationship between the minimum of lead-dust loading estimates (PbD) from individual samples, and the PbD values obtained from composites.

III. RESULTS

A. ANALYSIS OF THE COMPOSITE VS. WEIGHTED-AVERAGE RELATIONSHIP

The correlations (r) between weighted averages and composites for lead (Pb) loadings (mass of lead/unit area) and Pb concentrations (mass of lead/mass of sample) were .92 and .85, respectively. To examine the relationship between the composite sample and the weighted average of Pb loadings and Pb concentrations, regression models of the following form were used:

$$\ln(PbWAC) = \beta_0 + \beta_1 \ln(PbC) + \ln(E) \tag{1}$$
$$\ln(PbWAL) = \beta_0 + \beta_1 \ln(PbL) + \ln(E) \tag{2}$$

where

PbWAC = lead concentration based on a weighted average of individual samples,
PbWAL = lead loading based on a weighted average of individual samples,
PbC = lead concentration from the composite sample,
PbL = lead loading from the composite sample,

and ln(E) is normally distributed.

These models yielded the following results:

For Pb loading:
estimate of β_0 = .31, with a standard error of .58;
estimate of β_1 = 1.00 (.996), with a standard error of .11.
For Pb concentration:
estimate of β_0 = 2.04, with a standard error of 1.15;
estimate of β_1 = .72, with a standard error of .15.

B. ANALYSIS OF THE MAXIMUM INDIVIDUAL READING–COMPOSITE RELATIONSHIP

The maximum lead loadings and concentrations for each room were determined and compared to the composite values for that room. The maximum of the individual samples was highly correlated with the composite value for Pb loadings and for Pb concentrations (r = .93 and .77, respectively).

To examine the relationship between the composite samples and the maximum of the individual Pb loadings and concentrations, regression models of the following form were used:

$$\ln(PbmaxC) = \beta_0 + \beta_1 \ln(PbC) + \ln(E) \tag{3}$$
$$\ln(PbmaxL) = \beta_0 + \beta_1 \ln(PbL) + \ln(E) \tag{4}$$

where

PbmaxC = maximum lead concentration in the four individual samples,
PbmaxL = maximum lead loading in the four individual samples,
PbC = lead concentration from the composite sample,
PbL = lead loading from the composite sample, and
ln(E) = normally distributed.

These models yielded the following results:

For Pb loading:
estimate of β_0 = 1.02, with a standard error of .69;
estimate of β_1 = 1.02, with a standard error of .14.
For Pb concentration:
estimate of β_0 = 2.93, with a standard error of 1.5;
estimate of β_1 = .71, with a standard error of .19.

C. ANALYSIS OF THE MINIMUM INDIVIDUAL READING–COMPOSITE RELATIONSHIP

The minimum lead loadings and concentrations for each room were determined and compared to the composite values for that room. The correlations of the minimum of the individual samples with the composite value for Pb loadings and Pb concentrations were r = .84 and .56, respectively.

To examine the relationship between the composite samples and the maximum of the individual Pb loadings and Pb concentrations, regression models of the following form were used:

$$\ln(PbminC) = \beta_0 + \beta_1 \ln(PbC) + \ln(E) \tag{5}$$
$$\ln(PbminL) = \beta_0 + \beta_1 \ln(PbL) + \ln(E) \tag{6}$$

where

PbminC = minimum lead concentration in the four individual samples,
PbminL = minimum lead loading in the four individual samples,
PbC = lead concentration from the composite sample,
PbL = lead loading from the composite sample,

and ln(E) is normally distributed.

These models yielded the following results:

For Pb loading:
estimate of $\beta_0 = -.61$, with a standard error of .85;
estimate of $\beta_1 = .78$, with a standard error of .17.
For Pb concentration:
estimate of $\beta_0 = 3.71$, with a standard error of 1.44;
estimate of $\beta_1 = .37$, with a standard error of .18.

IV. DISCUSSION

The results of this preliminary study suggest that floor dust composites can be used as a practical means of determining floor dust-lead levels in a room and that compositing can be used to reduce the number of samples per house, without sacrificing information on total lead loadings. In particular, the three hypotheses (Section II.D) were not supported by the data. (The implication of this finding for widespread screening of houses is obvious.) It should be pointed out, however, that this was a small pilot study using a fixed number of individual samples (n = 4) per room. Results of a larger study will allow determination of the regression estimates, with higher precision, and, hence, will better determine the relationship between composites and individual samples. Overall, there appears to be a stronger relationship between weighted averages of individual samples and composite values for Pb loadings (r = .92) than for Pb concentrations (r = .85).

One aspect of sampling for environmental contaminants, using composite samples, that may be lost is the detection of "hot spots." This data set provided an opportunity to investigate this problem. Surprisingly, the maximum of the individual samples was highly correlated (r = .94 for loadings, r = .77 for concentrations) with the composite sample value, indicating that information on "hot spots" is not compromised using composite sampling. Minimum values of the individual samples did not correlate as well with the composite values (r = .84 for loadings, r = .56 for concentrations), indicating that compositing may not be very effective in determining minimum levels of lead contamination, particularly using concentrations.

It is recommended that further work on the properties of compositing be undertaken. In particular it is important to determine the optimum area to sample with a composite sample. One of the authors has developed a theory of the variability associated with composite sampling.[5] In a subsequent paper the authors will address the issue of variability in composite sampling and relate this variability to the estimates obtained from this dataset. This investigation will shed further light on the relationship between composite samples and individual samples. Investigations of alternative sampling devices should also be undertaken to determine whether they exhibit desirable compositing properties.

ACKNOWLEDGMENTS

This work was supported by the EPA (Contract No. 68-DO-0126), Office of Pollution Prevention and Toxics, Design and Development (DDB) and Field Studies (FSB) Branches, Washington, D.C. We thank the following persons for their assistance: Desmond Bannon, Chief Technician of the Kennedy Krieger Institute (KKI) Trace Metals Laboratory; Ruth Quinn, KKI Outreach Coordinator; Peter Lees, PhD, Associate Professor in the Department of Environmental Health Sciences, Quality Assurance Officer; Susan Dillman, EPA DDB Task Manager; and Ben Lim, EPA FSB Chemistry Consultant.

Although the information described in this chapter has been funded by the U.S. Environmental Protection Agency, it does not necessarily reflect the views of the Agency, and no official endorsement should be inferred. Findings appear, in part, in the "Draft Final Report for the Lead-Based Paint Abatement and Repair and Maintenance Pilot Study," submitted to EPA Office of Pollution Prevention and Toxics, May 5, 1992, Washington, D.C.

REFERENCES

1. The nature and extent of lead poisoning in United States children: a report to Congress, U.S. Agency for Toxic Substances and Disease Registry, Washington, D.C., 1988.
2. Preventing Lead Poisoning in Young Children, a statement by the Centers for Disease Control, U.S. Department of Health and Human Services, Public Health Service, Centers for Disease Control, Atlanta, 1991.

3. Comprehensive and workable plan for the abatement of lead-based paint in privately owned housing: report to Congress. U.S. Department of Housing and Urban Development, Washington, D.C., 1990.

4. Kennedy Krieger Institute, Quality Assurance Project Plan for the Lead Paint Abatement and Repair and Maintenance Pilot Study in Baltimore. Submitted to Battelle under Subcontract No. Y-6839(1938)-1749, EPA Contract No. 68-DO-4294, Office of Pollution Prevention and Toxics, Design and Development Branch, Washington, D.C., 1991.

5. Rohde, C. A., Composite sampling, *Biometrics*, 32, 273–282, 1976.

6. Numerical Algorithms Group, The Generalized Interactive Modeling System: The GLIM system, Release 3.77, Royal Statistical Society, London, 1987.

Chapter 27

Sampling Methodology and Decision Strategy for Testing for Lead-Based Paint in Public Housing

D. C. Cox and J. G. Schwemberger

CONTENTS

I. INTRODUCTION

In the federal government responsibility for addressing the problem of lead-based paint in housing in the U.S. has been assigned to the Department of Housing and Urban Development (HUD). The Environmental Protection Agency (EPA) has signed a Memorandum of Understanding with HUD, to provide technical support on lead-based paint issues.

In December, 1989 HUD was under pressure from Congress to release, by April 1, 1990, guidelines on testing and abating lead-based paint in public and Indian housing developments. HUD had been working on a set of guidelines, in conjunction with the National Institute of Building Sciences, for approximately 2 years. EPA agreed to give HUD the technical advice and support necessary to complete the guidelines and publish them by the required deadline.

This paper presents the statistical approach to testing for lead-based paint in public housing, developed by EPA in support of HUD. The rationale for the selection of a sample of units to be tested is presented, followed by a discussion (with examples) of statistical decision rules to be used by public housing authorities to make decisions on abatement of units and components within units.

II. THE LEGAL REQUIREMENTS

There are about 1.5 million public housing units in the country. About half of these, or 750,000, are units where children are likely to reside. (Some units are restricted to the elderly.) There are about 4000 public housing authorities that manage the public housing units. The federal government does not own any public housing, but has considerable leverage over the management of the units, because it controls the money that is used for improvements.

Congressional legislation requires all public and Indian housing authorities to conduct a random sample of all public housing developments where children live, or are expected to live, and to test these units for lead-based paint. All developments are to be tested by December, 1994.

III. THE MEASUREMENT INSTRUMENT

Federal statutes and HUD regulations specify the use of an X-ray fluorescence (XRF) analyzer for measuring the levels of lead-based paint. A level of 1.0 mg/cm^2 or greater is, by statute, a positive finding of lead-based paint. An XRF instrument is portable and provides results quickly. Public and Indian housing

1-56670-113-9/95/$0.00+$.50

authorities may collect paint chip samples and send these samples to laboratories for atomic absorption spectroscopy (AAS) analysis, either as a confirmation of XRF testing or in place of XRF testing.

There are currently two types of XRF devices available: direct readers and spectrum analyzers. Direct readers are more common and less expensive than spectrum analyzers. A direct-reading instrument will display an apparent lead concentration per unit area. This reading includes lead from both the paint and the material, or substrate, behind the paint, as well as the impact of interferences from the substrate. Because direct readers are quite variable, a *measurement* consists of the average of three readings. When using a direct reader, it is necessary to take a measurement on the bare substrate from which the paint has been removed, to obtain a measurement that will be subtracted from the measurement on the paint and substrate combined. The difference of the combined measurement and the bare substrate measurement is the measurement of lead in the paint per square centimeter.

The spectrum analyzer is relatively new. It is more expensive than the direct readers, but is thought to be more accurate. The spectrum analyzer has proprietary software that corrects the XRF reading for interference from a variety of common substrates, including aluminum, wood, plaster, sheetrock, brick, plywood, steel, and concrete. Because of this automatic correction, the manufacturer claims that the operator does not need to correct the spectrum analyzer measurements, by scraping the paint from the surface and taking a bare substrate measurement, as is required for direct readers. This claim was borne out in a limited study conducted in the laboratory by the National Institute of Standards and Technology (NIST).[1] However, more recent field data[2] suggests that the spectrum analyzer's built-in substrate correction is by no means perfect. Extensive validation data collected as part of HUD's national survey of lead-based paint in housing demonstrates that the spectrum analyzer tends to underestimate lead levels in paint on wood, drywall, and concrete (by as much as 70%), and to overestimate lead levels in paint on steel (by 60%). Moreover, the degree of bias observed depends on the particular instrument used. Thus, although the HUD Guidelines do not require substrate correction for the spectrum analyzer, it is strongly recommended.

IV. THE TESTING PROBLEM

Public housing authorities are faced with the problem of having to test a random sample of their developments and then making a decision on which components to abate. Lead-based paint must be abated under the guidelines for public and Indian housing authorities, published by HUD.

The XRF analyzers are permitted by statute and are convenient to use, but are not considered reliable, especially near the legislative threshold of 1.0 mg/cm^2. Collection of paint chips, and subsequent laboratory analysis, while regarded as reliable, is cumbersome, expensive, takes time, and might surpass the available laboratory capacity.

V. THE TESTING SOLUTION

The solution is based on the uniformity in construction across dwelling units in multiunit buildings and developments. It is expected that dwelling units in a public housing development have similar construction. Although all units were probably painted in a similar fashion shortly after construction, later paintings are assumed to be nonuniform.

To test for lead-based paint, the Guidelines direct a public housing authority to identify a sampling frame of units with a common construction history. These units may be in one building or in a group of buildings. A random sample of units is selected from this frame of units. Each painted or varnished component with a unique painting history in each of the selected units is to be tested with an XRF instrument. For a direct reader a few units are set aside for scraping of paint from components so substrate adjustments can be made. As mentioned above, substrate corrections are now recommended for the spectrum analyzer also, although not required by the HUD Guidelines. Decisions on the presence of lead-based paint at, or above, the legislative threshold are made for each component. For each component all the XRF measurements are classified as positive, inconclusive, or negative. Next, for each component the percentages of positive, inconclusive, and negative results are computed. These percentages are compared to threshold percentages derived from simulations, and decisions are made based on the comparison of percentages from the samples to the threshold percentages. There is one set of threshold percentages for direct-reading XRFs and another set for spectrum-analyzer XRFs.

The solution, as described above, relies on the XRF device as the primary testing mechanism. A set of decision rules was derived to interpret the set of XRF readings by making comparisons to the threshold percentages. If the application of the decision rules to the XRF readings leads to no decision, the Guidelines direct the public or Indian housing authority to collect paint-chip samples for laboratory analysis by atomic AAS, or a similar method. In this case the final decision is based on both XRF and AAS results.

VI. STATISTICAL CRITERIA

There are two primary sources of error to control: sampling error and measurement error. Sampling error is a factor because a random sample of dwelling units is permitted by law, and the sizes of many of the public and Indian housing developments suggest that sampling of units is appropriate. As noted above, measurement error in the direct reader is a major issue. The spectrum analyzer, while considered more accurate than the direct reader, also is prone to measurement error.

To control the sampling error, the following criterion was developed and accepted: the number of units to be tested will be chosen so that if no lead paint is found in any unit, there is 95% confidence that no more than 50 units or 5% of the units, whichever is less, have lead-based paint. Given the number of units in the housing development, say N, calculate $\max(50, .05 \times N) = K$. Then for a given sample size n, the probability that no leaded units will be sampled is given by the hypergeometric probability $H(N, K, n, 0)$. To meet the 95% confidence criterion, determine the smallest value of the sample size n for which $H(N, K, n, 0) < 0.05$. This value of n is the required number of units to sample in a development with N units. A program was written in the Statistical Analysis System (SAS), to compute the exact sample size for each development size from 20 to 5000 units.

To control the measurement error in the XRFs, the decision rules contain threshold percentages that meet certain criteria. The threshold percentages were chosen so that for each component, they are highly unlikely to be exceeded when testing paint that is below the 1.0-mg/cm² threshold. Thus, if the decision rule percentages are exceeded, one may be highly confident that lead truly is present on at least one of the tested components above the threshold. The percentages were derived through a simulation approach.

VII. THE SIMULATION APPROACH

The actual distribution of lead levels across a public housing building component is unknown. To obtain the threshold percentages, a uniform, a normal, and two lognormal distributions were used, in turn, for the distribution of the true values of lead in paint, assuming that a very high percentage (95 to 100%) of the true lead levels were below the 1.0-mg/cm² threshold. Measurement error for both types of XRF devices was assumed to be additive and to follow a normal distribution. The simulation process[3] involved first taking a random value from one of the distributions of the true values of lead in paint and then adding a random value from the measurement error distribution. Using more than one distribution for the true values gave some perspective to the results. From the simulation approach, it was possible to derive estimates of the percentage of XRF readings that would fall into the categories "positive," "negative," and "inconclusive," under the scenario of very little lead-based paint present at, or above, the threshold. After review of the range of percentages from the simulations, threshold percentages were chosen for the decision rules, to achieve the high confidence desired.

VIII. THE DECISION RULES

There are four decision rules.[4] First, if a sufficient percentage of XRF readings for a component are positive, all such components in all units in the frame of units must either be abated for lead-based paint or all should be tested to determine which should be abated and which should not. Second, if no XRF results are positive, and no more than a threshold percentage are inconclusive, then the component does not have lead-based paint. Third, if the percentages from the sample are not covered by the first two rules, then confirmatory testing is required. All positive XRF results must be confirmed, and all inconclusive readings above 1.0 mg/cm² must be confirmed. Fourth, if confirmatory testing confirms the presence of lead-based paint above the regulatory level on a sampled surface, then all components of the given type must either be abated or exhaustively tested to determine which ones must be abated.

Table 1 Hypothetical XRF Testing Data from a 55 Unit Development

Component	Number tested	% Positive	% Inconclusive	% Negative
Baseboards	275	30	10	60
Walls	275	0	5	95
Doors	55	13	15	82
Shelves	110	2	20	78

A. HYPOTHETICAL EXAMPLE

Assume we are using a direct reader. The substrate corrected readings for a direct reader are classified as follows:

Results of 1.6 mg/cm^2 or greater are classified as positive;
Results of less than 0.5 mg/cm^2 are classified as negative;
Results between 0.5 and 1.5, including 0.5 and 1.5, are classified as inconclusive.

Assume we sample in 55 units in a public housing development. There are five rooms per unit, and we collect the data shown in Table 1. Decision rule 1 for direct readers says that if more than 15% of the results are positive, then lead-based paint is present on the component, at a level of 1.0 mg/cm^2 or greater. Decision rule 2 for direct readers says if no results are positive and if no more than 17% of the results are inconclusive, then the component does not have lead-based paint at the level 1.0 mg/cm^2 or greater. Decision rule 3 says if there are positive or inconclusive results, with the following percentages, then confirmatory testing must be done: less than or equal to 15% of the results are positive, or more than 17% are inconclusive.

Applying the decision rules to the example above, we see that rule 1 leads to the decision that baseboards have lead-based paint. Because of the high percentage of negatives and the low percentage of inconclusives, testing all components, rather than abating all components, may be the preferred strategy. Rule 2 leads to the decision that walls do not have lead-based paint. Rule 3 leads to the decision that doors and shelves require confirmatory testing.

B. EXAMPLE WITH FIELD DATA

The data was collected by a spectrum analyzer XRF from a public housing development, by a contractor. For a spectrum analyzer, results of 1.3 mg/cm^2 or greater are classified positive; results below .8 are classified as negative; and results between .8 and 1.2, including .8 and 1.2, are classified as inconclusive.

Decision rule 1 for a spectrum analyzer says lead-based paint is present if more than 11% of the readings are classified as positive. Decision rule 2 says if no positives and no inconclusives are found, then the component does not have lead-based paint. Decision rule 3 says that confirmatory testing is required if either the percentage of positives is no more than 11% or if there are some inconclusive readings.

The field data are shown in Table 2. Applying the decision rules, the following components do not have lead-based paint at, or above, 1.0 mg/cm^2: lower walls, baseboards, chair rails, upper walls, and bedroom doors. The following components do have lead-based paint at, or above, 1.0: living room doors, kitchen doors, window sills, window sashes, and window wells. Door casings and window casings are inconclusive, and confirmatory testing is required.

IX. POTENTIAL IMPROVEMENTS

The sampling and testing aspects of the Guidelines were developed in a 4-month period from December 1989 to March 1990. Some aspects related to sampling and testing had to necessarily receive less attention to meet the project deadlines. For example, substrate corrections are an area that could benefit from additional statistical design work. In addition, as more is done on lead sampling and analysis, improvements to the Guidelines will become obvious. Such improvements are particularly important in testing so-called scattered-site housing; i.e., housing where each unit is unique and has a different painting history. The approach outlined in this paper is not valid for such housing. Each unit must be tested separately. Nevertheless, initial indications are that the approach in the Guidelines is workable for most public housing and will uncover many of the cases of lead-based paint.

Table 2 Actual XRF Field Testing Data (Interior Samples Only)

Component	Positive		Inconclusive		Negative	
	Number	Percent	Number	Percent	Number	Percent
Lower walls	0	0	0	0	236	100
Baseboards	0	0	0	0	216	100
Chair rails	0	0	0	0	45	100
Upper walls	0	0	0	0	29	100
Living room doors	24	55	12	27	8	18
Kitchen doors	30	83	2	6	4	11
Bedroom doors	0	0	0	0	35	100
Door casings	1	1	1	1	139	99
Window sills	28	15	26	13	139	72
Window casings	1	2	3	5	53	93
Window sashes	137	72	25	13	29	15
Window wells	44	98	0	0	1	2

REFERENCES

1. McKnight, M. E., Byrd, W. E., Roberts, W. E., and Lagergren, E. S., Methods for Measuring Lead Concentrations in Paint Films, Report NISTIR 89–4209, National Institute of Standards and Technology, Washington, D.C., 1989.
2. Westat, Inc., Comprehensive Technical Report on the National Survey of Lead-Based Paint in Housing, Part II, EPA Contract No. 68-D9–0174, U.S. Environmental Protection Agency, Washington, D.C., 1992.
3. Cox, D. C., Statistical Methods Used in Hazard Identification and Abatement of Lead-Based Paint in Public and Indian Housing, EPA Contract No. 68-D0–0061, U.S. Environmental Protection Agency, Washington, D.C., 1992.
4. Lead-Based Paint: Interim Guidelines for Hazard Identification and Abatement in Public and Indian Housing, U.S. Department of Housing and Urban Development, Washington, D.C., 1990.

Chapter 28

Efficient Methods of Testing Lead-Based Paint in Single-Family Homes

A. Greenland, D. C. Cox, J. G. Schwemberger, and C. Foster

CONTENTS

I. INTRODUCTION

The presence of lead-based paint (LBP) on both interior and exterior building components of dwellings has been shown to be one of the major contributors to lead poisoning, especially among young children. Because of this hazard, the U.S. Department of Housing and Urban Development (HUD) was given the responsibility to lead the government's effort to investigate the problem and suggest guidelines for its solution. The Environmental Protection Agency (EPA) is providing technical support to HUD. With the publication of the HUD "Interim Guidelines,"[1] a first major step in the process was completed. The Guidelines focused on the problem of lead-based paint in public housing, touching only briefly on the parallel situation for single-family homes. This document considers the problem of sampling for LBP in such homes.

A. SAMPLING IN SINGLE-FAMILY HOMES

Chapter 4 of the HUD Interim Guidelines suggested methods of testing for LBP in public housing developments. Those methods employ statistical procedures to specify the selection of a sample of units to be tested. Within each of those sampled units, the Guidelines require testing of all surfaces in all rooms. This approach is efficient in that it limits testing in developments that are free of LBP, thus avoiding a possibly costly and wasteful requirement to test all surfaces.

There is a clear difference in the sampling problem for multifamily housing and that for single-family homes and scattered-site housing. The basic statistical assumption for sampling within multifamily developments is that they are exposed to central management and, therefore, a consistent and similar painting history. Selecting units at random within those developments is justified. The case for single-family homes is much different. The decision as to whether the house is leaded or not (has surfaces at, or above, the 1.0 mg/cm^2 threshold) must be made using measurements taken within that house alone.

This report describes the findings of a project to study the use of sequential sampling of LBP in single-family and scattered-site housing. The Interim Guidelines currently require 100% testing of all surfaces in all rooms of such housing, to determine the presence of LBP. Sequential sampling could provide a rationale that reduces the number of X-ray fluorescence (XRF) tests that must be taken. This is accomplished by examining the test results, one measurement at a time, to see if the collection of samples taken up to that point provides enough information to decide about the levels of lead in that dwelling unit (or subset of that unit).

Sequential sampling is often used in situations in which the cost of sampling is an important factor. Standard sampling methods specify the sample size in advance. The sample is selected and analyzed in support of the decision-making task of interest. Sequential sampling differs from standard sampling in that the sample size is unknown at the beginning of the sampling process. In fact, the sample size is one of the random results of the process.

Sequential sampling for LBP in single-family homes is implemented as follows: begin sampling components within a group of components in the dwelling of interest, by selecting a surface at random. After each test is completed, a computation is performed adding the new measurement to those already taken. If the computed "test" value is above, or below, a prespecified range, respectively, the group of components is classified as having lead or not. If the test value is between the two limits, the classification is inconclusive, and another surface is selected at random for testing. The process continues until either a decision is reached or all surfaces are measured.

B. THE RESEARCH PLAN

Prior to beginning this research project, there was speculation that sequential statistical methods could be of value for testing of LBP in single-family homes; however, there was no evidence to substantiate that belief. Therefore, the following research plan was defined:

- Investigate the underlying lead distributions and select a theoretical model to use in the analysis;
- Derive the sequential procedure;
- Test that procedure, using existing testing databases.

The first and third of these bullets require the examination of data, and we had two basic sources: the HUD Demonstration Database and the data collected by Westat for the HUD National Survey of Lead-Based Paint. Both of these databases were examined during this study.

Although we considered options that would have used both of these data sources, we settled on using the HUD Demo Database to estimate parameters of the underlying distribution and to use as a source of data to test the sequential procedure. The main reason for this decision was that the HUD Demo Database represents a set of measurements that is very similar to the data that will be experienced by lead inspectors in the field, and we want the sequential procedure to be tested in a realistic environment.

Section II below discusses the sequential procedure, including general information about the methods, as well as specific information about how it was applied in this case. Section III discusses the use of the HUD Demo Database as a test bed against which to measure the effectiveness of the sequential procedure.

C. SUMMARY OF THE KEY FINDINGS

The sequential sampling procedure proved to be very successful in both reducing the costs associated with testing and maintaining accurate sampling for the presence of Lead-Based Paint.

In particular we found that the number of XRF tests required to classify surfaces as leaded were reduced in the range of 30 to 70 percent (depending on the sampling strategy used). In addition, the costs associated with testing for LBP could be reduced by an amount in the range of 17 to 50 percent.

While reducing the costs as mentioned above, we found that the sequential procedure was very accurate in classifying surfaces as leaded or not with a misclassification of less than 1 percent of surfaces. This finding is considered to be very encouraging.

Section III below provides a comprehensive discussion of the methodology used in evaluating the procedure and determining the key findings. We conclude that more work needs to be done, both in fine tuning the sequential procedure to better balance the competing cost and accuracy constraints, and to consider other applications for sequential procedures to lead sampling, for example clearance testing.

II. DEVELOPMENT OF THE SEQUENTIAL SAMPLING PROCEDURE

A. SOME BACKGROUND ON SEQUENTIAL METHODS

Sequential sampling methods were developed to deal with situations in which the cost of sampling could be prohibitive. If a decision that needs to be made can be reached easily, for example, by sampling only a few cases, a substantial portion of the cost associated with testing can be saved. All sequential procedures are characterized by a ''stopping rule'' and a ''decision rule.'' The stopping rule determines when to stop selecting more cases to be included in the sample. The decision rule tells the researcher what action to take (or what decision to make). In this application, the decision rule must determine whether the group of components of interest should be classified as having lead-based paint at or above the HUD standard or not.

The most basic and well-known sequential procedure is the Sequential Probability Ratio Test (SPRT). This method derives its name from the ratio of likelihood functions, which is often used to develop test statistics. It was developed in the 1940's by A. Wald [1947]. The SPRT was selected for this application because the ratio of likelihood functions, which is often used to develop test statistics. The Sequential Probability Ratio Test (SPRT) was selected for this application because the theory of this method is well developed and because of the ease of computing the test statistics.

In its most basic form the SPRT is used to distinguish between two simple statistical hypotheses of the form:

$$H_1: \theta = \theta_1 \text{ vs.}$$
$$H_2: \theta = \theta_2,$$

where θ represents a vector of values that define the decision reached by the procedure. In applications, one must select values for θ_1 and θ_2 that allow for deciding between the two real-world situations. For example, in the application of interest, we want to distinguish between whether a group of surfaces have lead at, or above, the HUD standard or not. To do that we must select a specific value for θ_1 that represents one of those options (for example, that the group of surfaces are ''not leaded'') and a value for θ_2 that represents the group being ''leaded.'' We will discuss below how these values are selected.

B. THE UNDERLYING MODEL

In order to apply statistical procedures, the researcher must have a specific target population in mind and a set of parameters defined to use in the decision-making process. Conforming to definitions used in the HUD Guidelines, our target population is a collection of building components that are related to each other by their painting histories. Based upon prior work on sampling of LBP in the HUD National Survey and based upon the availability of supporting data, we further focused attention on groups of building components that we call *cells*. The cells are defined as follows:

Room types:

1. Wet (interior rooms having plumbing)
2. Dry (interior rooms with no plumbing)
3. Exterior

Building component types:

1. Walls, ceilings, and floors
2. Nonmetal substrates (baseboards, windows, etc.)
3. Metal substrates (baseboards, windows, etc.)
4. Other substrates (cabinets, shelves, etc.)

The three room types and four component types are consistent with the definitions used in the HUD National Survey; they were selected not only because of their expected homogeneity, but also because summaries found in the report for that survey could be used in this work. Since there are three room types and four component types, there are a possible 12 cells in each dwelling unit.

Within each dwelling unit of interest, all of the building components are grouped into cells. The sequential procedure was applied to the collection of measurements within each cell separately. The decision rules, discussed later in this section, are defined so as to classify each of these cells as being ''leaded'' or ''not leaded.'' This method relies upon the concept that the above-defined cells are sufficiently related in their painting histories that making a ''group'' decision is reasonable, and the findings of the study, which are described in Section III of this report, bear out that concept.

C. DEFINITION OF THE STOPPING AND DECISION RULES

The classic formulation of the SPRT for normal distributions (see for example, Chernoff[3]) is given by:

1. Stop sampling and accept H_1 if $\overline{X} \le L_n$.
2. Stop sampling and accept H_2 if $\overline{X} \ge U_n$.
3. Continue sampling if $L_n < \overline{X} < U_n$.

where

$$L_n = \overline{\Theta} - [\ (\log A)\ \sigma^2]\ /\ n\ (\Theta_2 - \Theta_1)$$
$$U_n = \overline{\Theta} - [\ (-\log B)\ \sigma^2]\ /\ n\ (\Theta_2 - \Theta_1)$$

Consider next the computation of the values A and B. These two quantities rely only on the probabilities of Type I and Type II error, α and β, respectively. The researcher decides in advance what levels are required to control both Type I and Type II error, and the values of A and B can be estimated. One of the key results in the theory of sequential probability ratio tests (also due to Wald) is that the values A and B can be approximated as follows (see Chernoff[3]) :

$$B = \frac{(1-\beta)}{\alpha} \text{ and } A = \frac{\beta}{(1-\alpha)}$$

In this study, we used $\alpha = \beta = 0.05$. Under those circumstances the test values are computed as follows:

$$\log A = 2.944 \text{ and } \log B = -2.944.$$

The final aspect of the decision and stopping rule is practical in nature. Further, it requires modification of the strict theoretical formulation of the sequential procedure. When a surface was selected on which the XRF value in the HUD Demo Database was at, or above, the threshold of 1.0 mg/cm^2, that entire group of surfaces was designated as "leaded," and the sequential procedure is terminated. This is consistent with the HUD Guidelines, which indicate that any surface found to have a lead concentration at, or above, the threshold value must be abated. By that definition the group of surfaces in which the single leaded surface is found makes that collection of surfaces "leaded."

The requirement described in the prior paragraph results in a very clear departure from the pure sequential procedure and results in a modified set of stopping rules. The new rules are

1. Stop sampling and accept H_1 if $\overline{X} \le L_n$.
2. Stop sampling and accept H_2 if $\overline{X} \ge U_n$ or any $X \ge 1$.
3. Continue sampling if $L_n < \overline{X} < U_n$ and all $X_1, \ldots, X_n < 1$.

D. SELECTION OF THE PARAMETERS OF THE SEQUENTIAL PROCEDURE

As discussed above, SPRT requires the researcher to use only simple hypotheses in the sequential decision process. In the particular application of the presence of LBP in a collection of components, the actual goal is to classify the collection as "leaded" if any one of the components has a concentration at, or above, 1 mg/cm^2, and to classify the group as "not leaded" if none are above that value.

The strategy used for determining the two parameters θ_1 and θ_2 was developed by trial and error. One basic principle used was that the value associated with the "leaded" decision should use a mean value, θ_2, over the HUD limit of 1 mg/cm^2, and similarly the value θ_1, associated with a decision that the group of surfaces is "not leaded," should be a value less than 1 mg/cm^2. Also, we spaced the two values equally from the central value of 1 mg/cm^2. From experimentation it was clear that as the values for θ_1 and θ_2 were moved farther away from 1 mg/cm^2, there was a tendency for the procedure to terminate prematurely (often with the wrong answer). Similarly, as the two values got closer to 1, the procedure tended to be unable to terminate with a decision before all of the measurements were taken.

After examining several alternatives, we settled on the following definitions for the two key parameters:

$$\theta_1 = 0.8 \text{ mg/cm}^2$$
$$\theta_2 = 1.2 \text{ mg/cm}^2$$

As will be discussed in Section III, these values resulted in accurate decisions; however, we do not claim that these are the optimal values. For example, there may be a different set of values, θ_1 and θ_2, which provide a better balance between accuracy of classification and cost of measurement. One of the areas that we believe demands additional research is to "tune" these parameters to produce the most efficient procedure.

In addition to the mean values along the lead distribution axis, the Normal model sequential procedure defined in Section II.C requires a known standard deviation, σ. To determine this value we looked at the empirical lead distribution obtained from the HUD Demo Database. These values are displayed in Table 1. There are some generalizations that one can make by looking at these values. In general, XRF measurements on exteriors are more variable than those in wet rooms, which are, in turn, more variable than those in dry rooms. The walls/ceilings/floors substrate groups are generally less variable than the nonmetal substrate groups, but the metal and other substrate groups do not tend to follow a pattern.

In order to simplify the programming process, it was decided to select a single standard deviation, σ, for all applications of the sequential procedure. There is a similar tradeoff in the selection of σ as in the selection of the two means. Ordinarily for SPRT procedures, these are very large or very small values in the data that direct the decision-making algorithm to one or the other of the hypotheses. However, when the value of σ is large, it is difficult to distinguish large or small values from variability in the data. This explains why when the assumed standard deviation is large, more iterations of the sequential procedure are expected. Also, when the standard deviation is large, the sequential procedure will not be as likely to classify a cell that has some higher XRF values (those near to, but not greater than, 1 mg/cm^2) as "leaded." Therefore, the number of misclassifications should drop.

A standard deviation of 1.0 mg/cm^2 was selected for this analysis. Many values in Table 1 exceed that value; however, compared to the HUD standard of 1 mg/cm^2, such a standard is quite large (100% of the size of the standard). Also, we took an exploratory approach to the selection of the standard deviation. We started out with $\sigma = 1.0$ mg/cm^2 and expected to reconsider it only if the results were discouraging. In fact, the results of that first test were very encouraging (fewer than 5% of cells misclassified and fewer than 1% of surfaces misclassified), so we did not reconsider the assumed level of σ.

Based on the definitions of the SPRT, it should be noted that the test for normal distributions relies solely upon the two functions: L_n and U_n. Also, based upon the formulas for these quantities, it is clear that doubling the value of σ^2 and also doubling the distance between θ_1 and θ_2 leaves the operation of the sequential test unchanged.

Further research on this methodology should also consider alternative values of σ and, in addition, the use of different levels of σ for different groups (for example interior vs. exterior surfaces).

III. EVALUATION OF THE EFFECTIVENESS OF THE SEQUENTIAL PROCEDURE

A. THE METHOD OF EVALUATION

The plan to evaluate the two proposed sequential procedures (one for each of the distributions discussed in Section II.C) included the following steps:

1. Prepare the database of XRF measurements;
2. Apply the sequential procedure to those houses using Monte Carlo simulation;
3. Use the results of the simulation to:

- measure the number of tests that were done as compared to the number that would have been under the HUD Guidelines
- estimate the number of misclassifications that resulted from the application of those methods

4. Interpret the results in terms of costs and benefits.

Each of the next four sections of the chapter contain discussion of one of these bullets.

B. PREPARATION OF THE HUD DEMONSTRATION DATABASE

The main reason for selecting the HUD Demonstration Database for use in this study is that it is a collection of actual direct-reader XRF measurements taken in private housing units, similar to the types of measurements that will be encountered in the field, by inspectors. Moreover, the database is available in a computer readable form for ease of access. Since it has 306 dwelling units and measurements on over 30,000 surfaces,

Table 1 Estimated Standard Deviations in mg/cm² from HUD Demonstration Database by City, and Room Type within Component Types

Substrate room type	City from HUD demonstration project				
	Baltimore	Birmingham	Denver	Indianapolis	Seattle
Walls, ceilings, and floors					
Wet	3.39	2.63	2.17	2.05	3.55
Dry	1.88	1.13	1.10	0.70	1.65
Exterior	5.82	7.49	4.16	5.23	7.88
Metal substrate					
Wet	4.74	0.26	2.63	1.53	1.13
Dry	2.47	0.56	1.08	0.94	2.82
Exterior	5.40	0.76	—	3.09	3.03
Nonmetal substrate					
Wet	4.56	2.54	2.87	2.15	3.82
Dry	4.67	2.39	3.46	2.20	3.61
Exterior	6.77	5.39	5.23	6.53	8.20
Other substrate					
Wet	2.72	1.84	0.80	1.73	1.80
Dry	2.01	2.71	1.55	0.67	2.26
Exterior	1.75	—	—	—	—

Table 2 Accounting of Records Useable for the Sequential Sampling Test Among Those in the HUD Demonstration Database

City	HUD Demo Database records available for sequential analysis			
	Total number records	Number having room and component	Number having XRF measurements	Number with room, component, and XRF
Baltimore/DC	8,728	8,647	7,367	7,323
Birmingham	4,720	4,617	4,408	4,352
Denver	10,623	10,521	9,485	9,427
Indianapolis	5,689	5,615	5,120	5,080
Seattle/Tacoma	5,037	4,984	4,753	4,722
Total	34,797	34,384	31,133	30,904

it is a formidable collection of information. The number of records in each dataset by city is shown in Table 2.

Further details of the data preparation, including the definition of cells and the treatment of missing and negative XRF values, are available in the EPA report[4] prepared concurrently with this study.

C. IMPLEMENTATION OF THE SEQUENTIAL PROCEDURE

The sequential sampling procedure was performed within each cell within dwelling unit and city. Within each of those cells, surfaces were selected one at a time for inclusion in the procedure. Since in the field the surfaces must be selected purely at random, not favoring any type of surface in the cell, the surfaces in this simulation were selected from the total within the cell, using a random number generator. A cell is defined as a combination of unit, room, and component type.

The computer program that implements the sequential sampling procedure had five distinct steps. They were to

- Preprocess data to handle missing and negative XRF values
- Assign values for the θ_1, θ_2, and σ parameters
- Randomly sort the XRF measurements within the cell
- Calculate an overall number, S_n, for each record sampled until the cell is defined as "leaded" or "not leaded"
- Save a dataset that includes the records involved in the procedure

Table 3 Total Number of Surfaces and Tested Surfaces in HUD Demo Database Under Sequential Simulation (Underlying Normal Distribution)

City	Number of units	Total number of surfaces	Number of tests required to classify	Number of tests required to find all leaded surfaces
Baltimore	73	8,647	2,171	6,513
Birmingham	37	4,617	1,533	2,913
Denver	97	10,521	3,735	7,712
Indianapolis	53	5,615	1,880	3,399
Seattle	46	4,984	1,666	3,637
Total (percent of total surfaces)	306	34,384	10,985 (32.0%)	24,174 (70.3%)

Further details of this implementation is found in Greenland.[4]

D. DATA ANALYSIS

During the simulation phase, information was collected that permits evaluation of both the reduction in cost using the proposed sequential procedure, and the level of accuracy of the method. Both cost and accuracy are related to information collected about the surfaces in dwelling units, in the HUD Demo Database. The useful information collected during the simulation included the following:

- For each cell in each dwelling unit, we have the XRF measurements on all of the surfaces in the cell: both those selected randomly for inclusion in the sequential procedure and those not selected (because a decision was already reached in that cell);
- We know the total number of surfaces that were required for use in the sequential procedures, n, and the total number of surfaces, m, that are in that cell.

Using simple rules about how XRF testing is done according to the Interim Guidelines, we can determine cost and accuracy. The next two sections deal, respectively, with cost and accuracy.

Cost Analysis

The findings related to cost are shown in Table 3. That table is organized by total number of surfaces. The first column in Table 3 contains the number of units in each city in the HUD Demo Database. The next column has the total number of surfaces in the database whether or not there was a XRF measurement value for that surface in the database. Because the Interim Guidelines require that all surfaces in all rooms must be tested, we interpret the number of "Total Number of Surfaces" to be equal to the number of XRF tests that would have to be done following the HUD Guidelines.

The next column contains the total number of surfaces that were tested when simulating the sequential procedure, as defined in Section II.C of this report. The column is titled "Number of Tests Required to Classify." It represents exactly the number of XRF measurements that, according to the sequential procedure, were used to decide if there was lead present in a group of surfaces or not. However, it may not represent the totality of tests that must be taken when using the sequential procedure. Consider two key groups: (1) cells that are classified as "leaded" and (2) cells classified as "inconclusive." One assumption we used in analyzing the data is that when the sequential procedure determines that a particular cell is "leaded," then the Interim Guidelines become applicable to that cell. Therefore, we infer that XRF measurements will have to be taken on the remaining surfaces of the cell. It is useful to give an example of this situation. Suppose we are considering the cell defined for all nonmetal surfaces in all wet rooms. This includes baseboards, window frames, doors, etc., that are in bathrooms and kitchens and other rooms with plumbing. Also assume there are 30 such surfaces in a dwelling unit. If the sequential procedure determines that this cell is leaded after examining 10 surfaces, one scenario for reaching abatement decisions requires that all of the other 20 surfaces be tested to determine exactly which are leaded and which not. The use of this and other scenarios for defining abatement strategies will be further discussed below.

Consider the group of cells that are defined as "inconclusive." By definition, all of the surfaces in these cells must have been tested, but the sequential procedure did not determine that they were leaded or not. In fact, they must be not leaded. The sequential procedure was not able to decide from the observations alone that they were, but since all surfaces were tested and none had lead at, or above, the HUD standard (otherwise the procedure would have classified the cell as "leaded"), we can conclude that the cell has no lead. Since all surfaces were included in the sequential procedure, they are counted (in Table 3) among those required to classify leaded surfaces.

The difference between the number of surfaces required to *classify* the cells and the number of surfaces *required to find all leaded surfaces* is captured in the two columns, of Table 3, with those titles. As can be seen in Table 3, the number of XRF measurements required to find all leaded surfaces is much larger than the number needed to classify the cells. The table shows that 32.0% of the surfaces need be tested to classify cells, while 70.3% of the surfaces need to be tested to uncover exactly which surfaces have lead in the cells classified as "leaded."

Having both levels of testing is useful. Policy makers may want to consider suggesting different abatement strategies for homeowners. One strategy is to abate a cell completely when it is classified as leaded (i.e., without any further testing). Another is to continue testing to determine exactly which surfaces have lead at, or above, the HUD standard in those cells classified as "leaded." The economics of choosing one of those strategies will depend on the particular type of building components involved (walls will be different from window frames, for example) and the abatement strategy (removal will be different from encapsulation). Those issues are beyond the scope of this project. However, we do suggest that tabulating and considering both levels of reduction in needed XRF measurements is beneficial for future policy analysis.

A Simple Economic Model of Cost

The information contained in Table 3 relates to numbers of XRF measurements that must be made. However, it cannot be translated directly into cost savings. In order to understand the true costs better, we developed a simple economic model.

The cost of XRF testing is determined by many factors. In addition to the obvious per hour labor charge for inspectors, the total cost includes the transportation costs related to getting a test crew to each dwelling unit, capital costs associated with owning the XRF equipment, maintenance of the equipment, and other standard overhead costs.

We also make some assumptions about costs associated with testing under the current Interim Guidelines, as compared to that using the sequential procedure. Under HUD Guidelines for a single-family dwelling unit, an inspector would be required to obtain XRF measurements on every surface in every room of a dwelling unit. Using the sequential procedure, that number would be, according to Table 3, either 70.3% of total surfaces (if all surfaces in "leaded" cells are tested) or 32.0% of total surfaces (if we are concerned only with the cost of *classifying* cells as leaded or not).

In preparing this report we interviewed several inspection contractors and other knowledgeable professionals in the lead-based paint field, to gather information about such costs. Our cost model incorporates the cost structure in use by one of the major LBP inspection firms in the Massachusetts area,[5] but applied to federal guidelines rather than Massachusetts testing requirements. The model also ignores costs associated with managing a random sample rather than sampling all surfaces in a unit. Based upon information obtained, we assume that following the HUD Guidelines exactly, a contractor would charge approximately $250 to test a six-room dwelling unit. Estimating 7.2 as the average number of rooms per house, we conclude that the average cost per unit in the HUD Demo Database is $300. Also, based upon contractor discussions, we assume that using a two-person crew, one "HUD Guideline" inspection would be completed per day. The total labor and equipment cost to inspect the 306 HUD Demo houses without the sequential procedure would be

$$(306 \text{ units}) \times (\$300 \text{ per unit}) = \$91,800$$

Note, this figure includes the cost of the inspectors' time in transit to and from the job site, but not the travel cost. Again, using information obtained from contractor interviews, we estimate the average daily round trip mileage at 60 miles. Using a $0.28 cost per mile, the travel cost is estimated at

$$(60 \text{ mi/day}) \times (306 \text{ days}) \times (\$0.28/\text{mi}) = \$5,141$$

The total estimated cost to inspect the 306 HUD Demo Database units under the Interim Guidelines is $96,941. These results are displayed as the first row of Table 4.

Table 4 Summary of Estimated Cost Reductions Using the Sequential Procedure for LBP Testing

	Number of days	Number of hours	Daily cost ($)	Travel cost ($)	Total ($)	Percent cost decrease
Using procedure in HUD Interim Guidelines	306	2,142	91,800	5,141	96,941	
Using sequential procedure to classify cells only	137	685	41,100	5,141	46,241	52.3
Using sequential procedure to determine all leaded surfaces	251	1,506	75,300	5,141	80,411	17.1

Consider now the impact on the cost of using the sequential procedure, rather than the HUD Guidelines approach. We estimate the costs under the two scenarios described above. They are (1) the sequential procedure used to classify cells only (i.e., the abatement decision is made from that information alone) or (2) the sequential procedure used to identify all leaded surfaces (i.e., all surfaces in cells classified as "leaded" will be tested).

Table 4 contains all the information needed to understand the cost model. The first row of that table is dedicated to the cost for sampling under the current HUD Interim Guidelines, while the second and third rows are for the two sequential-sampling scenarios described above.

There are six columns in the table. The first column indicates the number of sampling days that would be required by a two-person crew to complete the 306 inspections. The second column in the table has the associated number of inspection hours. The latter columns have dollar cost estimates that are derived from the first two columns and unit cost assumptions.

To begin, consider HUD Guidelines sampling (the first row of Table 4). We assume that being required to complete only one house per day, the crew would spend 7 h of an 8-h shift inspecting and 1 h in transit. Therefore, the total number of inspection hours for the 306 units is 2142. The other numbers in that row were derived above.

Consider now the scenario of using the sequential procedure to classify cells as leaded or not leaded. As noted in Table 3, this will require 32.0% of the surfaces to be tested. Compute the total number of inspection hours required for this scenario by multiplying the 2142 hours required for HUD Guidelines sampling by 32.0%, obtaining 685 hours of inspection for this scenario. Next, estimate the number of days of inspection required by making an assumption about how many inspections could be done in a single day. Since there is roughly a two-thirds reduction in the number of surfaces to be inspected, we assume that two to three units can be inspected per day; but much more of the day will be spent traveling. To accommodate the increased travel, we assume that only 5 h day of useful inspection time will result. Therefore, the 685 inspection hours will require 137 days of work. Multiply the 137 days by $300 per day to determine the value shown in the column labeled "Daily Cost" ($41,100), and add the travel cost to obtain the total. Thus, the sequential procedure to classify cells results in a total cost of $46,241 and represents an overall reduction from the HUD Guidelines cost of 52.3%

Note, in the above analysis we assumed the same travel cost for inspecting the 306 dwelling units as that obtained for the HUD Guidelines sampling protocol. This assumption is reasonable because having more than one unit per day provides both some travel cost savings and some possible additional travel costs. If you have one inspection per day, there is always the distance both to and from the site. Having more than one unit to inspect in a day means that you can plan the route so as to have a travel leg to the first site, to the second site, etc., and one "return" leg. This tends to reduce the total travel cost. However, in some cases the inspection crew may have to return to the same site over a 2 day period, thus increasing the total travel cost. For the sake of simplicity we assume the travel costs would be about the same for all three scenarios.

Consider next the scenario in which we use the sequential procedure to determine *all* leaded surfaces (the third row of Table 4). In this case all surfaces in cells defined as "leaded" must be tested with the XRF, so the total number of surfaces requiring testing is larger than the previous scenario. From Table 3 we use the percent of total surfaces to be inspected (70.3%) applied to the 2142 inspection hours, to obtain 1506 total inspection hours required. Next, we assume that between one and two inspections can be done

Table 5 Total Number of Cells Misclassified in Testing the Sequential Procedure on the HUD Demo Database (Underlying Normal Distribution)

City	Total number of cells	Number of cells misclassified	Percent misclassified
Baltimore	593	31	5.2
Birmingham	349	25	7.2
Denver	894	29	3.2
Indianapolis	449	30	6.7
Seattle	410	10	2.4
Total	2,695	125	4.6

in a day. For simplicity we assume that there would be 6 h of inspection out of an 8-h day (leaving 2 h for travel time between inspections). Therefore, it would require $1506 \div 6 = 251$ days to inspect all 306 units. Estimate the total daily cost at $251 \times \$300 = \$75,300$; add the travel cost of $5,141 to obtain a total cost of $80,411. This translates to a 17.1% cost reduction when using the second inspection scenario.

We propose that these two scenarios be viewed as upper and lower bounds on the cost savings derived from employing the sequential sampling procedure defined in Section II of this paper. This is reasonable because it may be the case that some contractors and homeowners will adapt various strategies that are mixtures of the two "pure" strategies described above. Under this concept we conclude that the cost savings from using the sequential procedure described in Section II of this paper is between 17 and 50%.

Accuracy Analysis

The cost analysis of the prior section cannot be considered without carefully examining the accuracy of the sequential procedure in finding the lead in a house. It must be emphasized, also, that this method is statistical in nature, as it looks at only a group of components and attempts to classify the collection of those components as leaded or not. As with any statistical procedure, there will be some level of misclassification. In this particular case the level of misclassification is quite small and well within the-less-than 5% misclassification criteria mandated in the Interim Guidelines.

To quantify this aspect of the sequential sampling process, we examined the results of the simulated tests on the HUD Demo Database, in detail. Our accuracy analysis focuses on both cells and surfaces, to determine the accuracy of the classification process. Recall that a cell is a room/component type within a dwelling unit. The collection of surfaces within a cell is the collection of components on which a single application of the sequential procedure is applied. As Table 5 indicates, the HUD Demo Database has a total of 2695 cells. Thus, we had 2695 times to classify a group of components as leaded, not leaded, or inconclusive. We assume that once the group of cells is defined as leaded or not, then sampling stops, and no other XRF measurements are observed. However, since we were using a set of data that had XRF measurements on almost all of the surfaces in the database, we had the capability to determine whether any of the measurements that were not part of the sequential procedure had lead at, or above, the HUD limit or not.

It is important to comment on the use of XRF measurements to classify surfaces. This study is based upon the supposition that XRF measurements classify the presence or absence of lead, without error. We know that supposition is not strictly correct, because of the existence of measurement error. However, we needed a standard from which to work, and we believe that the basic findings of the study are not affected by this assumption. We considered methods to incorporate measurement error in the analysis, and although modifications to the study were possible, they were outside the scope and resources of this study. Therefore, we suggest that incorporating the impact of XRF instrument measurement error be one of the enhancements considered for further work on sequential procedures applied to LBP testing.

The results of the classification process for cells are displayed in Table 5. A cell was determined to be "misclassified" in two cases. The first case is if the sequential procedure said a cell was "not leaded," but one of the subsequent observations (the ones not included in the sequential procedure) had a lead value at, or above, the HUD limit. Conversely, a cell would be misclassified if the sequential procedure determined it was "leaded," but in fact there were no surfaces with lead at, or above, the HUD standard. In this simulation we encountered no cells that were classified as "leaded," but were really "not leaded."

Also as noted above, any cell that was determined to be inconclusive by the sequential procedure is a cell in which all surfaces were tested and no lead was found. So, from the point of view of classification,

Table 6 Total Number of Surfaces Misclassified in Testing the Sequential Procedure on the HUD Demo Database (Underlying Normal Distribution)

City	Total number of surfaces	Number of surfaces misclassified	Percent misclassified
Baltimore	8,503	75	0.88
Birmingham	4,608	46	1.00
Denver	10,349	69	0.77
Indianapolis	5,578	60	1.08
Seattle	4,974	13	0.02
Total	34,012	263	0.77

we assume that such cells are correctly classified. The only cells that are considered misclassified are those that are classified as "not leaded," by the sequential procedure, but which have at least one surface that is at, or above, the HUD standard, among those not tested.

Under those definitions Table 5 indicates that 125 out of 2695 were misclassified (4.6%). This is a very good record indeed. But this method is actually very stringent as a measure of accuracy, because there may be only one or two leaded surfaces among many unleaded surfaces in the "misclassified" cell. Therefore, we also tabulated the misclassifications by surface, and these results are shown in Table 6.

The surfaces are classified as "truly leaded" if the XRF measurement is at, or above, 1 mg/cm^2. If the surface is included in the sequential procedure (either as leaded or not leaded), we consider that it to be classified correctly. Also, if the surface was in a cell that was labeled inconclusive by the sequential procedure, we also assume that it is correctly classified, because all surfaces are tested. It is only those surfaces that were in a cell that is classified as "not leaded," but has an XRF value at, or above, the HUD limit, that are counted as a misclassification. As noted in Table 6, of over 34,000 surfaces in the database, only 263 (less than 1%) were misclassified by the sequential procedure. We consider this level of accuracy to be outstanding, as it is well within the 5% misclassification requirement in the HUD Guidelines.

Note, the total number of surfaces shown in Table 6 is smaller than that listed in Table 2. This is because in some cells all the records contained missing values for the XRF measurement, and these cells were excluded from the sequential procedure. It resulted in a smaller set from which to do the accuracy analysis.

E. CONCLUSIONS

The sequential procedure provides a very useful alternative to complete testing in single-family homes. The specific procedure developed in Chapter 2 of this document may still need fine tuning to be workable; however, we consider the results quoted in the prior section to be very encouraging. The analysis earlier in this chapter revealed that the number of surfaces required to decide whether there is a lead hazard, can be reduced to between 30 and 70%. The simple economic model we formulated indicates that the associated dollar cost reduction is in the range of 17 to 50%.

To counterbalance the cost benefit, we must consider the loss in accuracy of classification of groups of surfaces as leaded or not. The findings here are dramatic. There is a less than 1% misclassification of surfaces. This is well within the requirements of the HUD regulatory mandate, and we consider it to be excellent.

We further believe that the findings of this study justify further research on sequential procedures for LBP testing. In particular we suggest further work on tuning the parameters of the sequential procedure. The findings of the previous sections indicate that the level of misclassification is very low, substantially lower than the 5% level required by HUD Guidelines. In contrast, the cost reduction, especially for the scenario in which all leaded surfaces are to be identified, is not as dramatic as one would want. It is likely that by tuning the parameters of the sequential procedure, we can bring these two competing measures into a better balance.

A second suggestion for further work involves investigating the application of sequential procedures for sampling of other media. Consider in particular the application to clearance testing. In this case one must sample for the presence of lead in dust in a dwelling unit in which abatement was done. It is likely that if abatement was done properly, there will be very small amounts of lead in the dust. This situation is one in which sequential procedures are particularly useful, and we believe that some research into that application would be beneficial to the lead program.

Finally, we suggest that future work with the HUD Demo Database include a method of dealing with measurement error in the XRF instruments.

ACKNOWLEDGMENT

Support for this work was funded by the U.S. Environmental Protection Agency, Office of Pollution Prevention and Toxics, Exposure Evaluation Division, under contract number 68-D0–0099. Opinions expressed in this paper are those of the authors and do not represent official policy of the EPA or any other public agency.

The authors also gratefully acknowledge the comments and suggestions of reviewers of this paper.

REFERENCES

1. Lead Based Paint: Interim Guidelines for Hazard Identification and Abatement in Public and Indian Housing, U.S. Department of Housing and Urban Development, Washington, D.C., 1990.
2. Wald, A., *Sequential Analysis*, John Wiley & Sons, New York, 1947.
3. Chernoff, H., *Sequential Analysis and Optimal Design*, Society for Industrial and Applied Mathematics, Philadelphia, 1972.
4. Greenland, A., Development and Testing of a Sequential Procedure for Sampling of Lead Based Paint in Single Family or Scattered-Site Housing. Final Report for Task 2–2 under contract No. 68-D0–0099, September 30, 1992.
5. Weydt, J., personal communication, 1992.

Chapter 29

Relationships Among Lead Levels in Blood, Dust, and Soil

D. A. Burgoon, S. W. Rust, and K. A. Hogan

CONTENTS

I. INTRODUCTION

Significant and permanent detrimental health effects have been shown to result from elevated blood lead levels in children. Ten micrograms of lead (Pb) per 100 ml of blood is the new, lower level of concern adopted for lead poisoning, by both the Centers for Disease Control (CDC) and the U.S. Environmental Protection Agency (EPA). Lead has been found in environmental media such as soil, surface dust (indoors and outdoors), water, and air. Sources of this lead include crumbling and chalking lead-based paint, airborne lead particulates released by industrial or waste elimination emissions, and lead plumbing materials, and a large reservoir of lead is in the surrounding soil, from the fallout of decades of leaded gasoline emissions. The pathways by which children are exposed to lead are even more varied: unintentional soil and dust consumption (usually via their hand-to-mouth activity), dustfall on food, ingestion of lead in tap water, and inhalation of airborne lead particles. Today tighter regulations govern industrial and automotive emissions, and lead pipes are gradually being eliminated. The existing reservoir of lead in paint, soil, and dust may well present the greatest hazard.

Public health policy requires reduction of exposure to these sources, either by removing them or by reducing contact with them. Lead control and abatement techniques, such as soil removal, lead-based paint removal or encapsulation, reduction of water corrosiveness, and dust-lead control via wet mopping, are among the options available. One measure of the effectiveness of these procedures is their ability to reduce blood lead levels. Analysis of the relationships among blood, soil, and dust lead levels may be useful for predicting the effectiveness of abating particular lead sources. Any examination should begin with a consideration of the body of information already in existence. Lead-contaminated environmental media, and their impact on blood lead levels, have been the focus of debate and study, for a number of years. The available information can be examined in two ways: (1) as a review of existing results and conclusions within the literature; and (2) in the form of a reanalysis of existing datasets, aimed at addressing specific issues and concerns. This article documents the results of such a two-pronged examination. Section II reviews and summarizes eleven relevant studies identified from the literature. A reanalysis of selected datasets is documented in Section III. Section IV summarizes the implications of the review and reanalysis

Table 1 Linearized Slope Estimates of Relationships Among Blood, Dust, and Soil Lead Levels

Study	Slope estimate of dust to soil	Slope estimate of blood to soil	Slope estimate of blood to dust
Omaha	ND[a]	14.4 µg/dl per 1000 ppm	10.6 µg/dl per 1000 ppm
New Haven	ND	2.2 µg/dl per 1000 ppm	2.9 µg/dl per 1000 ppm
Boston	ND	2.2 µg/dl per 1000 ppm	ND
Cincinnati	177.7 ppm per 1000 ppm	6.2 µg/dl per 1000 ppm[b]	3.4 µg/dl per 1000 ppm
Edinburgh	ND	ND	1.8 µg/dl per 1000 ppm
Helena Valley	ND	ND	5.0 µg/dl per 1000 ppm
Kellogg	ND	0.6 µg/dl per 1000 ppm	1.9 µg/dl per 1000 ppm
Telluride	636.1 ppm per 1000 ppm	ND	8.1 µg/dl per 1000 ppm
Leadville	96.5 ppm per 1000 ppm	2.8 µg/dl per 1000 ppm	ND
Midvale	734.5 ppm per 1000 ppm	1.9 µg/dl per 1000 ppm	ND
Butte-Silver Bow	461.7 ppm per 1000 ppm	1.8 µg/dl per 1000 ppm[b]	ND

[a] ND = estimate could not be satisfactorily determined from the published results.
[b] Estimate provided in published results.

on the perceived relationships among dust, soil, and blood. Finally, Section V discusses the conclusions developed from this examination.

II. REVIEW OF RELEVANT STUDIES

The relationships among lead levels in environmental and body burden media have been examined in a number of field and laboratory studies. These studies are, in turn, documented within the scientific and governmental literature. This section contains brief reviews of some of the relevant studies described in the literature. For each study considered, a summary of the study design and its pertinent results is presented. The studies are summarized in chronological order. In addition, whenever possible linearized estimates of the relationships among dust, soil, and blood were determined. These estimates were developed directly from the regression analysis results reported in the literature and were not obtained via reanalysis of the study's datasets. The derivative (with respect to soil or dust) of an estimated regression equation predicting the parameter of interest (either dust or blood) was calculated in order to predict the desired slope coefficient. This transformed equation was then estimated at the geometric mean values for the parameters in the equation. This procedure effectively estimates the dust-to-blood and soil-to-blood slope coefficients at the geometric mean blood lead concentration. The linearized estimates that could be determined for each study are displayed in Table 1.

A. OMAHA LEAD STUDY[1]

Children 1 to 18 years of age were recruited from three areas of interest in Omaha, Nebraska: an urban-commercial area in the vicinity of a small battery plant; an urban-mixed residential area contiguous with downtown Omaha; and a strictly suburban area. The blood and environmental media sampling occurred over a 7-year period, 1970 to 1977. Soil, dust, and water were among the environmental media sampled. No attempt was made to randomize, nor obtain equal sampling distribution, among the three areas. Adequate data for analysis was available from a total of 831 children. The study's authors concluded that community-wide changes in blood lead concentration were multifactorial. Air lead, soil lead, water lead, housing, and socioeconomic shifts all produced an additive, or possibly even synergistic, effect on blood lead. In addition, there was some evidence of seasonal variation in blood lead concentration. It should be noted that soil and dust samples were collected from the residences of only 37 children; all others were assigned the sampling values taken from their school. There was less variability in the assigned values than in the individual residential values. The geometric mean blood lead concentration measured across the study population was 21.47 µg/dl. Linearized estimates of the blood-to-dust and blood-to-soil uptake slopes were 10.6 and 14.4 µg/dl per 1000 ppm, respectively.

B. NEW HAVEN, CT, LEAD STUDY[2]

The study examined a sample of "at-risk" children, age 1 to 6 years, residing in New Haven, CT, and enrolled in its blood lead screening program. A subset, 377 out of 8289 screened, was selected in 1977 to have environmental measurements taken. The environmental media sampled included dust, soil, air, paint, and water. The children had resided at their address for at least 1 year, and their measured blood lead concentration was at least 30 µg/dl. Substantial levels of lead were present in the soil, paint, and house dust, throughout New Haven. The most important variables impacting blood lead levels were determined to be soil lead concentration and exterior paint-lead loading. Elevated levels of lead in the proximate environment, however, explained only a small portion of the variation in blood lead levels. The geometric mean blood lead concentration across the 377 children was 26.94 µg/dl. The linearized estimate of blood-to-soil uptake was 2.2 µg/dl per 1000 ppm. An estimate of 2.9 µg/dl per 1000 ppm for blood-to-dust was also developed.

C. BOSTON HOSPITAL FOR WOMEN LEAD STUDY[3]

This longitudinal study traced the relationship between infant blood lead levels and various environmental factors, from late pregnancy to 2 years of age. From a base population consisting of 11,837 consecutive births at the Boston Hospital for Women (now named Brigham and Women's Hospital) between April 1979 and April 1981, 249 infants were enrolled. The infants were categorized into highest (> 10 µg/dl), lowest (< 3 µg/dl), and middle (6–7 µg/dl) deciles of umbilical cord blood lead concentration. Nearly equal samples were obtained from each of the three distinct cord blood lead levels. Follow-up blood samples were also collected at 6, 12, 18, and 24 months of age. Paint, soil, dust, and water samples were also collected whenever possible, at 1, 6, 18, and 24 months of age. Only 195 children had soil samples (frozen ground prevented sampling) collected, and only 91 children had complete paint samples. Soil lead, dust lead, and past refinishing activities were highly related to blood lead levels. In addition, there is evidence of seasonal variation in blood lead concentration, peaking during the summer months. The families enrolled in this study were more affluent than families in other urban lead exposure studies. Mean blood lead levels at 12 and 24 months of age were 7.7 and 6.8 µg/dl, respectively. A linearized slope estimate of 2.2 µg/dl per 1000 ppm was calculated for blood-to-soil uptake at 24 months of age.

D. CINCINNATI LONGITUDINAL STUDY[4,5]

This study, conducted from 1980 to 1987, systematically monitored blood-lead levels in children from birth through 5 years of age. Residential dust, soil, and paint were among the environmental media sampled. The study enrolled expectant mothers residing in one of a group of census tracts identified as having a long history of producing children with elevated blood lead levels. The mothers were patients at one of three prenatal clinics in the area, intending to deliver at Cincinnati General Hospital. In all, approximately 250 women were involved in the study. The analyses reported here focused on the children at 18 months of age. Environmental lead in inner-city residences was found at potentially hazardous levels. Further, an important interplay existed between environmental sources of lead and social factors, in the determination of hand lead and blood lead levels in very young children. The authors also described a dust lead gradient, with highest levels outside the residence, lower levels at its entrances, and still lower levels within the dwelling. The authors concluded that there was a general movement of lead from the outside of the residence to the inside and finally to the child via the pathway: soil lead (PbS) → interior dust lead (PbD) → hand lead (PbH) → blood lead (PbB). This regression model suggested that blood lead levels would rise 6.2 µg/dl for each 1000 ppm increase in soil lead. In contrast, the developed linearized estimate for dust suggested blood lead concentrations would increase 3.4 µg/dl per 1000 ppm rise in dust lead levels. A linearized estimate of the dust-to-soil slope was 177.7 ppm per 1000 ppm. The geometric mean blood lead concentration at 18 months of age was 16.91 µg/dl.

E. EDINBURGH LEAD STUDY[6]

This 1983–1985 cross-sectional study sought to identify those primary sources of environmental lead exposure for children attending primary classes 3 and 4 from 20 schools in the older, central areas of

Edinburgh, Scotland. Overall, 898 children were sampled. Following the blood sampling, a subsample of the children was identified. In each school all children in the top quartile of the blood lead distribution were selected, and a one-in-three random subsample of the remaining children was obtained. The subsample consisted of 520 children. Blood samples for each child were taken at the school, and environmental samples, including water and dust lead levels, were taken approximately 2 months after the blood sample. A median blood lead concentration of 10.1 µg/dl was measured. The reported analyses emphasized the data collected for the subsample of 520 children. Dust lead concentrations were significantly related to children's blood lead levels. The authors cite a 1.9 µg/dl increase in blood lead concentration per 1000 ppm rise in dust lead levels. A comparable linearized estimate, 1.8 µg/dl, was calculated. The estimate of the dust lead/blood lead relationship may have been impacted, however, by the median 2-month lag between the collection of the blood samples and the environmental samples.

F. HELENA VALLEY CHILD LEAD STUDY[7]

This cross-sectional study was conducted in the summer of 1983 in the vicinity of an operating lead smelter in East Helena, Montana. All households, with children aged 1 to 5 years, located in the vicinity of the smelter were sought for a study of residential exposures to smelter-associated lead and other heavy metals. Three study areas, concentric rings emanating from the smelter, were identified (Area 1: <1.0 mile from the smelter; Area 2: 1.0–2.25 miles; Area 3: >5 miles). In total, 396 of the 437 eligible children were enrolled, and blood and hand-wipe samples were collected. Paint, soil, dust, and air measures were collected from each enrolled residence. Children residing close to the smelter had higher blood lead levels than did children living farther away. Dust lead, soil lead, and air lead all varied significantly between the areas. Multiple regression identified dust lead concentration, air lead concentration, residence near the smelter (indicator), and household member smokers (indicator) as major contributors to blood lead levels. Soil lead was dropped from the regression equation, after residence location was incorporated. The linearized estimate for blood-to-dust uptake was 5.0 µg/dl per 1000 ppm. The geometric mean blood lead concentration among the children was 8.76 µg/dl.

G. KELLOGG REVISITED LEAD STUDY[8]

This study examined the relationship between children's blood lead levels and associated environmental lead levels among 1- to 9-year-old children residing in the vicinity of an inactive smelter in Kellogg, Idaho. As part of the cross-sectional study in the summer of 1983, three study areas were identified (Area 1: <1.0 mile from the smelter; Area 2: 1.0–2.5 miles; Area 3: 2.5–6 miles). Area 3 was specified to be the control area. All families residing in Area 1, with children between the ages of 1 and 9 years, were asked to participate in the study. Every other eligible family in Areas 2 and 3 were sought for the study. In total, 364 of the 400 recruited children were enrolled, and blood and hand-wipe samples were collected. Paint, soil, dust, and air measures were collected from each enrolled residence. Children residing close to the smelter had higher blood lead levels than did children living farther away. A multiple linear regression calculated low, though statistically significant, slope estimates for the contribution of soil and dust lead to blood lead concentration. Very little lead was found in other environmental media, suggesting that the positive association between house dust lead contamination and children's blood lead levels were the result of soil lead contamination from atmospheric fallout of emissions from the now-inactive smelter. Low linearized-slope estimates of soil and dust uptake were determined, however. Blood lead levels rose 0.6 µg/dl per 1000 ppm increase in soil lead concentration, and 1.9 µg/dl per 1000 ppm climb in dust lead level. The geometric mean blood lead concentration 15.18 µg/dl.

H. TELLURIDE LEAD STUDY[9]

This 1986 study investigated the threat from residual lead to residents of a former lead mining and milling community. Young and older children, young adults, pregnant women, and occupationally exposed (to lead) adults were enrolled, and their blood lead levels sampled. In total there were 258 individuals and 45 residences included in the study. Paint, soil, dust, and water lead levels were measured at each residence. The authors hypothesized the following exposure model of an indirect pathway for dust lead: PbS → PbD → PbH → PbB. Using structural equations models, this exposure pathway was examined and supported. The geometric mean blood lead concentration was 6.1 µg/dl. Blood lead levels were estimated to increase between 2 and 4 µg/dl for each 1000 ppm increase in soil lead. A higher, 8.1 µg/dl, linearized estimate was calculated for the blood-to-dust slope. The linearized slope estimate of dust to soil was 636.1 ppm per 1000 ppm.

I. LEADVILLE METALS EXPOSURE STUDY[10]

The entire population of the community of Leadville, Colorado was the focus of this cross-sectional study in 1988. All households with children aged 6 to 71 months, residing in the Leadville census tract were asked to participate in the study. Additional samples of 25 individuals from each sex were randomly selected in each of three age groups (6–14, 15–44, 45–65 years). The final sample sizes were as follows: 150, 6–71 months old; 29, 6–14 years old; 28, 15–44 years old; 26, 45–65 years old. Blood samples were collected from each participant, and soil, dust, paint, and water measures were collected from each residence. Soil lead and dust lead levels were correlated. Blood lead levels were correlated with soil lead samples, but not with indoor dust samples. A geometric mean blood lead concentration of 8.7 µg/dl was measured. The linearized estimate of the blood-to-soil slope was 2.8 µg/dl per 1000 ppm. The study's authors suspected the lack of a dust lead/blood lead relationship was due to household cleaning before the visit of the sampling team, and problems experienced in collecting dust samples. This suspicion is supported by the dust-to-soil linearized slope estimate of 96.5 ppm per 1000 ppm.

J. MIDVALE COMMUNITY LEAD STUDY[11]

This cross-sectional study focused on children aged 6 to 72 months, living within close proximity to a former smelter site in Midvale, Utah. In September 1989, families with eligible children within the community were selected at random and asked to participate in the study. Blood lead samples were obtained for 291 individuals: 181 from children less than 6 years of age. Complete blood lead and environmental samples (dust, soil, paint, and water) were collected from a random sample of 122 eligible children in the area. The geometric mean blood lead concentration was 4.93 µg/dl. Lead-based house paint and lead-contaminated soil were identified as the primary contributors to blood lead. Based upon a structural equations model, blood lead levels were predicted to increase 1.25 µg/dl for each 1000 ppm increase in soil lead. The linearized estimate, in contrast, was 1.9 µg/dl per 1000 ppm. The dust-to-soil linearized slope estimate was a 734.5 ppm increase in dust lead per 1000 ppm rise in soil lead.

K. BUTTE-SILVER BOW ENVIRONMENTAL HEALTH LEAD STUDY[12]

The sampling frame of this study emphasized children less than 72 months of age, residing in one of seven study neighborhoods in Butte, Montana. The neighborhoods vary in their proximity to mining wastes and age of neighborhood (thereby, in their likelihood of having leaded water pipes and lead-based paint). The cross-sectional study entailed blood lead measurements from each child volunteered for the study, and his/her residential environment (paint, dust, soil, and water). The study was conducted during August to September 1990. A total of 430 individuals had blood samples taken (294 of which were less than 72 months of age), while the environmental media was sampled at 217 residences. Using structural equations modeling, the researchers concluded that lead-based paint contributed lead to the surrounding soil, which, in turn, contributed lead to the interior house dust. It was noted that 40% of the variability in soil lead was attributable to lead-based paint. Only lead in house dust was determined to be directly related to blood lead. A geometric mean blood lead concentration of only 3.56 µg/dl was measured. The effect of soil lead on blood lead was described by the authors as small. However, blood lead was found to increase 1.8 µg/dl per 1000 ppm increase in lead in soil. Only 5.4% of the variance in blood lead concentration was attributable to lead in soil. The linearized estimate of the uptake slope of dust to soil was calculated to be 461.7 ppm per 1000 ppm.

III. REANALYSIS OF SELECTED DATASETS

The reanalysis of the individual field study datasets emphasized an investigation of the relationships among blood, dust, and soil lead levels. Though the literature provides reference to a number of field studies that collected environmental media and body-burden lead levels, only five were available to us:

- Boston Hospital for Women Lead Study (1979–1983)
- Helena Valley Child Lead Study (1983)
- Kellogg Revisited Lead Study (1983)
- Midvale Community Lead Study (1989)
- Butte-Silver Bow Environmental Health Lead Study (1990)

These five datasets are typical of the types of studies available in the literature. They do not, however, reflect entirely the variety of environmental lead exposure in the U.S. For example, urban environment

Table 2 Results of Nonlinear Regression Fit of Dust to Soil

Study	No. of samples	Intercept (β_0)	Slope (β_1)	R^2
Boston	130	638 ppm[a]	253 ppm per 1000 ppm	0.03
Helena Valley	171	288 ppm[a]	1350 ppm per 1000 ppm[a]	0.48
Kellogg	103	1211 ppm[a]	285 ppm per 1000 ppm[a]	0.26
Midvale	113	168 ppm[a]	671 ppm per 1000 ppm[a]	0.60
Butte	273	239 ppm[a]	371 ppm per 1000 ppm[a]	0.44

[a] Statistically significant ($p < 0.05$).

studies should be better represented. The gradual decline in air lead levels during the 1980s also suggests more recent studies may be informative. Additional datasets should be reanalyzed to enhance the extent to which the reanalysis results are representative.

The reanalysis consisted of fitting a sequence of simple regression models to the blood, dust, and soil lead levels obtained in each study. The reported regression analyses were specific to the particular study. For example, one study included measures of age of house (as a surrogate for lead-based paint assessment), while another emphasized distance to the smelter site. These study-specific analyses address the hypotheses of the authors, but may obscure the larger relationships, through confounding among the variables. The reanalysis sought to, directly and simply, address the association between dust, soil, and blood lead levels. Four regression models were considered:

1. Dust lead levels (PbD) on soil lead levels (PbS)
2. Blood lead levels (PbB) on PbD
3. PbB on PbS
4. PbB on PbS and PbDr, where PbDr are the residuals from PbD on PbS

The skewed nature of lead exposure levels in environmental and body-burden media has been extensively noted within the literature. Log transformations of the lead exposure variables are used predominantly to adjust for this. Furthermore, the impact on blood lead levels of varied sources of lead exposure has been shown to be additive. Regression models that fit log-transformed blood lead levels to an additive function of log-transformed lead exposure levels (e.g., $\ln(PbB) = \beta_0 + \beta_1 \cdot \ln(PbD) + \beta_2 \cdot \ln(PbS) + \varepsilon$), therefore, are not ideal. A better model may be the nonlinear regression models employed in the reanalysis. Specifically, the four regression equations fitted were,

$$\ln(PbD) = \ln(\beta_{10} + \beta_{11} \cdot PbS) + \varepsilon_1 \tag{1}$$
$$\ln(PbB) = \ln(\beta_{20} + \beta_{21} \cdot PbD) + \varepsilon_2 \tag{2}$$
$$\ln(PbB) = \ln(\beta_{30} + \beta_{31} \cdot PbS) + \varepsilon_3 \tag{3}$$
$$\ln(PbB) = \ln(\beta_{40} + \beta_{41} \cdot PbS + \beta_{42} \cdot PbDr) + \varepsilon_4 \tag{4}$$

These models are consistent with the "Log Total Exposure Model" employed in the EPA's "Air Quality Criteria for Lead" report.[13]

The reanalysis results for the nonlinear regression fit of dust lead levels to soil lead levels are presented in Table 2. Soil lead concentration was found to be a significant predictor of interior dust lead concentration, except in the Boston study. The lack of correlation in Boston may be a function of problems in the sample collection procedures. In Midvale, soil lead levels explained 60% of the variation in dust lead levels; only 26% of the variation is explained by Kellogg. The regression slope coefficients suggest that for a 1000-µg/g (ppm) increase in soil lead levels, dust lead levels may rise between 285 and 1350 ppm.

Table 3 displays the nonlinear regression fits of blood to dust for each of the five reanalyzed studies. Dust lead concentration was a significant predictor of blood lead levels in the Butte, Helena Valley, and Kellogg studies. The R-squared values of these relationships are low, though as is evident within the literature, blood lead levels are usually difficult to predict. Where significant, only about 20% of the variation in blood lead levels can be attributed to concentration of dust lead in the interior of the residence.

Table 3 Results of Nonlinear Regression Fit of Blood to Dust

Study	No. of samples	Intercept (β₀)	Slope (β₁)	R²
Boston	138	3.994 μg/dl[a]	0.4 μg/dl per 1000 ppm	0.03
Helena Valley	171	6.608 μg/dl[a]	2.3 μg/dl per 1000 ppm[a]	0.21
Kellogg	103	9.593 μg/dl[a]	2.5 μg/dl per 1000 ppm[a]	0.20
Midvale	113	4.377 μg/dl[a]	1.0 μg/dl per 1000 ppm	0.03
Butte	202	2.329 μg/dl[a]	2.2 μg/dl per 1000 ppm[a]	0.18

[a] Statistically significant ($p < 0.05$).

Table 4 Results of Nonlinear Regression Fit of Blood to Soil

Study	No. of samples	Intercept (β₀)	Slope (β₁)	R²
Boston	138	3.106 μg/dl[a]	2.7 μg/dl per 1000 ppm[a]	0.08
Helena Valley	171	7.368 μg/dl[a]	3.4 μg/dl per 1000 ppm[a]	0.15
Kellogg	103	12.430 μg/dl[a]	1.0 μg/dl per 1000 ppm[a]	0.14
Midvale	113	3.775 μg/dl[a]	2.5 μg/dl per 1000 ppm[a]	0.10
Butte	273	2.996 μg/dl[a]	0.8 μg/dl per 1000 ppm[a]	0.06

[a] Statistically significant ($p < 0.05$).

The blood lead level of a child is a function of both his recent and long-term lead exposure. Lead stored in the child's bone tissue may be activated into the bloodstream, and a myriad of factors impact the rate of lead absorption. Though these limitations suggest caution, the similarity in the slope coefficients is remarkable. In the three communities with a significant blood-to-dust relationship, these results suggest that a 1000 ppm increase in dust lead concentration would produce an approximately 2 μg/dl rise in blood lead concentration. This is somewhat surprising, since the Butte and Kellogg studies had the lowest and highest, respectively, median blood lead levels among the reanalyzed studies. This average blood-to-dust slope is also considerably less than the 5 μg/dl slope suggested by Duggan and Inskip[14] and the 5- to 10-μg/dl range calculated by Brunekreef.[15]

The results of fitting blood lead levels to soil lead levels, for each reanalyzed study, are shown in Table 4. Unfortunately, the R-squared values are even lower than for the blood-to-dust fits. This is not altogether surprising, especially since soil is not usually cited as a direct pathway of blood contamination. The lead in soil may stem from a variety of sources, including leaded gasoline emissions, peeling or chalking lead-based paint, and point source emissions from a smelter. The slope coefficients are all significantly different from zero, but vary to some degree. For a 1000 ppm increase in soil lead concentration, blood lead levels would be suggested to rise between 0.8 and 3.4 μg/dl. Even the four smelter communities exhibit variation in the blood-to-soil lead slope coefficients. These estimates are, however, less than 8.57 μg/dl, the 95% upper confidence bound on the median estimated uptake slope, as determined by Madhavan et al.[16]

The final reanalyzed regression model fitted blood lead to soil lead and dust lead. Since dust and soil are usually correlated, the residuals of the nonlinear regression fit of dust lead to soil lead levels were employed, instead of the dust lead levels themselves. The two independent variables, therefore, were orthogonal. The results of this reanalysis are presented in Table 5. Again, the R-squared values are disturbingly low, suggesting that the regression explained only between 9 and 23% of the variation in the blood lead levels. The slope coefficients, however, do hint at some interesting results. The two communities where the dust lead residual coefficient was not significantly different from zero, Boston and Midvale, have comparable coefficients for soil lead concentration. Their coefficients imply that if soil lead levels rose

Table 5 Results of Nonlinear Regression Fit of Blood to Soil and Dust Residuals

Study	No. of samples	Intercept (β_0)	PbS coeff. (β_1)	PbDr coeff. (β_2)	R^2
Boston	139	3.029 µg/dl[a]	2.6 µg/dl per 1000 ppm[a]	0.3 µg/dl per 1000 ppm	0.09
Helena Valley	171	7.088 µg/dl[a]	3.6 µg/dl per 1000 ppm[a]	1.8 µg/dl per 1000 ppm[a]	0.22
Kellogg	103	11.807 µg/dl[a]	1.0 µg/dl per 1000 ppm[a]	1.8 µg/dl per 1000 ppm[a]	0.23
Midvale	113	3.786 µg/dl[a]	2.7 µg/dl per 1000 ppm[a]	−1.4 µg/dl per 1000 ppm	0.13
Butte	273	2.882 µd/dl[a]	0.8 µg/dl per 1000 ppm[a]	2.0 µg/dl per 1000 ppm[a]	0.15

[a] Statistically significant ($p < 0.05$).

1000 ppm, blood lead concentration in the residents would elevate by 2.6 µg/dl. In the other three communities there appear to be two sources for the lead in the bloodstreams of the residents. Both the soil lead and dust lead residual slope coefficients were significantly different from zero in the Butte, Helena Valley, and Kellogg studies. Furthermore, though the contribution to blood lead levels from soil lead varies — between 0.8 and 3.6 µg/dl increase in blood lead per 1000 ppm increase in soil lead concentration — the contribution from dust lead remains remarkably consistent. Recall from Table 3 that the contribution to blood from dust was very similar in these three communities. This consistency continues when examining only that contribution from dust lead that does not stem from the surrounding soil. These results suggest a 2-µg/dl increase in blood lead concentration per 1000 ppm increase in dust lead levels stemming from sources of lead other than soil. The Air Quality Criteria for Lead noted a similar result.

These results are remarkably consistent given the differences among the studies. The Butte study was conducted in 1990; the Boston study, in the early 1980s. Different sampling and laboratory analysis techniques were employed. Still, the reanalyses results hint at some intriguing global conclusions. The low R-squared values are of some concern, but may stem, in large part, from the basic uncertainty in predicting blood lead levels. The reported, study-specific analyses did not produce much larger R-squared values:

- Butte-reported, 0.20; reanalysis, 0.15
- Helena Valley-reported, 0.28; reanalysis, 0.22
- Kellogg-reported, 0.33; reanalysis, 0.23,

despite incorporating other relevant variables such as air lead levels or a measure of a child's tendency to mouth nonfood objects. Since the circulating blood lead level of a child is a function of both recent and long-term exposure, it may be extremely difficult to explain a large percentage of its variation.

IV. SUMMARY OF STUDY FINDINGS

Though only a small number of studies were reanalyzed, their results are reasonably consistent with those developed in the review. Altogether, these results reaffirm the soil-to-dust-to-blood pathway said to represent the dominant mechanism of childhood lead exposure. Furthermore, they provide estimates of the uptake slopes along this pathway.

The linearized and reanalysis estimates of the dust-to-soil uptake slope suggest that interior dust lead levels elevate between 96.5 and 1350 ppm per 1000 ppm increase in soil lead concentration. The extremes of this range may have been heightened by extenuating conditions. The authors of the Leadville study noted the possibility that the examined residences may have been cleaned in preparation for the arrival of the sampling teams. This would certainly lower the dust-to-soil uptake slope. In a similar vein, the East Helena smelter was still operational when the environmental sampling occurred. The impact of fallout from the elevated air lead levels, therefore, may be included in the 1350-ppm estimate.

The blood-to-soil uptake slope estimates ranged from 0.8 to 14.4 µg/dl per 1000 ppm. Though soil lead is often described as not directly impacting blood lead levels, it may still be responsible for significant elevations in children's blood lead concentrations. The 14.4-µg/dl estimate from Omaha may have been exaggerated by the elevated air lead levels identified in the study. In fact, a reanalysis of the Omaha data suggests this to be the case.[17] The reanalysis examined the simultaneous impact of dust lead, soil lead, and air lead levels on blood lead levels, using a nonlinear regression model such as those employed in our reanalysis. The resulting estimated blood-to-soil slope coefficient was 6.8 µg/dl per 1000 ppm. This is much closer to the range identified among the other studies. Given the tremendous differences in the timing of each study, the sampling and analysis methods employed, and the nature of the communities examined, it is striking to encounter such consistency in the uptake slope estimates.

The blood-to-dust slope estimates are perhaps the most remarkable. When their relationship was statistically significant, the uptake slope estimates ranged from 1.8 µg/dl to 10.6 µg/dl per 1000 ppm. This is surprising, since dust lead levels exhibit significant variation and were collected using a wide variety of techniques. Furthermore, in the two communities (Boston and Midvale) where blood and dust were not significantly associated, the blood-to-soil slope coefficient is approximately 2.5 µg/dl per 1000 ppm. These estimates are well within the identified range for blood-to-dust uptake. Elevated dust lead levels appear to prospectively elevate blood lead levels at least 2 µg/dl per 1000 ppm increase.

V. CONCLUSIONS AND RECOMMENDATIONS

As the summary above indicates, the review and reanalysis did develop information regarding the relationships among blood, dust, and soil. This information can be used to gain some insight into the possible efficacy of various abatement approaches. The reanalysis, however, provides perhaps the most valuable information in this regard; the nonlinear regression equation results of fitting soil and dust residuals to blood. The blood-to-dust uptake slopes do not distinguish between the sources of the lead found in the dust. The blood-to-soil uptake slopes do not include additional independent variables that may impact the estimated coefficient for soil. One alternative is the simple, nonlinear regression model of soil lead levels and dust lead residuals predicting blood lead concentration. By providing a readily interpretable model that partitions the sources of dust lead, the prospective effectiveness of abatement interventions may be considered.

Consider the reanalysis results portrayed in Table 5. These suggest that the efficacy of soil abatement may produce between a 0.8 and 3.6 µg/dl reduction in blood lead levels per 1000 ppm decrease in soil lead levels. The abatement of sources of lead in dust other than soil, in turn, is estimated to reduce blood lead levels 2 µg/dl per 1000 ppm. Interestingly, the Boston Soil Abatement Project reports reductions on the order of 2 µg/dl per 1000 ppm decreases in surrounding soil lead concentrations.[18]

Perhaps additional reanalyses of this sort can provide insight into the effectiveness of abatement interventions, and the relationships among blood, dust, and soil. There are a myriad of reasons not to combine the results from different studies, including differences in study design, sampling and analytical design, location and timing of the study, and the population under examination. This review suggests, however, that such a combination may be more successful than initial expectations.

REFERENCES

1. Angle, C. R. and McIntire, M. S., Environmental lead and children: the Omaha study, *J. Toxicol. Environ. Health*, 5, 855, 1979.
2. Stark, A. D., Quah, R. F., Meigs, J. W., and DeLouise, E. R., The relationship of environmental lead to blood-lead levels in children, *Environ. Res.*, 27, 372, 1982.
3. Rabinowitz, M., Leviton, A., Needleman, H., Bellinger, D., and Waternaux, C., Environmental correlates of infant blood lead levels in Boston, *Environ. Res.*, 38, 96, 1985.
4. Bornschein, R. L., Succop, P. A., Dietrich, R. N., Clark, C. S., Que Hee, S., and Hammond, P. B., The influence of social and environmental factors on dust lead, hand lead, and blood lead levels in young children, *Environ. Res.*, 38, 108, 1985.
5. Bornschein, R. L., Succop, P. A., Krafft, K. M., Clark, C. S., Peace, B., and Hammond, P. B., Exterior surface dust lead, interior house dust lead and childhood lead exposure in an urban environment, in *Trace Substances in Environmental Health, II, 1986. A Symposium*, Hemphill, D. D., Ed., University of Missouri, Columbia, 1986, 322.
6. Laxen, D. P., Raab, G. M., and Fulton, M., Children's blood lead and exposure to lead in household dust and water — a basis for an environmental standard for lead in dust, *Sci. Total Environ.*, 66, 235, 1987.
7. Lewis and Clark County Health Department, Centers for Disease Control and U.S. Environmental Protection Agency, East Helena, Montana child lead study, Final Report, Centers for Disease Control and U.S. Environmental Protection Agency, U.S. Public Health Service, Washington, D.C., 1986.
8. Panhandle District Health Department, Idaho Department of Health and Welfare, Centers for Disease Control, and U. S. Environmental Protection Agency, Kellogg revisited–1983: childhood blood lead and environmental status report, Final Report, U.S. Public Health Service,Washington, D.C., 1986.

9. Bornschein, R. L., Clark, C. S., Grote, J., Peace, B., Roda, S., and Succop, P., Soil lead-blood lead relationship in a former lead mining town, in *Lead in Soil: Issues and Guidelines, Supplement to Volume 9 of Environmental Geochemistry and Health,* Davis, B. E. and Wixson, B. G., Eds., 1989, 149.

10. Colorado Department of Health, University of Colorado at Denver, and Agency for Toxic Substances and Disease Registry, Leadville metals exposure study, Final Report, April 1990.

11. University of Cincinnati, Midvale community lead-study, Final Report, July 1990.

12. Butte-Silver Bow Department of Health, and Department of Environmental Health, University of Cincinnati, The Butte–Silver Bow environmental health lead study, Draft Final Report, June 1991.

13. U.S. Environmental Protection Agency, Air Quality Criteria for Lead Volumes I–IV, EPA Report No. EPA-600/8–83–028CF, Environmental Criteria and Assessment Office, Office of Research and Development, Research Triangle Park, NC.

14. Duggan, M. J. and Inskip, M. J., Childhood exposure to lead in surface dust and soil: a community health problem, *Public Health Rev.,* 13, 1, 1985.

15. Brunekreef, B., *Exposure of Children to Lead,* University of London, Monitoring Assessment Research Center, London, 1986.

16. Madhavan, S., Rosenman, K. D, and Shehata, T., Lead in oil: recommended maximum permissible levels, *Environ. Res.,* 49, 136, 1989.

17. Angle, C. R., Marcus, A., Cheng, I., and McIntire, M. S., Omaha childhood blood lead and environmental lead: a linear total exposure model, *Environ. Res.,* 35, 160, 1984.

18. Weitzman, M., Aschengrau, A., Bellinger, D., Jones, R., Hamlin, J. S., and Beiser, A., Lead-contaminated soil abatement and urban children's blood lead levels, *JAMA* 269, 1647, 1993.

Chapter 30

Application of a Novel Slurry Furnace AAS Protocol for Rapid Assessment of Lead Environmental Contamination

Michael S. Epstein, Sarah M. Smith, and Joseph J. Breen

CONTENTS

I. INTRODUCTION

The rapid and accurate assessment of environmental hazards is the cornerstone in any legitimate program to protect both the environment and the public. Threats to public health must be identified and quickly eliminated by officials who have the proper information to make educated decisions. Instances of environmental injustice, where minority and low-income populations are subject to disproportionately high and adverse human health or environmental risks, need to be identified and corrected.[1] Analytical science plays the pivotal role in providing data for environmental decision-making, from site hazard assessment, through evaluation of remediation efforts, and finally in the appraisal of pollution prevention technology. Analytical scientists have been most successful in developing fast and accurate methods in the area of inorganic elemental analysis.

Public awareness of lead toxicity is not, as many think, a product of the environmental consciousness of the 1960s and 1970s. Lead poisoning in adults was first described in the second century B.C. Benjamin Franklin spoke of the "bad effects of lead taken inwardly" in 1786, and the first cases of lead poisoning in children were reported in Australia over 100 years ago. Unfortunately, the findings of investigators in the latter case that "painted walls and railings were the source of the lead and that biting of finger nails and sucking of fingers were the means of conveyance" were ignored by those in authority. When organic compounds of lead were introduced as an anti-knock ingredient for gasoline in 1923, similar concerns about public health were expressed by some in the scientific community. In 1925, the Surgeon General called experts from business, labor and public health to assess the hazard of "ethyl gas," with the subsequent conclusion that there were no good grounds for prohibiting its use, provided that its distribution and use were controlled by proper regulations. The danger signs were ignored since there were no adequate measurement methods and standards to prove the case for lead toxicity at low concentration levels. Without hard scientific evidence, economic concerns prevailed. However, by the late 1950s evidence had built up, and paint manufacturers voluntarily limited the lead content of paint to 1%. But it was not until the 1970s that the U.S. government began to take an active role. There was a great deal of ground to cover, since pollution from lead had become widespread in the environment, arising from a number of sources, such as lead-based paint, leaded gasoline, and lead-based solder. Today, we face a three-fold challenge: (1) to identify environmentally hazardous sources of lead and eliminate them; (2) to characterize lead toxicity and reduce lead levels in the general population; and (3) to recognize and rectify environmental injustice to minority and low-income populations. The need for rapid and accurate analytical technology has never been greater.[2]

The ideal analytical method for the assessment of environmental contamination by toxic elements such as lead should be rapid and cost-effective, while retaining enough accuracy and precision to allow conclusions to be drawn from the data. Most analytical techniques do not meet these criteria. To obtain reasonably accurate results, they require the sample to be leached or dissolved in an acid media or fused at high temperature into a soluble form. Such sample preparation demands decrease sample throughput and thus lengthen the response time for environmental remediation, as well as requiring dedicated and expensive technician time. The few analytical methods that can be modified for direct elemental analysis of solids without pretreatment, such as X-ray fluorescence spectrometry, are limited by cost, matrix interferences, elemental coverage or sample size.

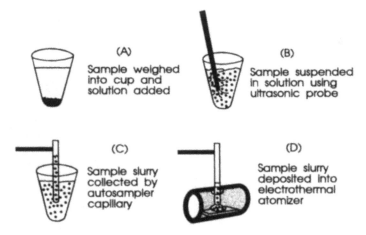

Figure 1. Steps in the use of slurry sampling for ETAAS.

The slurry method of sample introduction for electrothermal atomic absorption spectrometry (slurry-ETAAS) provides a unique combination of minimal sample preparation, proven accuracy, low cost instrumentation, and rapid and unattended sample throughput that makes it ideal for the evaluation of large numbers of samples for toxic element contamination.[3] Slurry-ETAAS has been used successfully by a number of researchers for the determination of toxic elements in soils and sediments[4-6] and paints.[7] Lead is one of the elements most easily determined by the slurry method, since it readily extracts into the solvent phase, thus maximizing precision and accuracy. Lead is also one of the most pervasive and toxic elements in the environment.

In this investigation, slurry-ETAAS was used to determine lead in samples of paint and soil from 53 parks and playgrounds in Arlington County, Virginia. The analysis results indicated a significant number of sites at which further evaluation is needed. The locations of high lead levels were compared to demographic information on income level to look for evidence of environmental injustice.

II. EXPERIMENTAL

Soil was collected with a plastic utensil from several locations within each park and individual samples were combined to form a composite sample representing each park. Similarly, soil samples near streets adjacent to the parks were also collected and combined. The paint samples were taken from jungle gyms, slides, swings, benches, and posts within the parks and then combined to form a composite paint sample. Paint samples were taken with a stainless steel blade only from those playgrounds where equipment or structures had paint that was already chipped or peeling. Sampling tools were wiped with a lead-free baby-wipe between collections. A total of three composite samples (paint, park soil, and street soil) represented each playground or park, for a total of $\sim$ 150 composite samples. All samples were transferred to plastic bags and taken to the laboratory where they were crushed and mixed with a mortar and pestle. After homogenization, two subsamples weighing between 10 and 20 mg each were taken from each sample bag and placed into plastic autosampler cups, with the subsequent addition of 1 mL of a diluent solution consisting of 5% (v/v) HNO_3 (high-purity, sub-boiling distilled) and 0.05% (v/v) Triton X-100 surfactant (Rohm and Haas registered trademark for octyl phenoxy polyethoxyethanol). Samples were analyzed using an atomic absorption spectrometer with Zeeman-effect background correction, a transversely heated graphite furnace atomizer, and an ultrasonic slurry mixing device. Peak area absorbance measurements were employed. Figure 1 illustrates the steps involved in ultrasonic slurry sample introduction for ETAAS. The samples were suspended in the diluent solution using an ultrasonic probe and 20 µL of slurry was removed and introduced into an electrothermal atomizer for analysis. During the suspension, which is accomplished by ultrasonically disrupting the solution for 10 seconds, some of the sample matrix is dissolved in the acid, while the rest of the sample is suspended as small particles. For some elements, such as Pb, a very high fraction of the element will be dissolved during ultrasonic agitation in the acid media. But even if a very small fraction of the element is dissolved, as long as the particles are not larger than the diameter of the sampling capillary or too dense (so that they do not remain suspended for a few seconds

Table 1. Instrumental Parameters

Instrument:	Perkin-Elmer 5100ZL Atomic Absorption Spectrometer with USS-100 ultrasonic mixing device
Wavelength:	261.4 nm
Source:	Hollow Cathode Lamp at 10 mA

Furnace Program:

Step	Temp	Ramp	Hold	Gas flow	Other
1	110	1	20	250	Dry (stage 1)
2	130	5	30	250	Dry (stage 2)
3	400	10	20	250	Ashing
4	20	1	5	250	Cooldown
5	1600	0	10	0	Measure absorbance
6	2500	1	2	250	Cleanup

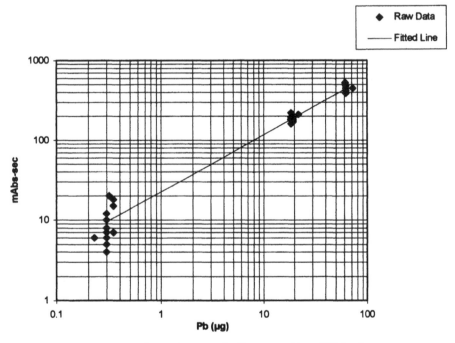

Figure 2. Calibration curve for Pb prepared from SRM soils.

after the ultrasonic disruption), both solution and particles are quantitatively transferred to the electrothermal atomizer by the micropipetor. Rather than using the most sensitive lead line at 283.3 nm, a lower sensitivity spectral line of lead at 261.4 nm was used to provide a working range that would cover the concentrations of lead in soil and paint considered hazardous: $> \sim 0.02\%$ for lead in soil and $> 0.5\%$ for lead in paint. Instrumental parameters for the analyses are shown in Table 1. A calibration curve was prepared from NIST Standard Reference Material Soils with concentrations of 18.9 ± 0.5 μg Pb/g (SRM 2709), 1162 ± 31 μg Pb/g (SRM 2711), and 5532 ± 80 μg Pb/g (SRM 2710) that were run before and after each sample set. Sample throughput (duplicate measurements per sample) was approximately 15 samples per hour. The entire dedicated analysis time for ~ 400 samples, standards, and controls was ~ 30 hours (10 separate runs of 40 samples per run). After all analyses had been completed, Pb concentrations in the samples were calculated from a log-log plot of absorbance versus concentration, prepared from the pooled data from all analyses of the SRM soils and shown in Figure 2. Since both paint and soil samples from the parks and playgrounds were compared to the calibration curve prepared from the SRM soil samples, reference paint samples were also run blind as controls.

Table 2. Results for Control Samples of Paint

Analysis Designation	Identity or Source	Concentration (%Pb)	
		Reference	Determined
A	SRM 2582	0.0209 ± 0.0005	0.01
B	RTI[a]	0.14	0.16
C	RTI[a]	1.05	1.06
D	RTI[a]	0.85	0.98

[a]Standards prepared by Research Triangle Institute, Research Triangle Park, NC.

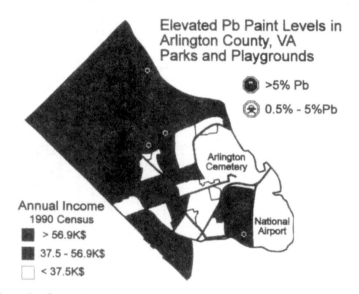

Figure 3. Comparison of annual income level with elevated lead levels in paint.

III. RESULTS

Analysis results for the control paint samples are shown in Table 2. The method appears to provide adequate accuracy for survey analyses at lead concentrations greater than 0.01%, which is the range of primary concern. Analysis results for the samples of paint and soil from the parks and playgrounds are shown in Table 3. In general, there was sufficient agreement between subsamples to draw conclusions, except in a very few cases at lower lead concentrations. Sample inhomgeneity probably accounts for these few results, since the samples were pooled from several locations within each park. No significant contamination from preparation or handling of the samples was detected. As noted in Table 3, soil lead concentrations were found to vary between 0 and ~700 μg Pb/g, and paint lead concentrations varied from 0 to ~8 % Pb.

The concentrations of lead were compared to location and income level in the vicinity of the parks and playgrounds. Figures 3 and 4 show the distribution of sites where elevated lead in paint and lead in soil were found, as a function of annual income based on demographics from the 1990 census. In neither case does there appear to be evidence of environmental injustice. The majority of high lead level parks are in the medium income areas.

This survey was intended primarily as a test of the speed, accuracy, and precision of the slurry-ETAAS method when applied to a real analytical problem, and as a preliminary evaluation of lead poisoning dangers in public parks. Significant lead concentrations were found in some locations, and more thorough evaluation appears to be in order. The only significant limitation of the slurry-ETAAS method, as we used it, is the sequential nature of the atomic absorption measurement, which limits the range and speed of multielement analysis on selected samples. It is quite adequate for rapid surveys when a limited number of known toxic elements are being investigated.

The accuracy base for slurry sampling combined with electrothermal atomization for atomic absorption is currently being established through international round-robin measurements.[8] Preliminary results show agreement within 20% of most participating laboratories on soil reference materials. While our study was

Table 3. Summary of Results for Survey of Lead Contamination in Arlington County, Virginia Parks and Playgrounds. Results Separated by a "–" are the Average Values for the Two Subsamples from Each Composite Sample of Soil or Paint

Park Name	Income Level	Park Soil Pb (ppm)	Street Soil Pb (ppm)	Park Paint Pb (%)
Arlington Arts Center	high	11–14	100–430	0
Glencarlyn Park	high	7	110–150	0.1
Woodmont Center	high	2–11	70–110	–
East Falls Church Park	high	4–22	41–130	–
Edison Minipark	high	9–21	37–110	0.2
Stewart Park	high	22–30	47–56	–
Glebe Road Park	high	3–20	43–45	0.005–0.006
Fort Scott Park	high	7–8	11–58	2.5–3.7
Nottingham Elementary	high	6–25	23–28	0.004–0.006
Madison Manor Park	high	2–7	23–25	0.1–0.2
Madison Recreational Center	high	14–110	15–30	0.1–0.4
Greenbrier Park	high	17–56	14–20	0.0004–0.0008
Jamestown Playfield	high	7–10	4–11	0.5–0.6
Marcey Road Recreational Center	high	27–120	9–12	0.9–2.1
Taylor Park	high	5–31	3–7	4.4–8.1
Langston Community Center	middle	18–180	320–580	0.005–0.008
Quincy Street Playground	middle	36–66	340–490	0.5–0.7
Clarendon Playground	middle	4–5	90–320	0.7
Dawson Recreation Center	middle	1	32–270	5.4–6.2
Lacey Woods Park	middle	10–220	96–210	0.0005–0.0007
Barcroft Park and Playground	middle	21–31	100–180	–
Jackson School	middle	56–170	110	0.2–0.6
Alcova Heights Park	middle	9–26	80–110	0
Lee Recreation Center	middle	7–9	86	0.001–0.006
Lyon Village Playground	middle	27–38	40–130	–
Abingdon School	middle	<1	15–140	–
High View Park	middle	4–62	58–88	4.5–7.8
Hayes Playground	middle	11–14	50–92	–
Eades Playground	middle	12	25–120	–
Parkhurst Playground	middle	7	35–74	0.6–0.7
Woodlawn Park	middle	20–27	31–65	6.5–7.8
Virginia Highlands Park	middle	4–15	19–56	–
Utah Field Park	middle	7	22–37	–
Bon Air Park	middle	7–9	15–30	4.9–7.3
McKinley School	middle	6–25	17–21	
Westover Park	middle	12–27	11–23	0.0002–0.0007
Butler Holmes Playground	middle	6–19	7–16	0.1–0.3
Wilson Adult Center	middle	28–40	5–11	0.0008–0.0009
Doctor's Run Park	low	9–660	230–260	0.3
Carver Community Park	low	2–5	180–200	0.0001–0.001
Clay Playground	low	1–28	95–280	0.0001–0.0002
Fort Bernard Park	low	4–9	130–160	0.02
Walter Reed Recreational Center	low	2–3	120–160	–
Troy Playground	low	7–9	94	0.1–0.7
Patrick Henry Playground	low	7–12	60–86	0.1
Shirley Park	low	5	28–110	–
Nauck Playground	low	19–89	21–88	–
Lubber Run Community Center	low	4	34–60	0.009–0.02
Arlington Heights Park	low	5–51	32–37	–
Towers Park	low	27–50	20–46	0
Rocky Run Playground	low	6–11	29–40	–
Long Branch Elementary	low	<1	17–28	–
Drew Community Center	low	1	9–26	–

270

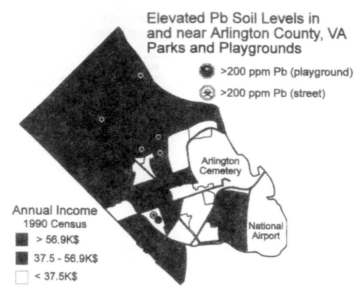

Elevated Pb Soil Levels in
and near Arlington County, VA
Parks and Playgrounds

● >200 ppm Pb (playground)

◉ >200 ppm Pb (street)

Arlington
Cemetery

Annual Income
1990 Census
■ > 56.9K$
■ 37.5 - 56.9K$
□ < 37.5K$

National
Airport

Figure 4. Comparison of annual income level with elevated lead levels in soil.

conducted with a single-element AA instrument, multielement AA spectrometers are now available. The future of the slurry-ETAAS method lies with such instrumentation as well as in its combination with sensitive multielement techniques, such as inductively-coupled plasma mass spectrometry, for rapidly and accurately surveying toxic element contamination in the environment.[9] Not only can large numbers of elements be rapidly surveyed, but such methods allow for source identification through multiple element or isotope pattern recognition,[10] providing policymakers with even more solid grounding to make just decisions on environmental issues.

REFERENCES

1. Clinton, W. J., Executive Order 12898, Federal Actions to Address Environmental Justice in Minority Populations and Low-Income Populations, Feb. 11, 1994.
2. Semerjian, H. G., Statement before the Subcommittee on Technology, Environment and Aviation, U.S. House of Representatives, on the Role of the NIST in U.S. Technology Policy, July 26, 1993.
3. Miller-Ihli, N. J., *Anal. Chem.*, 64, 964A, 1992.
4. Epstein, M. S., Carnrick, G. R., Slavin, W., and Miller-Ihli, N. J., *Anal. Chem.*, 61, 1414, 1989.
5. Hoenig, M., Regnier, P., and Wollast, R., *J. Anal. Atom. Spectrom.*, 4, 631, 1989.
6. Hinds, M. and Jackson, K., *J. Anal. Atom. Spectrom.*, 5, 199, 1990.
7. Garcia, I. L., and Cordoba, M. H., *J. Anal. Atom. Spectrom.*, 4, 701, 1989.
8. Miller-Ihli, N. J., Slurry sampling GFAAS: a preliminary examination of results from an international collaborative study, *Spectrochim. Acta*, in press, 1994.
9. Gregoire, D. C., Miller-Ihli, N. J., and Sturgeon, R. E., *J. Anal. Atom. Spectrom.*, 9, 605, 1994.
10. Lord, C. J., III, *J. Anal. Atom. Spectrom.*, 9, 599, 1994.

ACKNOWLEDGMENTS

The authors thank the Perkin-Elmer Corporation for the use of the USS-100 ultrasonic mixing device. Mention of commercial equipment in this manuscript does not imply endorsement by the National Institute of Standards and Technology, nor does it imply that the equipment is the best suited for such use.

Epilogue

Check Our Kids for Lead:
Empowering Employees to Make a Difference

*J. J. Breen, C. R. Stroup, S. Wooten, V. R. Anderson, K. A. Benjamin,
G. H. Bergeison, S. F. Brown, B. T. Cook, G. Cooper, L. M. Harris, B. S. Lim,
D. G. Lynch, and B. A. Myrick*

Over the past couple of years, a number of us have been challenged by the federal government's growing involvement in the lead-based-paint area. In 1992 the level of activities exploded with involvement in a number of major across-government and private-sector activities, all geared toward protecting children from lead poisoning. At EPA we have had many reasons to feel good about what we, the Office of Pollution Prevention and Toxics (OPPT), are doing to address lead poisoning in children.

Feel good, that is, until we heard, at the national conference Lead Tech 92, comments to the effect, ". . . all these government people ever talk about are their programs. They never talk about the kids as someone's child. It's as if the kids were more a thing, a commodity, than my kid. . . . What does it mean to my sister that 200,000 of 'em are at risk? She wants to know about her boy."

This comment prompted OPPT staff to ask, "What are we, as an organization of some 500 people, doing to ensure our kids under the age of 7 years have their blood-lead levels checked? It's all well and good to develop and implement programs for the greater public audience "out there," but we also need to know what is going on with our own kids, back here at OPPT, to ensure those at increased risk are being helped and their parents, our colleagues and fellow workers, are educated and counseled to prevent or reduce that risk." Answering those questions might then allow us to formulate a program to export to other organizations, to allow them to put a human face on this important environmental and public health issue.

We believed OPPT management should look at the lead-in-kids issue as more than some remote, inanimate problem and bring it home. OPPT needs, as leaders in the lead paint area, to put a strong personal element in our OPPT management program and serve as a model to other organizations, both in and out of government.

We felt the Agency's strong spirit of volunteerism, together with the special technical talents OPPT has in the area of lead poisoning in kids, could be brought to focus on the question of testing OPPT's own kids. Toward that end we obtained the support of OPPT management to call for volunteers from across OPPT, to plan a program focused on OPPT's kids. The response was excellent! Twenty-five volunteers formed a group called the "OPPT **Check Our Kids for Lead** Workgroup." We solicited and obtained support and participation from across EPA, including the Office of Human Resources Management: the Agency's Safety, Health, and Environmental Management Division; and two employee unions: National Federation of Federal Employees and Association of Federal Government Employees.

OPPT **Check Our Kids for Lead** Program

MISSION STATEMENT

Develop a model program to educate OPPT staff and the local community on lead health issues as they pertain to young children. Encourage and facilitate the testing of all OPPT and local community children under 7 years of age for blood lead levels.

OPPT **Check Our Kids for Lead** Program

PROGRAM OBJECTIVES

1. To develop a program, information packet, and program implementation plan, for use by OPPT and the local community, and ultimately other EPA program offices
2. To educate OPPT staff on lead health issues
3. To educate the local community on lead health issues
4. To encourage all OPPT children under 7 years be tested for blood lead levels

5. To encourage all local community children under 7 years be tested for blood lead levels
6. To explore on-site EPA (e.g., Wellness Center, Day Care Center) screening possibilities for OPPT, other EPA, and local community children
7. To develop an information package on lead in day care centers
8. To encourage EPA to conduct environmental monitoring at the EPA Day Care Center
9. To develop a model program for export to other EPA offices, other federal agencies, day care centers, schools, and private-sector and public institutions

The volunteer workgroup developed and distributed officewide a lead information package and developed a Washington-area screening brochure. While the primary goal was to educate OPPT staff and to encourage childhood blood screening, we were also interested in pursuing the export of the program; that is, once there was a meaningful, programmatically sound, and successful OPPT, **Check Our Kids for Lead** program, the workgroup planned to make it available to the other EPA program offices. The program can be modified as necessary, and used by the regional offices, ORD laboratories, other federal agencies, and the private sector.

The volunteer workgroup sponsored an agency-wide seminar on lead poisoning, with presentations by the U.S. Public Health Service, HUD, EPA, and the DC Lead Poisoning Prevention Program. The seminar was videotaped and copies are available. They also held a seminar at the Howard University College of Nursing, as part of their summer school program. The workgroup worked with Howard University to incorporate aspects of lead poisoning issues into their curriculum development for the College of Nursing. The OPPT **Check Our Kids for Lead** Program Workgroup and the College of Nursing collaborated with Howard's School for Communications, on developing a videotape suitable for distribution to others.

The OPPT **Check Our Kids for Lead** Program Workgroup made presentations before the Leadership–Washington Group, the Baltimore Innovative Teams for the Environment (BITE, part of the EPA-Morgan State University's environment equity program), and the Purdue University Chemistry Department Industry Associates Annual Meeting on Environmental Chemistry. The workgroup is currently engaged in a dialogue with several private-sector companies, regarding the establishment of employee-focused programs for those companies. Companies that have expressed an interest include Midwest Research Institute, Hallmark Cards, and Westat, Inc. We also worked with a local D.C. area school to prepare a 1995 calendar with the lead-poisoning theme expressed through children's artwork.

The OPPT **Check Our Kids for Lead** Workgroup received the EPA Gold Medal for superior achievement in 1994. Information on OPPT's **Check Our Kids for Lead** Program and starting your own program is available from Cindy Stroup (202–260–3889) or Joe Breen (202–260–1573).

Check Our Kids for Lead!
It's the right thing to do.

INDEX

A

AAS, see Atomic absorption spectrometry
Abatement demonstrations
  FHA
    lead-based paint abatement methods, 31
    PHA abatement demonstration and, comparison,
      32
  HUD, 49
    abatement methods, 39–41
    database, 38, 41–44, 44–47, 244, 246–247
    interior vs. exterior room, 40
    of lead-based paint, 77–78
    participants, 37–38
    sampling and analytical methods, 39
    square footage abated, 40
  PHA
    FHA abatement demonstration and, comparison, 32
    lead-based paint abatement project, 31–32
    waste disposal, 34
    work force, 33
Abatement methods
  effect on metals in household dust and soil, 225
  EPA strategy, 71–72
    nonregulatory activities, 73
    objectives, 86
  lead-based paint, 39–41, 82
    ASTM subcommittee, 101–102
    FHA, 31
    HUD plan, 77–78
    modernization and, relationship, 32
    PHA, 31–32
  lead dust and, 42, 45
  national implementation plan objectives, 87
  pilot CAP study, 51–54
  soil lead concentrations and, 41–44
Adhesive lift sampler, surface particle, procedure, 157
Agency for Toxic Substances and Disease Registry
    (ATSDR), 143
  lead prevalence in homes, 135
Alliance To End Childhood Lead Poisoning, 97
American Society for Testing and Materials (ASTM),
    lead hazards subcommittee, 101–102
ASTM, see American Society for Testing and Materials
Atomic absorption spectrometry (AAS)
  flame, 121
  graphite furnace, sample data, 210–213
  ICP and, comparison, 180
  lead measurements, 144, 161, 172
ATSDR, see Agency for Toxic Substances and Disease
    Registry

B

Backscattered electron imaging
  dust particles, 148–150
  soil particles, 153
Battery recycling, lead emissions, 63
Bias, see also Error
  classification, national survey data, 26–28
  definition, 21–22
  measurement, of XRF spectrum analyzers, 26–28
  recommended procedures to limit, 29
  surface, 22–23
Blood lead
  dust and soil levels, relationship
    linearized slope estimates, 256–259
    reanalysis of selected datasets, 259–262
  levels in children, studies, 256–259
  screening, 93
Blood level, measurement methods, delves cup method,
    121–126
Blue-nozzle dust collector, collection efficiency
  protocol, 193
  test results, 197–198, 200

C

Cadmium, in household dust and soil, 217–220
  abatement effects, 225
  correlation with other metals, 223, 226–227
  renovation effects, 224
CAP study, see Comprehensive Abatement Performance
    study
CCSEM, see Computer-controlled scanning electron
    microscopy
CDC, see Centers for Disease Control
Cells, 245
Centers for Disease Control (CDC), lead poisoning in
    children
  current prevention program, 80
  history of involvement, 79
  public statements, 71
Check Our Kids for Lead program
  mission statement, 271
  objectives, 271–272
Childhood Lead Poisoning Prevention Program (CLPPP),
    health education programs, 105
Children, lead poisoning, 8, 75, 127, 135, 143, 161, 183,
    231, 243
  CDC preventive actions, 79–81
  characteristics, 90–91
  dust sources, 9